PBVS-273:
An Introduction to the Theory of Subvolition
(Primary Psychology) – PART I

VOLUME TWO

PBVS-273:
AN INTRODUCTION TO THE THEORY OF SUBVOLITION
(PRIMARY PSYCHOLOGY) – PART I

VOLUME TWO

BY

ANDREW J. GALAMBOS

SPACELAND PUBLICATIONS

FIRST EDITION

ISBN: 978-1-968146-47-4

The designation "P.I." stands for post-integration, the time elapsing since Isaac Newton's publication of *Principia Mathematica* in year 1687 of the so-called Christian era. This positive designation by Professor Galambos is intended to replace the mystical basis and nomenclature of the past history of man and to establish a rational calendar, with each new year beginning on the vernal equinox. It should also be noted that the naming of days, weeks and months, etc., will follow the same rational reasoning.

Typography and Index by Erik G. Falvey

CONTENTS

SESSION 6

SESSION 7

SESSION 8

SESSION 9

SESSION 10

SESSION 6

I would like to discuss here a little bit of the type of thing I'm doing here in the early sessions of this course. Clearly, in the very first few sessions of this course, what I tried to do is to show you what the subject is that I'm trying to talk about which I've called psychology—with some reservations—because the word has been previously used by other people and I explained why I chose to use it anyhow, because it covers the same domain of knowledge.

A New Approach to the Subject of Psychology with Classical Terminology

I am unwilling, in this case, to take the position that I did in social science where I also covered the same domain of territory as social science, but I called it volitional science. I changed the name because I consider my antecedents in that domain total quacks. Here I cannot make that statement. I don't know enough about psychology in the classical sense and I do not certainly make any such claim. Some are, and some aren't. That's true in every profession. Some physicists are quacks but the profession isn't quackery. Well, obviously there are quacks amongst psychologists and psychiatrists but that doesn't make the profession false and therefore I make no such claim and I think the word is probably quite a respectable term.

I am covering the same domain, so instead of calling it by a different name I'm calling it by the same name but making it clear there's a new approach to it, and I started out with discussing that. I took about two sessions to do that, to tie it in with physics and volition and I identified the principal terms of the subject.

The Principal Terms of Psychology: Ego and Ego Minor

The principal terms of the subject are ego and ego minor. Ego is the basic concept. Ego is to psychology what property is to volition. Ego minor is to the ego what coercion is to property. It's a delusion concerning the proper handling of the subject, and it's a delusion which leads to false results. And so, just as in volition, we have a beautiful, clean, clear handling of what seemingly is a confusion of concepts in so-called social science handling of the subject. In volition it comes out very simple, totally integrated, and everything can be explained in terms of one central concept called property, and malfunctions of the handling of property which produces flatland. Coercion is the source of flatland. In other words, the reason the world is flat is because coercion is the tool of handling property. It's a malfunction. It's a disease.

Subvolition Is Galambos' Term for Psychology

Quite similarly, I've made a parallel structure here for a subvolitional science. I've explained why I call it subvolitional and it's certainly tied in with volition, but inwardly directed to the individual rather than outwardly directed from the individual to other individuals, and is therefore both related and meshed with volition, properly, the way I've done it. I'm also not using a new vocabulary, I'm using the same vocabulary but with new meanings or, more accurately, expanded meanings to apply to the subvolitional domain of psychology. I have introduced one new term principally: ego. Other than that, I'm still talking about property. Only property. And before that, life, and before that, physics and therefore mass, length and time and the supreme unifier of physics, energy. These are all in the backflow of this subject. And I've introduced one principal new term, the ego. So far, that's the only major new term I've introduced, and yet I've made a lot of headway with the old terms, plus the ego, which is a form of property, incidentally, I tied that in too.

The Beginning of the Unification of All Knowledge

The ego minor is a supreme simplifier to explain all of the malfunctions of the ego, exactly the same as the single term coercion is the supreme unifier of what is ailing the world in terms of the malfunctioning of the handling of property. I hope you recognize both the correlation and the simplicity of both. *Ja oder nein*? That means, "Yes or no?" [*Some laughter in the audience and some "Yeses."*] Is this to you what it seems to me? A clean sweep. Now you may have at the moment little sensitivity to what's yet to come, but I'm trying to prepare the groundwork for it, so I brought in a number of other topics that may or may not immediately be apparent to you that are related to the subvolitional subject of psychology.

Galambos' Definition of Cure is Applicable to All Science

For example, I discussed the word *cure* from V-201. Cure, of course, is ultimately a volitional concept which can be applied to biology which is upstream of volition, and also can be applied to psychology which is downstream of volition. Same word, same definition, three sciences. I hope you consider that what I consider it to be. The word is beautiful. The fact that it brings three different subjects into a consecutive flowstream. This is even more potent when you recognize that all three flow from physics, and this is the beginning of the beginning of the unification of all knowledge, but it is a beginning.

Entropy Is the Measure of the Disorder of the Universe

I don't know how far it will go at the moment, myself, but it will go to the end but not necessarily through me and the end is no limit. You know, every time you get smart enough to know the answer to something you get smarter to ask more questions, and the end is as far as the cosmic time scale goes. And man shall ride with it provided we don't botch it up. But he will ride with it not as a freeloader, but as an upstream swimmer. Upstream against the disorder. Functioning of the universe, the disorder growth of the universe which is called entropy, incidentally. Entropy is the measure of the disorder. It also does not go without effort. It goes with great effort but it's the only way to go, to fly, to swim. The only way to do any of these things.

Both the Quality and Quantity of Happiness

I've also brought in the subject of the quality as well as the quantity of happiness, which is the extension of what I have previously stated in V-50 and V-201 on the word *happiness*, and shown that there are two different ways of measuring happiness. The total amount that you have a capacity for vs. the total percentage of what you have a capacity for

3

that you have attained or can find attainable. And the larger the capacity the smaller the relative attainability, and the smaller the capacity the larger the attainability, which favors the high-capacity seekers because it gives them a higher incentive. Incentive is born from that concept. The concept of incentive in psychology is born from the drive to learn more and achieve more. That's the source of that very important builder of civilization.

All of Galambos' Courses Are Interrelated to Each Other

I also brought in the subject of long term and short term which appears in all my courses someplace in some measure and I'm going to continue doing this for a while, bringing up subjects from various courses. I remind you: All courses that I've ever taught—without exception—all the courses I ever taught at FEI are interrelated to each other. There are no such things as a disjointed product, which is somewhat of a loose piece that you haven't found to fit into the same picture. Somewhere it fits into the total structure. There are no two courses I have ever put forth that are unrelated to each other or to any other course. And the reason they're put in the different courses is for different administrative units for production, or more accurately, product sales so that you don't get to the point where you say, "Well, I don't even know what the hell I'm buying." There's only one course that I have like that, it's called the open-end course where you don't know what you're buying because I don't know what I'm selling yet. [*Some chuckles in the audience*.] I make it up as I go along and every time I bring in a new subject, it comes in and then I find a place to put it in where it belongs. That's possible too, but that requires a more select determination of who's willing and able to take the effort and the time to be there.

Difficulty of Marketing the Open-End Courses

Also, it's not an easy thing to market. At least it wasn't. It's easier now than in the beginning. If I had not had V-50 or 201 it would have been hopeless. Can you imagine without V-50 and V-201 going up to somebody and telling him, "Would you like to take a course where I don't know what I'm going to be talking about each time and I can't tell you in advance because I won't know until I get there myself and it'll be always on a different subject and sometimes it'll be a short one and sometimes it'll be a long one? Sometimes it'll overflow, sometimes it will not fill up even ten minutes and I have to find a new topic. I don't know how long the sessions will be and the course will last as long as I live." [*Chuckles in the audience*.] Well, as I say, the marketing of that is rather unique also. [*AJG chuckles*.]

Difficulty of Marketing V-50

That's essentially not unrelated to the courses where I do tell you what it's about. I give you administrative units. This one is on the basic course of volitional science. Now of course you don't know what that is until you come in. That's V-50. Now that really does have a name, but we don't tell you because we found it was dangerous to tell you before you come into V-50. Its name was originally "Capitalism—The Key to Survival!" and I got a lot of yo-yos who thought that that means to save your little hides from the revenuers and run up to the hills. I got a lot of weird types then. I still do. But they're weird in different ways now. The weirdness has altered. So later the name of the course was upgraded to "Capitalism—The Liberal Revolution" because *the key to survival* referred to the survival of the species and most people took it to be the survival of their own little hides and it got, therefore, a dual ambiguous application.

I meant actually both but with the emphasis on the whole species, but I got a lot of little short-term nobody's, getting all hot and then getting all upset because they didn't get what they thought they were getting and then those who found out that it was the right thing then they dropped their short-term activities and then all hell broke loose with the John Birch Society and everywhere else. Because they lost all of their flock followers. Not all of them, just some. And then of course, the John Birch Society decided, I guess, that the real enemy is not communism but Galambos. [*Audience laughs with AJG*.] And they concentrated on keeping everybody out of my course, which is fine because you always get your prestige in terms of who doesn't like you. It's an honor not to be liked by some people. [*Chuckles in the audience*.]

The Index of Forbidden Books

You know, speaking about that, I was greatly disappointed some years back when I found out the Catholic Church had abolished its Index, the Index of Forbidden Books. That really shook me up [*audience laughter*] because I fully expected to make that Index [*audience and AJG*

5

laugh] and have the pleasure of being on the same list as Giordano Bruno, Galileo Galilei and all these other heretics. [*AJG chuckles.*] That's second only to being banned in Boston. [*Audience laughs.*] Imagine, I got a course in Boston and they haven't banned it yet. [*AJG chuckles.*] Ironically, that's the first eastern city which we have a sustained market in, thanks to a very qualified contractor.

Galambos' Courses Are Pro, Not Anti

Well anyhow, I'd like to point out that the entire nature of V-50 is productive but by calling it "Capitalism—The key to survival!" originally I got a whole bunch of negative yo-yos who were super hell bent on hating something or other, like communism or anti-this or anti-that and I wasn't anti-this or anti-that, I was pro-capitalist. I tried to make a clear distinction between the importance of pro-capitalism vs. anti-communism which is a disease. Communism is a disease and anti-communism is a disease because anti-anything is a disease. A negative posture is an unhealthy way of handling anything. Anti-cancer is not a way to cure cancer. Doctors may someday learn that and might make some headway after that. You're not going to eliminate it by hating it or by fighting it. Neither is appropriate to man's dignity nor his intellectual prowess.

Hoodlums and Hooligans Are Not Revolutionaries

Well, so I changed the name of the course to "Capitalism—The Liberal Revolution" and that was fine for a while until Snelson started reporting some feedback. He was getting some problems that the word *liberal* turned off the conservatives; the word *capitalism* turned off the pseudo-liberals; the word *revolution* turned off everybody except rabble rousers whom we don't want. The rabble rousers are the last people we want to get but they're the only ones who like the word *revolution* these days. They're not revolutionaries, they're rebels and hoodlums. Hoodlum is actually the better word. Rebel isn't good enough, it isn't strong enough to describe the hooligans that are called revolutionaries these days. They're hooligans! Assassins, terrorists! They're not revolutionaries. The word *revolutionary* is a clean, sweet, beautiful word. It means you're changing the world for the better, you're turning it around from barbarism to civilization.

The Naming of Course V-50

"Capitalism—The Liberal Revolution" I think still is the best title for V-50 and I haven't changed it. But since it doesn't sell well, because it produces all kinds of image distortions because of the false usages of words these days. We both gulped—Snelson and I—and said, "Okay, we'll sell the course without a name." "You mean just V-50?" What do you put down there these days? What's that other slogan he puts on? Nothing? [*Audience chuckles.*] You used to have something on there. What was it that used to be on there? [*Snelson says, "I don't recall."*] You don't remember either? Something like "the basic course of volition" or something like that. It doesn't matter, it's just to attract attention to the fact that there are words there so you don't feel it's empty. Apparently he dropped it. Snelson changed it from "Course V-50" to "The V-50 Lectures." I can't say I approve, but what the hell, he's selling the course. [*AJG laughs.*] As long as he's not distorting the subject matter, it's okay. But that basically deals with something called property.

The Naming of Course V-201

Then the second major course which is actually far more important, but we have to get some early, introductory, elementary concepts to make it possible to understand, is V-201. That subject is called primary property. That course has any number of names you can give it, none of them matter because it's not sold by name either. It's just called V-201 and it has several names, one of which is "The Nature and Protection of Primary Property" or the theory of primary property; this is arbitrary. It's on the subject of primary property and civilization, freedom, durability, and so forth. All of these subjects, and then quite a bit more.

Other Courses Are Farther Developments of V-201

These are the two basic courses in tandem, one after the other. And everything else is other parts of the jigsaw puzzle to fill it out. This is, you might say, the backbone of the subject. And there are historical components, "Positive History" for one. Step three component. Step three means the ideological program, step three, the distribution, the advertising, the image creation, image transference, sales, communications, lumped under the title "Positive Journalism" because basically it's the way to go from the producer to the consumer. That's a very impor-

7

tant subject and actually, it meshes at both ends with the other steps of the ideological program—second, first and fourth. In other words, the third step must mesh with both the first and second steps in proper sequence, it also must mesh with the fourth. That's basically what that subject is about, it's an integration of that. Actually, it's a much larger topic than you think just from hearing that.

As a matter of fact, that's one of the broadest scoped courses I've ever put on the market. Then of course, I have topical subjects in V-212, which deals with extensions in the domain of contractualism in V-201, stressing the significance and importance of contracts and long term, and integrity and sensitivity. These are the subjects that V-212 deals with. Basically, it's a strengthening of the understanding of the integrity machine. It's a farther development of V-201.

This Primary Psychology Course Draws
from All Other Volitional Science Courses

The reason I'm mentioning this is that this course now, which is chronologically after all of these others, is drawing from every one of them. It's drawing from V-30 also which is on investments and insurance for defective flatland products and also from that part of V-201 itself, which is also on investments and insurance for non-defective products. It's drawing upon even historical topics such as the American Revolution, which is interesting, has durable components and has erroneous components which are nondurable and which produce destruction. The durable components of the American Revolution are covered in V-76. Most people think that it's all hot and good. There are nondurable and damaging components to the American Revolution which have led to the witchcraft state which is worse than the tribal chief state. That's covered in V-111. And most recently, last spring, I put on a course which shows the significance of Adolf Hitler and the mess that he created in Germany and by contagion it spread to the rest of the world. That course though seemingly trivial is not because it partially is a historical summary of what led to the theory of freedom and also has many historical inputs, a few samples of which I gave you in the preceding session on political states, political states in general like Germany, in particular. Germany also, but the general discussion of national socialism which is not a German invention, it's a German application of a worldwide disease. It's a worldwide volitional disease.

The Most Sophisticated Form of Attack on Property Ever Designed

It's the most sophisticated form of attack on property ever designed and which includes as a component of it, partial secondary capitalism which is a form of national socialism. Shortened, that's Nazism. Most people have the image, "Oh, that's Hitler. That's Germany." No, it isn't. It's any form of fraud that is perpetrated to gain control of the individual and transfer it to the state with the consent of the individual. The tribal chief at least gives you the dignity of not liking it. The tribal chief state gives you the dignity of letting you know you're enslaved when you are enslaved, and letting you not like your slavery. You have left that one dignity, you don't have to enjoy it. In national socialism, you think you're free when you're enslaved and it is not a German invention. It was simply applied in Germany with the full vigor of a megalomaniac of supreme proportions and major malignancy.

This Psychology Course Is the First Major Course Born in the Open-End Course

All of these have been drawn on, plus many individual topics from my open-end course which is not a secret that it exists, except that there are topics there which are a little more sensitive than normally put elsewhere. Ultimately, if I put something new in there it ultimately becomes a course in its own right. This course is the first major example. I've had some other extensions into other courses. This is the first major course that was born in the open-end course. Now, I'm using in this course every one of these inputs as an entry point, a further expansion of what I have done in the past. That's what the recent few sessions were about which is what I'm explaining here, in case you, in the various topics that I've covered, you might lose the thread of what I'm trying to explain here.

The New Approach to Psychology: Will Simplify, Rather than Complicate, the Subject

I'm showing you that all of these seemingly diverse subjects have a meeting ground in a new area which I've not talked about before and that is the motivation of human conduct, which is what I call psychology. The science that determines and influences the motivation of volitional action. In other words, it's the subvolitional driving mechanism of volition. That is my concept of psychology, let's put it that way. And I'm using all of these inputs to tie it in with other things I've already talked about which establishes a base of operations from which we will converge upon and use all of these as platforms to extend into the new science of psychology. New in the sense that it's a new approach to it, a new approach to an older subject. And I'm trying to show that the new approach will simplify, rather than complicate the subject; explain probably more with fewer inputs. At the very least, I say it might integrate well with part of what is now called classical psychology. But I'm not in a position to discuss that integration because I haven't studied classical psychology. To the extent that it explains human conduct, it has to integrate. Because one of the two has to be wrong if it doesn't integrate.

Psychological Maturity Deals with Emotional Growth

Okay. I've also discussed topics which I have not made a course out of directly yet but which I've mentioned in other courses here and there, such as for example, childish characteristics, maturity problems, or rather problems that have to be surmounted to become mature, which involves of course, the definition of maturity. And of course, you don't have a single definition for maturity in all subjects because we're not talking about the same topic. Biological maturity deals with biological growth. Psychological maturity deals with emotional growth to the point where the emotions, which are basically feelings, which are basically not rationally developed, are not in contradiction to rationality. As I said, emotions do not have to be irrational. They just usually are, but they don't have to be. And maturity, in a sense, is a condition wherein an individual who not only thinks but also emotes—feels—has not got feelings that are contradictory to rational conduct. Or that if he does have such feelings or emotions, he will subordinate the emotions to rational behavior and therefore he will be able to conduct himself in a responsible manner which is defined in the sense that he will take justice properly.

The Justice Apparatus

Justice is the proper compensation for any form of market action, any form of market behavior. Justice is the return flow, in the sense of V-201. Justice is the proper compensation. If a person does something that enriches another person's life and makes him profit therefrom, then he owes gratitude which is payment—primary and secondary in the sense of V-201—to the person from whom he received the value. If on the other hand he has been injured by someone else, either by error, negligence or coercion, then that person who has injured him, there's a justice apparatus there too. But the justice here, the return flow, means that the one who caused the damage does not get paid. He does the paying. The flow is in the other direction, because by damaging someone else he has subtracted from the value and the property that someone else has had and therefore he owes him compensation. He owes him restitution. Restitution is negative compensation. It is compensation going the other way from the one who subtracted value from another instead of added value to another. And that's also justice. And a person is mature when he's capable of recognizing justice to be justice and to abide by the justice mechanism instead of opposing it. And the thing that will cause him to oppose it if he does, which makes him immature, is the ego minor, the universal psychological disease.

Ego Minor Actions Are Childish Behaviors Which Have Not Been Outgrown

All ego minor actions are childish behaviors which have not been outgrown and the maturity has not been attained. And if it's in a child, more than overlook it, you must expect it. The smaller the child, the more certain you will get it. It is impossible for a child to be mature. You don't get born with maturity. You have to acquire both knowledge and successful applications of knowledge to have a positive base called an ego major to stand upon, that you have the strength to expect and understand the meaning of things so that you can be rational about what happens and expect both reward and restitution, as the case may be depending on what you have done. That then becomes maturity. When a person is incapable of it that's ego minor. As I say with a small child that's not only necessary to overlook it, it's expectable, it's inescapable. And a parent, or a prospective parent, must take that into account prior to the bringing of a new creature into the world called a child. The child doesn't ask to be born, the parents have much more to say about it and they must assume the responsibility of expecting immaturity from the child and shielding the world from the irresponsible child's misbehavior. And the parent must assume liability for the child's immaturity and his

11

potential damage to other people. And if the parent doesn't do it, then the parent is a child, psychologically. The parents have choice over whether or not they bring children into the world, the child doesn't.

But after the child experiences life and begins the operation of thinking, gradually a child can acquire self-responsibility in the sense that I've just described and gradually grow to assume a larger and larger percentage of his own life's output and assume both the expectancy of compensation in the sense of positive value rendered to others and also the expectancy of restitution he must pay should he misbehave and be damaging someone else's property. When a person cannot accept the latter—everybody finds it easy to accept payment but very few people are capable of doing the paying when they have misbehaved, which is defined as injuring someone else. And when they have misbehaved and injured someone else they usually get angry, not with themselves but with the one they have victimized. The victim is accused of having injured them!

The Ego Minor of Adolf Hitler

May I point out that Adolf Hitler, whom I will bring up very frequently in this course because he is a major misbehaver. [*AJG chuckles.*] He is one of the most malicious and also most childish people who have ever lived and with the enormous potency as a military conqueror, he is also one of the least grown-up people who have ever lived. When he attacked Norway in 1940 obviously the reason for the attack on Norway was to expand Germany's territorial and military potency and also to achieve a major seaboard base to launch a future impending attack upon England and also to have a further Atlantic coastline and have a base of operations for U-boats. They had many other sub-reasons, additional reasons, including I believe a heavy water plant for atomic research, atomic development, which fortunately did not succeed for them. Damn fortunately.

Anyhow, they had many reasons for taking Norway, all of it aggressive. Not one of them moral. There's no such thing as a moral invasion of someone else's property. How did Mr. Hitler explain, he and his propaganda minister Goebbels—who in Charlie Chaplin's movie *The Great Dictator* was renamed Garbage [*audience laughter*] most appropriately in the Charlie Chaplin caricature of Hitler. Where did Mr. Goebbels, the propaganda minister for Hitler, what did he use for an excuse to justify

Germany's invasion of Norway? He said that Germany was forced to move into Norway to protect Norway from an impending British invasion. How touching. How positively sweet. He was protecting the poor Norwegians against an impending British invasion. What does that tell you?

The Standard Response of a Criminal towards His Victim

That's exactly how a child behaves. A child attacks property either out of ignorance, or out of malice, or out of just insensitivity to the subject of property because the parents have brought the child up poorly, which is a damn near universal characteristic, because the parent doesn't know much about property either. The child injures property of another, he is accused of wrongdoing, and he screams and rants and raves and accuses the victim of having harmed him and screams his head off, goes into a temper tantrum. You ever heard of such things? Those of you who are parents have you observed this characteristic from misbehaving children? Hmm? [*"Yeses" from the audience.*] It's rather common, isn't it? It's even more nauseating when you see a grown human being doing this, isn't it? But they do it. That's the standard response of a criminal towards his victim. It's the victim that forced him to commit the crime and he doesn't call it a crime. He calls it defense. The victim is blamed for his shortcomings.

The Welfare State's General, Whining Philosophy

What do the standard rabble rousers of today say when they attack property in any way? It's society that has deprived them of their fair share and therefore they have simply come into their own to demand what belongs to them, and what other people have which is more than their fair share they have to give up to the poor, underprivileged people who have been robbed. That's the welfare state's general, whining philosophy, isn't it? The weaklings are accusing the producers of having deprived them. Who are the bad guys in today's society? The major producers, the profit mongers. The oil companies have conspired to take away your opportunity to get oil. They have raised the oil prices. This is the new line during the so-called energy crisis and the hoax that they have perpetrated on that. The oil companies are the culprits.

The Cause of the Rise of Oil Prices and
the Oil Shortages

Well, I have no particular liking for oil company management because their capibut, stupid, quasi—capibut means "I believe in capitalism, but." It's my wife's beautiful designation of people who are hypocrites. Who say they believe in capitalism while they denounce it. I have no friendship towards any PSC industrialist who is basically a socialist disguised as a hip-hip-hoorah for free enterprise Chamber of Commerce pseudo-capitalist. But with all the shortcomings of all of these hypocrites who are industrialists and captains of industry, who are basically not real capitalists, at their worst they cannot be as rotten as those who denounce them. Because the oil companies have done nothing of the sort, that they have conspired to raise your oil prices and hurt you and deprive you of your natural rights and all that garbage.

The cause of the rise of the oil prices and the oil shortages are purely, totally manufactured by the various states. Your own, in cahoots with the Arab states which have now blackmailed the whole world by seizing the property in the oil that exists on their geographical soil which they would never have even known was there had not the Western oil companies brought in the technological ability to find that there is oil there, drilled for it, found it, brought it up, shipped it out and made it available to the world.

And then these 12th century sheiks, these black age or Dark Age leftovers, they then seize the products of Western companies' achievements and take it as their own simply because it was on their soil with their knowledge with plenty of royalties to them along the way. How the hell do you think that barbarian the king of Saudi Arabia got to be so rich? You think he got rich by eating sand? [*Some audience laughter.*] Or selling it to his savages? They have one of the lowest cultures in the world. The whole series of Arab states. They have not progressed since their ancestors were around in the 12th century. They haven't had any civilization added to them except what was brought in from the West, which they have now seized by the socialist term, nationalization. That means theft by the state, in simple English. All right, and what do we do? We just allow this to happen. The United States of America, the mighty nation of the Western world, the victor of World War II, kowtows to a bunch of 12th century barbarians.

The Barbary Pirates

I'm interested in pointing out that in the first decade of the 19th century, there was an affair in the Mediterranean wherein some Barbary pirates, who are both ethnic and cultural ancestors of the present barbarians in the Arab countries, the Barbary pirates were sending out pirate vessels to seize the merchant fleet, the merchant ships of the Western countries. British ships, French ships, Italian ships, American ships that sailed the Mediterranean Sea. They used to capture these ships and hold them for ransom and demand tribute. And all these countries paid that tribute to these pirates, the Barbary pirates. The British state did. The British state which was the ruler of the waves, you know, Britannia rules the waves. That was the age of Horatio Nelson, the great British admiral. They paid the tribute. The French paid the tribute with Napoléon as the emperor. The Italians paid the tribute. All the other European nations paid the tribute. And President Jefferson, third president of an infant republic which had just barely got started and was considered to be a hopeless foundling which will disappear soon enough. This little countrylet, President Jefferson said we will not pay tribute and he sent some Marines over there and cleaned out this nest of pirates. That's where the Marine Corps gets its famous words for the song, "From the halls of Montezuma to the Shores of Tripoli."

Well, I don't like war and I don't like warlike action but I'll tell you, when you have pirates, you treat them as pirates. And I'm not talking about the natural republic. President Jefferson did not have access to that concept and therefore he cannot be blamed for not using it. He only had available to him what he knew how, and this little infant country with no major political or military strength would not tolerate this barbarism and this tribute from these pirates and cleaned them out. And today the mighty United States, nuclear power victor of World War II and all this, bows to these 12th century barbarians. That's where the high oil prices come from.

Victimized by Blackmail and the State Adds to It

We are being blackmailed and we are victimized by it and the state adds to it. The American state then compounds the problem by then blaming the oil companies and making it look as though they are depriving us of our oil. And to make matters worse, when the oil companies want to beat the crisis in the most natural market-like manner by looking for new oil which there's plenty of new oil in the world. You can forget the whole of the Arab world and let them drown in their oil. They haven't got the machines to use it. Tell 'em to go to hell. There's plenty of oil elsewhere.

The Venezuelans have joined the Arabs in this blackmail, you can tell them to go to hell. There's plenty of oil on this planet. There's oil in Alaska, there's oil in Canada and there's oil offshore. And there's shale oil. I'm not even an expert on this stuff, but this is just, you might say, everybody knows this after it's pointed out. [*Chuckles from AJG.*] I mean, everybody knows this except I have to add that somebody has to point it out. Now that it's pointed out, I know it too, without being a geologist.

There Is Plenty of Available Energy Sources on This Planet

And more important than that, supposing we did run out of oil, which is not going to happen in the near future. We are nowhere near running out of oil. This planet has centuries' worth of oil left with our present rate of consumption. All the propaganda to the contrary notwithstanding. We're not running out of oil yet, but let's say we did run out of oil in thirty years. So what? In the thirty years of transition time we will have other forms of energy which is not going to run out. We're not going to run out of solar energy and before we can get that, we can use nuclear energy. The solar energy is going to be around for a couple more billion years.

How many of you, aside from what I may have said in other courses you may have heard from me, are aware of the fact that the sun emits every second of time, it emits as much energy as the entire human species if it consumed energy at the present rate of consumption would consume in half a million years. Supposing we got a millionth of that energy for ourselves by trying hard, huh? This Avis thing, you know, trying harder, and by using our brains. There would be such an energy super-abundance we wouldn't know what to do with it. And the sun's not going to run out of radiation energy for a large number of more billions of years. And then we have plenty of time to think about what happens after that. [*Some chuckles in the audience.*]

The Energy Hoax Is a Delusion

This is so stupid a thing, this entire energy hoax. Why do I bring it up? Well, this is one of the delusions that people suffer on a grand scale. And where did they get this delusion from? Oh, "The oil companies are causing this crisis. Everything is caused by the profit mechanism. It's because of greedy profiteers that the earth is being polluted." I have courses in which I cover this. The physics course. V-111 extension #2. V-31 extension. We have quite a few places I've covered this topic. I don't want to go into it anymore, let it just suffice it to say that this is so much nonsense and I've discussed this elsewhere already. Also in various points in my open-end course. Also in the future, I plan to cover it much more deeply still elsewhere.

The Moral Quest for Happiness Is the Only Mechanism for Man's Survival

There is, however, the point that all of these things are blamed on the only mechanism that is capable of driving man forward and that is the pursuit of happiness via profit; moral quest for happiness. That's the only mechanism that is possible for man to survive at all and I can derive that not from economics but, as I say, from thermodynamics. The word *capitalism* is arbitrary but the mechanism of property protection is not arbitrary. That's an absolute concept. We can call it capitalism, or Ajax, or liberalism, or natural society, or spaceland. I think spaceland will be a stronger word later because flatland is a very powerful way of talking about the world of the past, which you are still living in, and spaceland is where we're going. A larger number of degrees of freedom, a larger number of dimensions with more imagination and a more proprietary mechanism.

I think that will be a stronger word than the quasi-political term of capitalism, which is basically an economic origin and I've used it for a larger concept. But since it has a political and economic background, I'm less enamored of the word than originally and I like newer, stronger terms like spaceland, natural society, better. The name is not important, that's an arbitrary semantic identification. As long as you specify what you mean by a word and communicate it properly it's alright. If you don't use a term that will essentially close someone's mind and turn them off image-wise, it's an acceptable term. It isn't the name that counts, it's the meaning of it and the proper communication of that meaning both in rational and image terms.

Subvolitional Motivations against Moral Property Acquisition

The proprietary mechanism which is property protecting and property expanding is the only thing that is capable of providing a propulsion mechanism for volitional apparatuses. This is of course fully explained in my early volitional courses V-50 and V-201. What has to be added to this discussion is what the inner subvolitional motivations are for people to be turned off against property and the proprietary mechanism, the mechanism of moral property acquisition, the mechanism of profit. What turns people off on this? Clearly, delusions which they have acquired. A delusion, I repeat, is something which is false which they, however, believe to be true. What makes someone believe something to be true, when in fact it is not corroborable? The explanation is not difficult. Most people do not use the scientific method to adopt conclusions and therefore they do not resort to corroboration apparatuses, the fourth step of the scientific method and they accept as true what has in their mind been adequately impregnated by numerous exposures regardless of what the content is, but if they've heard it often enough and it's deeply enough implanted in their mind they will believe it regardless of what it is.

The Basic Mechanism of Fraud

No man in history has ever practiced that mechanism more firmly, but he did not originate it, who more strongly and firmly succeeded with this, than Adolf Hitler. The bigger the lie and the more often it is repeated the more likely it will be believed. This is covered in the Hitler course, superficially. It is covered in my journalism course not superficially but very deeply. This is the basic mechanism of fraud. This is the basic mechanism of the creation, implantation of false ideas in the minds of people who will then believe it with fanatic religious fervor that these things are true. Because most people do not know what proof is in terms of observational corroboration. They believe what they hear more often, see more often, and are exposed to more often by both the magnitude and the variety of sources that they hear it from. If they hear it from many different angles, many different points of view, different mouths, different faces, different forms of communication in terms of, let's say, visual communication by reading; billboards, television, hearing it on the radio, gossip at the barber shop, gossip anyplace else, from the pulpit. All of these places, any of them, all of them, when they hear it often enough and it's the prevalent view it will be the majority view. And the majority, you can count on it, is almost always wrong, especially when the ideas are new.

Where the Masses Are Right

When the ideas are sufficiently long-held they gradually, you might say, return back to normalcy and they might accept even something which is right. After centuries of persecution the masses now believe that the earth moves around the sun, after centuries of persecution of those who held that view. After Bruno was burned and Galileo was threatened and Copernicus was put on the Index, and after all of these centuries of attacks and hatred against the people who developed these ideas finally the masses have come around to say, "Yeah, everybody knows the earth goes around the sun. That's news? Why, I got that in kiddie school. Everybody knows that." And so the masses, for a change, are right. But please notice where they are right. Where it's no longer controversial. When it has become part of everyday civilization, where you can't function without that knowledge.

The Majority Not Understanding Earth Is Not the Center of the Universe

Can you imagine what this world would look like if the majority still thought that the earth is the center of the universe? And yet they do. At the same time, they don't. What does that mean? It means they know damn well it isn't so because they've heard it from everybody. That doesn't mean they understand it. That means they heard it, and therefore, because they don't seek proof, they seek majority acclamation, they believe it. And yet, every time they personally tried to think, they think of themselves as the center of the universe and pompously strut about the stage and say, "Us creatures are made in God's image." People still say that, at the same time that they recognize the earth as a minor planet around a minor star in an ordinary galaxy 30,000 light-years from the center of the galaxy. [*AJG chuckles.*] And after they come to accept this—not understand it, accept it—they still have the audacity to think that the only God of the universe would send his only begotten son down to this miserable hole, this planet. [*Audience laughter.*] And that man is made in God's image. What a preposterous nonsense. They still don't believe it after they believe it! I hope the point got through. [*Audience laughs.*] I know it's not the most popular thing to say even in the 20th century. [*AJG chuckles.*]

Delusions of the Masses Come from Propaganda

Now, the delusions come from, in simple terminology, mass implantation of false statements otherwise known as propaganda. The source of the propaganda could be the state or just simply other deluded people. No shortage of that. There is absolutely no shortage of delusion. Again, it's important enough to restate: a *delusion* is an acceptance of something that is not true, observationally noncorroborable, a falsehood to accept that as truth. Is it done maliciously as a conspiracy? Probably not. There are times when it is. There's no question that the Hitler propaganda was malicious and was a deliberate attempt to spread lies for controlling the masses purposes. That is no question it was malicious and there was a deliberate Nazi-inspired conspiracy to delude everyone about what they wanted people to believe. But as far as American socialism, oh, there are individual organized groups of political pressure mechanisms that are spreading propaganda on the same basis the Nazis did, and they are the same caliber as the Nazis. But the majority of the delusions are spread by people who themselves are holding false ideas. They're not doing this deliberately knowing it to be not true. They themselves don't know it.

People Who Do Things for Mass Popularity
Are Dangerous

For example, take Ralph Nader. And I'd rather give him to you. [*Audience laughter.*] I mean, you can have him. [*AJG chuckles.*] Take him, for instance. I don't know if this guy is crazy or criminal. He could be both. It's absolutely a certainty he's not functioning intelligently. My opinion is he's probably sincere in thinking he's a holy crusader and he probably thinks he's saving the world from all kinds of enemies, like the cat, he's fighting imaginary enemies that he invents as he goes along. Unlike the cat, he's not cute [*loud audience laughter*] therefore he's not likeable and yet he's most popular. Incidentally, you gotta worry about popular people, they're dangerous. People who do things for mass popularity are dangerous always, because they won't ever be sincere. They'll say what they want you to hear for them to acquire popularity and through that, power.

It Doesn't Matter Whether It's Done by Conspiracy or by Stupidity, or by Sincere Fanaticism

Now, whether Ralph Nader really believes what he says or what he's doing is just to gain control over the masses, I cannot say. My personal impression is without ever meeting him, which is the way I'd like to keep it [*some chuckles in the audience*] is I think he believes what he says. He sounds like one of those crazy fanatics who thinks he's God's chosen savior for mankind. He's the new Jesus, perhaps, I don't know. Maybe he's trying to make himself deified, or president, or one or the other, or both. He probably thinks he's sincere. I mean, I think he probably is sincere if you define sincere that he really believes what he's saying. But whether he does or not is immaterial! That's one of the beauties of my theory, by the way. If people ask me, "Do you believe in a conspiracy?" I say, "I'm sure I don't know." The important thing is, it doesn't matter. Whether it's done by conspiracy or by stupidity, or by sincere fanaticism, which means that a guy who is a fanatic really believes what he's saying. He thinks he's God or he thinks he's a chosen appointee of God to deliver man from sin or from hell or from whatever. It doesn't matter what his motivation is, which is a psychological thing. The thing is, what makes you believe that jerk? That is what matters. And what is the mechanism to overcome this hazard? It does not matter whether he's sincere or not.

For example, take the most massive criminal I have mentioned on a political scale which was Hitler. You could ask the same question about him. Did he really believe he was the great savior of mankind and the builder of the master race? Builder of the master race, that scrawny looking worm? He was the exact opposite of what he described to be the master race. Blonde giants. He's a dark-haired shrimp. [*Audience laughter.*] He's a weasel and a weakling. A physical weakling and a moral weakling. That shows how clever the propaganda is if you can pass that guy off as the head of the master race. Did you ever see the rest of his gang? In V-113 I showed a rogue's gallery. The most beautiful specimen was Goebbels. He looked like an emaciated rat. [*Chuckles in the audience*.] Then there was Himmler. You have to see these to believe it and not one of them looked like the master race! [*Galambos laughs*.] But that didn't stop people from believing them. If you repeat a lie often enough, etc., you know, and it's big enough.

Who Do You Reach the Masses With?

By the way, if you worry about how to reach the masses, don't worry. That's easy. That's in V-50, Session 15. And if you need a bigger dosage, and you sure as hell do, it's the 28th Session of V-283. The reaching of the masses is the easiest thing in the world to do. That should not worry you five minutes. The hard thing is who do you reach the masses with? That's the second step of the ideological program, not the third. And the second step of the ideological program is secondary production. The people who do that don't know what they're doing because they're in the trivial time scale. They are basically short-term quick-buck artists. They're basically promoters, hucksters, peddlers, get-rich-quick boys, schemers, and in general their business ethics makes a skunk smell clean.

Technological Solution to the Entire World's Volitional Disease

Then there are the higher-class businessmen who might have some personal ethics, and there are some. But they have a perspective which in general is longer and therefore more commendable. But it doesn't penetrate beyond the personal time scale because they don't have the imagination for it. And they cannot identify with anything beyond their own lifespan. And so although they're more ethical, and therefore more durable, and therefore more productive, and therefore their net worth to civilization is considerably greater—that's the real backbone of our present flatland civilization. Those businessmen who develop businesses and industries in the personal time scale, these people still haven't got the potency and the strength and the dynamism and the imagination to carry them to the species time scale because they're not on the same time scale as the innovators. And of course 201 integrates that. The real problem is to develop the second step of the ideological program to mesh with the first and enter the species time scale. That's the solution. That's the technological solution to the entire world's volitional disease, and not how you reach the masses.

You Can Get Anyone to Believe Anything

But then still has to come the discussion, which is this course, of what is it that motivates people to believe what they do believe? Well, clearly, the masses will believe anything that's told to them often enough and with big enough noise and fanfare. They'll believe that the North Pole is in the Southern Hemisphere if you tell it to them and by the way, it is. Did you know that? It is. The North Pole is in the Southern Hemisphere and that's a fact. How many of you knew that? The South Pole is in the Northern Hemisphere and the North, oh nevermind. That's because the North Pole attracts the South Pole. It's in the physics course. Nevermind. That really is a digression which is not necessary here. But I'm just trying to say you can get anyone to believe anything, you can get 'em to believe black is white. That's a little harder to imagine. Or that a dark room is bright and a bright room is dark. They can prove that now. [*Galambos turns off the lights in the room.*] This is a bright room. [*Audience laughter.*] Now I have shut the lights off and it's dark. Now you laugh because it's the other way around. For benefit of those who hear it on tape it's the other way around. When I said it was bright I shut the lights off and when I said it was dark the lights were on.

But you see if I had the cleverness of Hitler, and I repeated that to you often enough, it was a preposterous statement. But if I really hammered that into you and you were not intellectuals, which I hope I might perhaps [*audience laughs loudly*] dream about the illusion that you might be perhaps, maybe? [*More laughter.*] But if you were all the average flatland types and I were a Hitler—two conditionals, which I hope don't happen—then I could convince you if I said that often enough and persuasively enough, that when you're sitting in the dark it's brightly illuminated, and vice versa. When it's brightly lit, I tell you it's dark and you can't see.

And you may say, "Well, that's ridiculous. Nobody would fall for that." No? No, indeed. Well, how come that people think that if they have Social Security they have been granted freedom and if they don't get Social Security their rights have been deprived and they are enslaved, when in fact, the exact reverse is the case. To get Social Security means you have to be robbed all your life of the money you could have invested and if you don't have Social Security, the real thing is, you have reacquired control over your investable money. And you get Social Security you have been deprived of it and yet people think it's the other

23

way around. That's no more preposterous than black is white and black is brightness and brightness is darkness. That's no more preposterous.

A Hitler could get every man in this room to believe he's a woman and every woman in this room to believe she's a man. They could get it completely backwards. Well maybe not you, I hope. I hope. I granted too much credit to F-201 once and I don't want to fall for that trap. So I just say I hope. [*AJG drinks.*] But with the mentality that accepted Nazism, yes, they could get you to believe that. You say that's ridiculous. No, it's no more ridiculous than what they did believe. They believed the most preposterous fairy tales in the world except that there were nightmares attached to those fairy tales. They were absolutely out of the realm of observational corroborability but that didn't prevent anyone from believing it.

Hermann Oberth Believed Hitler Was Defending Germany

Listen, there are intelligent Germans today that still think that Hitler was defending Germany against Allied aggression. I just talked with some a few weeks ago in Germany. You may not even believe who it might be. Well, this may shock you. Hermann Oberth, one of the pioneers of astronautics. I asked Professor Oberth, "How did you react to the fact that your intellectual property"—I didn't say primary property, that would have meant nothing to him—"How did you react to the fact that your ideas"—is, I think, the way I put it—"were used for mass murder?" Because it is his rocket concept that was used by Hitler to unleash the terror of the V-2 upon England. "How did you react to the use of your ideas for mass murder?"

And he said, unhappily, "They didn't ask me."

And I said, "I know that. I know that sir. They didn't ask you. I expected that answer. Nobody asked your permission." That's in 201, incidentally. That the innovators are not asked for their permission, for the misuse of their ideas. I said, "I know that. But now that it was done without your permission, how do you feel about it? That it was used for that?" And I got an answer I did not like.

He said, "Well, I suppose it had to be done. Hitler tried to win the war. He had to do that for Germany. We were at war. We had to defend ourselves."

I said, "But Germany wasn't defending itself. It was the attacker. Germany had started the war."

"Oh no, no, no, no—Germany was attacked by the Allies, and Germany was on the defensive."

I said, "What were you doing a thousand miles into Russia? [*Slight chuckle from AJG.*] And all the way into France?"

"Well, that was—we had to defend ourselves."

This is not a slob on the street. This is one of the geniuses of our century who believed this. I do not think he's a criminal. I'd like to add that. I may be in error, but I would like to give the great man the benefit of the doubt. I am not convinced he is a criminal. I don't believe he approves of murder. But I do believe that he was deluded as the lowest proletarian into the false belief that Germany was fighting a defensive war against Allied aggression, which I think he does believe, because that was a propaganda which was implanted on a mass basis into the German mentality.

I mentioned the destruction of six million Jews to him and he said, "That didn't happen."

I said, "That's preposterous. Of course it happened."

And he said, "That's Allied propaganda."

By the way, that's not the only intelligent German I've ever heard say that. I have spoken to several who believe that. Cultured, professional, intelligent Germans. This man is not intelligent. He's a genius. Intelligence is an understatement. This one, Oberth. He said that didn't happen. That's Allied propaganda.

And I said, "What about Dachau, which is not too far from where you are?"

"That was built by the Allies after the war to show the world what we're supposed to have done."

All right. Do you believe that it works? Now that's direct transmission of what I heard with my ears, not hearsay, directly from Hermann Oberth. I wish to safeguard his reputation for posterity by saying my opinion, which is my, not necessarily a fact, but my opinion. I don't believe he is a criminal. I don't think he even endorses criminals. I think he believes this. I think this is a propaganda which he swallowed hook line and sinker.

But it shows one of two things: either he's an accessory to the crime—which I choose not to believe—or else it shows that even such a brilliant man, such a genius, is capable of being deceived on such a grand scale.

25

The Potency of Propaganda

And this is not the day after World War II ended, you know. This is 1975. This is thirty years after the war ended. Thirty years after the war ended! A large number of the people in this room were born since then, and are younger than the period of time I just named. And thirty years later, this genius of the 20th century, this benefactor of mankind, one of the three major men who got man off the earth—the other two being Ziolkovsky and Goddard—believed this about his political master, Adolf Hitler.

Does that, I hope, for those of you who had V-113 last spring, give you an interesting supplement to my course? [*"Yeses" in the audience.*] Which has come about since my recent visit to Oberth. And adds an interesting additional input into everything else I've said about the potency of propaganda. And what I talked about in V-113 and the positive journalism course of V-283 of the potency of propaganda. Someone else, not me, said but it's correct: "Propaganda made the Third Reich." The Third Reich is the thousand-year Reich which fortunately culminated 988 years earlier, and twelve years too late. But for the duration that it existed, those twelve miserable horrible years which damn near wrecked Western civilization, almost killed the most important part of the Western world. That twelve years would have not been possible but for mass propaganda.

The Communists' Use of Propaganda Is the Supreme Witchcraft

And by the way, the communists know this just as well or better. The communists I think have by now exceeded Hitler. They are more subtle about it but even more potent. And the welfare state philosophy is even more insidious, because it pretends it's pro-capitalist. That's the highest form of witchcraft there is, where you take a watered-down facsimile and counterfeit of capitalism—which is in fact socialism—and make that look as though it's capitalism. And when it doesn't work you blame it on capitalism. And capitalism takes the blame for all the blunders committed in socialism by calling it capitalism! Now that's the supreme witchcraft. Of course, everything is going to hell in this country and what's it being blamed on? Socialism? No. On the profit motive. Simple, no? When you analyze it clearly.

V-201 is Short, Not Long, When Analyzed Correctly

Why is it that people think my lectures are too long when you hear thousands of political speeches which say nothing but lies and horror and falsehoods and empty statements, and people listen to it with rapt attention by the millions. And here in a few short hours I can explain all of the diseases of mankind. I once pointed out for those of you who are subjected to the delusion of how long V-201 is—that is a delusion by the way. It's a false statement you believe to be true. I pointed out that the whole of V-201 in its most recent presentation, which was naturally the longest—I believe it's 56 sessions—and say it's four hours long per session. That's an exaggeration because an intermission takes part of that. But let's say it's four hours long per session—56 x 4 are 224 hours.

Ladies and gentlemen, that's only nine days if you divide that by 24. It's only a little more than nine days of your life. When you consider that even the youngest of you is at least old enough to be two decades old, because that's our limit to let you into 201, nineteen, by the time you finish you're twenty. Even the youngest of you who takes the course is twenty years old by the time it finishes. What's nine days out of twenty years?

And those of you who are thirty years old, or forty years old, or fifty years old, what's nine days out of—look at how much you learned in your twenty to fifty or sixty years or whatever age you have. Look how much you learned in all those years, which is all an error about human civilization. And then in nine days you got it cleaned up, and you say that's too long. [*Chuckles in the audience.*] You see the injustice of that? [*More chuckles in the audience.*]

Now there's an example of delusion, amongst my own graduates who fancy themselves intellectuals, and I wish they were. And sometimes maybe they will be, some of them. Perhaps. I hope. Maybe? If I'm lucky? [*AJG chuckles.*] If you're lucky? [*AJG drinks.*] So don't get too cocky about this. Everybody's capable of being deluded about something. [*AJG takes another drink.*]

I'll continue after the intermission. [*Applause from the audience for fifteen seconds.*]

PART B

I have discussed the fact that this course is now drawing from a large number of other areas which I have covered in other courses, which are basically all the same theory but different components thereof. Or, to use the analogy of the forest, are different trees in the same forest and I'm combining the results from the various components of the forest. This is part of the bringing together of many inputs and coming out with a single unified subject from many different sources which then will in turn be used to create an understanding of a new subject. This is what I'm going through with the present phase of this course, now and for a while to come.

Ego Minor: A Continuation of Childhood

Well, let's now discuss, for example, the fact that since it is clear from what I have previously discussed that ego minor which is normal in a child and is a disease in a supposed adult, it means a continuation of childhood beyond reasonable expectation. You may say, "Well, that's all there is to it? You call that a disease that somebody remains a child? Children are nice." [*Audience laughter.*] Well, I don't think so. Children are not rational. You say, "Don't you like children?" [*AJG makes a face followed by loud audience laughter*.] Not particularly. They annoy me. Because it's hard to deal with people that are neither rational, nor have a proprietary interest in their upbringing.

Every Parent Should Have a Proprietary Interest in Their Child

Now if I were their parent, that's different. Then I would have a proprietary both responsibility and interest in what comes out of their future. Then that's my child. Not my property, but my flowstream. I want to make sure that I didn't bring someone into the world that I'm ashamed of. And that's different. Every parent should have a proprietary interest in his child or her child. But why should I take an interest in irrationality of other people's children? If I wanted to take that trouble, I would expend it on my own. That means everybody is the guardian of his own property, his own morality and his own children. But in a child, though I won't say I enjoy having to put up with the fact that they know little and I have to waste my time teaching them—you see, I'm not teaching children here, I hope. I don't think I would make a good

28

teacher of children unless they were my children, in which case I would make a superb teacher. The greatest teacher I ever had was my Father. It was a class of one. He had a good class. I had a great teacher. I'd feel the same way towards my child. But when I have to undertake this with someone else's child, that's not my problem. Nor is it my proprietary flowline. I have too much to do without that.

Everything Has a Market Function and a Market Value for Those Who Wish to Do It

Now some people enjoy doing that. Well, that's fine. That's what they get paid for. That's what teachers for children should do. I mean, after all, that's what they want to do for a living, they should be good at it. I mean, there are a lot of things I don't want to do, but I like the service. I mean, I really do not desire to fix my watches when they break but I'm glad there are people who do know how to do this and want to do it and I'll be happy to pay them rather than learn myself. Same goes for lens grinding, making of clothing, shoes, cooking of meals. [*AJG takes a drink.*] Everything has a market function and a market value for those who wish to do it.

The Parent Should Be Responsible for the Misdeeds of Their Children

But if you say, "Do children have nice characteristics?" Yes, in some cases. No, in more cases. "Do they behave nicely?" Sometimes they're very sweet and then they fall into an abhorrent mood of irrationality, temper outburst, pure insensitivity; but you must overlook it. How else would they know? If it's your own child you make sure they don't annoy someone else. If they do, the adult parent should be responsible for the misdeeds of their children. That's part of the responsibility of being adult enough to be a parent.

Almost All Adults Are Emotional Children

Unfortunately, most people who are parents are not any more than emotional children themselves. So you see, that's why the ego minor is a disease. Because it's people who shouldn't have this childish characteristic. And you may say, "Well how prevalent is this?" Almost universal. A truly mature and adult parent, or parents, plural, would rear their

29

children in such a way that they smoothly and hopefully soon grow up. Now you can't expect it to happen in two years or even five years. But you might begin to hope for an improvement by the teens, and by the time one enters around fifteen or sixteen there should be some semblance of growing up. And certainly by the time they approach biological maturity, around two decades, the process of growing up should be also terminating psychologically. The biological maturity will terminate, nature takes care of that. Emotional maturity does not necessarily terminate at age two decades, and for some people it never terminates at all. And they never grow up and they die as irresponsible brats in grown-up human bodies who demand something for nothing, who cannot take responsibility for their actions, will not admit error, will not understand the concept of property, unless it's convenient.

The Ego Minor Mentality Applied to Innovation

What is that supposed to mean? It's supposed to mean this: everybody loves property when he is the recipient, when he's on the receiving end. When he's on the receiving end. Remember the two pigs story? "But I have two pigs." Well, that's different. Then it's your property. When it's somebody else's property, that can be shared. When it's somebody else's, then of course he owes it to the world. Isn't that the prevalent view in the highest form of property generation, called primary? Isn't the world's attitude towards science and scientists that the discovery belongs to the world? If the scientist fails to disclose it he has harmed mankind. The scientist harms mankind if he refuses to disclose because he doesn't like to be plundered. In other words, Newton was a criminal because he went to the mint and refused to help the Linuses. That is the ego minor mentality applied to innovation.

The world is full of plunderers. That's the easiest thing in the world to do. Steal. It's easier to steal a car than to make one. It's easier to steal an idea than to have one. And what is more important, most thieves are so stupid they don't even know they have stolen. You say, "How can that be?" For example, with an automobile that's pretty hard to imagine. The fellow who doesn't own the automobile is driving along and somebody asks him, "Say, do you own this automobile?" "What do you mean, do I own this automobile?" "Well show me your registration." "I don't have one." "Well it's not your automobile." "Yes, it is." Well, you can see that this is a little hard to get away with this, even in flatland.

"Well I didn't know it wasn't my car." "What do you mean, you didn't know it wasn't your car?" [*AJG chuckles*.] That's not too easy to get across, is it? With an idea, that's different. "It belongs to everybody."

Most Scientists Are Altruists

And by the way, unfortunately—and this is the source of man's destruction so far, it's in 201—the majority of the scientists have been brainwashed to believe this. That's no more difficult to believe when you look upon it, than the business that I said about Professor Oberth. That he swallowed the Nazi propaganda. Don't forget, the Nazi propaganda was very potent and there was nothing else circulating. There was only one source of information in a country like that. Well, the prevalent view on science is that this is done for mankind. Most scientists think they're altruists. This is what Ayn Rand exposed, incidentally. The sham, the falseness of altruism; that's the greatest achievement as I see it. The hypocrisy, the sham, the falseness of altruism.

A Scientist Does Not Owe What He Does to Mankind— He Owes It to His Personal Flowstream

Now actually, I'm not the slightest bit offended by the expression *humanitarian*, when it's defined right. Just when it's defined in an altruistic way do I find it offensive. Unlike Ayn Rand, I do not consider it altruistic if one says, as a scientist, "I am doing this for mankind." Because, believe it or not, I am doing what I'm doing for mankind. And you may say, "What, are you an altruist too? Did you just switch colors?" No. I'm doing this for mankind, this is absolutely correct and I don't consider that injurious to what I said that I'm an egoist. "Well, aren't you doing this for yourself?" Yes, I am. "Now wait a minute. You're doing it for yourself and mankind?" Yes. "Well, aren't they in opposition to each other? How can you do it for both?" Because they're consistent. The true scientist does do it for mankind, whatever he does, but it does not mean he owes it to mankind. He doesn't do this as a servant or a slave of mankind. He doesn't do it because he owes it to society. Because a scientist, an innovator, an achiever—not necessarily a scientist, any achiever, any producer—does not owe what he does to mankind. He owes it to his personal flowstream. What does that mean? It means the V-201 concept of the ideological flowstream. I'm referring to that.

31

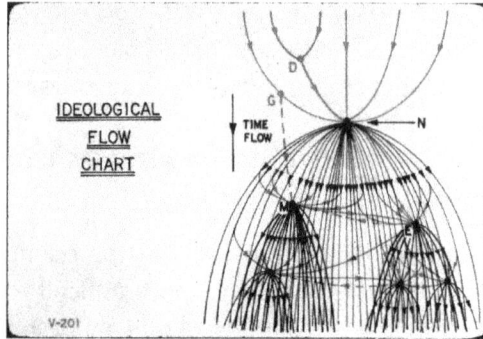

IDEOLOGICAL FLOW CHART

V-201

Compensation and Consent for Use of Ideas

Every innovator has a product, an intellectual product. Every secondary producer has a secondary product; this can apply to either. This can also apply to secondary production. The product is located here: N for now. The original way this drawing was interpreted is N for Newton. However, it can also mean now; the product that is now under consideration. This can represent the entire body of knowledge of mankind. It can also represent the body of knowledge that made it possible to produce a product at the present time, meaning now, whatever product is being created whether it's secondary or primary. And this is the outflow from that product which are its future derivatives. If it's a good product, it will have many derivatives called farther accounts.

But before it came into existence, the creator of that product—whether intellectual or secondary, whether primary product or secondary product—had to draw on his earlier inputs. Those who had knowledge before him that he used and integrated to create the product that he is now creating and putting forth as a larger product which incorporates knowledge from the past, adds knowledge that he created, and makes a larger and more significant product available for market development and distribution.

That person who puts that product out, whether it's an innovator who puts out intellectual products, or whether it is an entrepreneur who puts out market saleable products drawing upon past achievements of others, either way, that person who has used ideas and knowledge and intellectual or primary products from the past owes compensation to his predecessors. He also requires their consent. That's an important point in V-201, too. It's not adequate to pay a royalty. It is necessary to do it with permission.

ARD: Automatic Remoteness Dilution

If there has been no release to ARD, it is automatically not automatic. It is automatically negotiable. Property belongs to its owner. The most basic concept of the theory of volition: all property—primary, secondary and even primordial—belongs to its owner and to no one else and only the owner has jurisdiction over it. And if he has not said, "You may use it without consulting me," then that means you must consult me, which means negotiated discussion is necessary. If he finds that he has so many customers that he has safety in a market response to his product, and we have a better and stronger world where plunder is rarer and more easily discourageable and deterred, he will find it is to his interest to release it to a more general market so he doesn't have to be annoyed with individual separate negotiations, person by person, item by item, and that is called the release mechanism. That's the releasing it for automatic use either on a restricted or unrestricted basis. By unrestricted I mean to anyone at all, or on a restricted basis to certain kinds of people or certain groups of people either mentioned by name or by category or by some form of clear description as to who they would include.

The Two Standard Restrictions

Now for example, I have stated in V-201 that I recommend that anyone and everyone who releases any property for any use should make two standard restrictions. The two standard restrictions are, that there should be no coercive usage, and the other restriction should be that it should be for a positive, nonzero royalty which can be determined by market usage and that's the mechanism which leads to ARD. That's quite safe after there's a wide enough expansion of this mechanism. I pointed out in V-201 why that would be safe. Not immediately, but later.

NRD:
The Stage We Have to Go through to Get to ARD

There might come a day someday, not now certainly, when it will be proper, possibly proper to have every product released to ARD immediately because the world has become so civilized that plunder is so unlikely and so easily discourageable and deterred that plunder would not be a problem anymore. That's not in the near future, but it could be in the more remote future. I don't think that is at all impossible or even unlikely. But right now, NRD is the stage we have to go through to get to ARD. But if you have the consent of the owner, either by negotiation or because he released it to a category which does not exclude you, then that is consent. Without the release to ARD there is no consent, unless you have an individual negotiated contract.

Moral Usage of Ideas Requires Consent from the Owner

Many people are under the misconception that all you have to do is to pay a royalty and it's okay. That is not okay. That's no more right than somebody taking your automobile out of the garage and driving it off and then later sending you the blue book value of it. And he says, "Well, wait a minute. I didn't steal this car. Don't accuse me of being a thief. I paid you for it." I'm sorry you didn't pay me for it because I didn't sell it. I didn't offer it to you for sale. And then if you take my car and pay for it what you think is the proper value does not make it right because I wasn't offering it to you for sale. When a man says, "I do not sell" that means "I have set the price at infinity" which is above the payment capability of anybody. You understand that? When something is not for sale, that means the price is infinite.

The Haggling Mechanism: An Ego Minor Disease

Now of course if the guy's just haggling, and that's a very nasty characteristic of cheap type people. By the way, that's an ego minor disease too, the haggling mechanism. That's a very undignified way of conducting business and it will perish with flatland. That's a prediction. That will die out, not from one day to the next but it's a disgusting way of doing business. All right, let's skip that for now.

Using Ideas without Consent:
Same as the Statist Concept of Eminent Domain

If a person says I don't wish to sell, that's his property and that's his decision. If someone decides that he will take it anyhow, that's theft regardless of what he pays. That's the same as the state does with the concept of eminent domain. They say, "We condemned the property and take it for public use."

"Yeah, but I don't want to sell it," says the poor hapless owner.

"Well you don't have a choice in that matter. Your choice is either to accept the payment or not accept the payment but you're leaving the property."

That's clearly enough theft, isn't it? Even though it's permitted by the U.S. laws, even though the full weight and authority of the American legal system and the jurisprudence backs it. You recognize it: the owner has not consented to the sale.

Acceptance of Theft by the State Is Delusional

Now if he does and he says, "Okay, I accept the price and I accept the authority of the state to do it." Well, that's delusion but it's his property and by fraud it has been acquired from him because he thinks he's free. He may think it's okay and he will let it go but it's still a delusion. He did not have a free choice. Even if he says it's okay because he wasn't given the alternative of saying it's not okay. That's clear, I presume. Well, it's no different than intellectual property.

Secrecy the Only Tool Available to an Innovator in Flatland

And you say, "Why would a person hold something secret?" Well, that's a very good question. Secrecy is not a potent tool and yet that's the only tool available to an innovator in flatland, to withhold from an ungrateful motley crew of people who would use, without permission, ideas. And when an innovator gets tired enough of being plundered he does what Newton did or he does what Tesla did, or something different. The Wright brothers reacted still differently. Everyone has some slight variation in his response to this, but no innovator likes to be plundered.

People Cannot Identify with Primary Property

Why should that be unusual? Do you like your car to be stolen? Would you like it if you go home and you find someone has squatted in your home and is sleeping in your bed without your consent, and you can't get rid of him? Do you consider that a pleasant thing? Why should it be any different for primary property? It isn't. But the sensitivity is weaker because most people have little—I won't say no primary property—but little primary property. And they can therefore not identify with it, and they look upon the owner of primary property as some kind

35

of a screwball or an oddball or a neurotic and say, "Well what's the matter with you? After all, this belongs to everybody. You owe it to the world." That's what I'm saying. No, he doesn't.

The Bridge to Freedom:
The Very Difficult Transition Period

He owes it to no one except those he got value from and that's his upstream flow derivatives. Those whose ideas led to what he did. To those people he has a debt and to those people he should pay and from those people he should have consent. And you say, "How do you obtain consent from somebody who died two thousand years ago, or for that matter one hundred years ago, or died even yesterday? How do you get consent from such a person?" Well, in the future you will know that this mechanism will be available through the natural estate mechanism through trusteeship which is set up in one's lifetime. But how about the past? We're now in the very difficult transition period, the bridge to freedom period. We have entered upon the bridge to freedom, but it's a long bridge. It's an uphill bridge and it's a difficult bridge to traverse and it will be a long crossing, and an unpleasant crossing.

The Concept of Property Is Lost in Flatland

How do we do it now? How do you get permission from Isaac Newton or from Archimedes? That is very easy, ladies and gentlemen. That's in V-201 but this is to bring you up to date, refresh your memory and let us start from a new base to continue from. That's very easy. In the past everyone lived in flatland. Every human being who has ever lived has lived in nothing else than what I call flatland. The world where property is not considered sacrosanct, in general, and primary property in particular. Primary property isn't even recognized as property by most people at all, it's not even recognized as property. To a major intellectual, it's the higher things in life. To a nonintellectual, he owes it to the world. All right. So therefore, the concept of property is lost in flatland, in this quagmire of false conceptions and delusions.

In Flatland All Ideas Are in the Public Domain

Therefore, in flatland, there never has been a proprietary handling of primary property, and therefore it's very simple. Under flatland law, under flatland custom, in short, in the world of flatland, all ideas are in the public domain except those very few ideas which are subject to patent or copyright or trademark or service mark. And these are very minor ideas and the patents and copyrights have expiration periods which are relatively short duration in terms of the species time scale. The laws of various countries differ and in one country the copyright lasts this long, in another country it lasts that long, but in all countries, ultimately it expires and ultimately soon in the species time scale. If you wait long enough, everything is in the public domain.

Ideas and Concepts Not Protectable in Flatland

Most things never get into the realm of copyrights and patents and other so-called state protections. Ideas, per se, cannot be protected. Only applications of ideas. Concepts are not protectable. Cosmological concepts are uniformly and universally not protectable. Laws of nature fall in that category. That's only the most important things there are. There's nothing more important than man's repertoire of property than the laws of nature. So therefore you just lost the ball game. To use a disgusting vernacular, you just lost the game right there. [*Audience laughs and AJG chuckles.*] And no matter what happens after that you can't win. That's the hot end just being busted up by the cold end.

Well, that means in flatland everything that's important is in the public domain from the moment it gets disclosed and the less important things which are patentable or copyrightable, which are merely applications or specific illustrations of things, they are in the protectable domain only for a short time and the protection is dubious, illusory, expensive and untenable. Otherwise, it's fine.

Flatland Laws Protect Criminals Not Property Holders

If you are attacked, the thief has every advantage over you. The law is always structured to favor the dishonest. A man who goes bankrupt is better protected than his victim. A plunderer is better protected than his victim, whether the plunder is ideas or automobiles. There are laws to protect criminals but no laws to protect property holders. So that means even when there is protection, it's illusory, and if it weren't, if it were any good at all, if there were a miscarriage of injustice and the one who owns the property by some fluke won, in general, the party who stole it from him is insolvent and he could declare bankruptcy and get off the hook or escape into another state and flee, or ignore the judgment and the law will do nothing, in general. In other words, the protection is again illusory, even where there is any. Even if this were not true in some exceptional case, some lifeboat-squared or -cubed case, some very rare, extreme case where there might be some minuscule semblance of justice, the patent and copyright will expire someday, and *someday* is soon in species time scale.

Moral Innovator Treating Ideas as Property

Which means in the long run, which will set in relatively soon in terms of this discussion, everything, I mean everything, is ultimately in the public domain. And that's very simple for a moral person to handle then, since everything is in the public domain ultimately, if not immediately, and what's in the public domain, a moral entrepreneur or a moral innovator will treat as though it were ARD. In other words, he will voluntarily, even though it was not released to ARD, but he will treat it as property. In other words, the simple, beautiful, clean transition only for the hot-enders who know what they're doing and who are the major producers of the future, they will treat what other people have lost as property and not steal it and pay for it. And then negotiation is not necessary, but that's only because no negotiation was asked for because they didn't know the mechanism was available.

No Mechanism for Protection of Ideas Existed
when Newton Published His Book

For example, when Newton published his book, he did not publish in his book any restrictions on the usage. Why? Because, great man, illustrious genius, integrator of the knowledge of the world that he was, he was not God and he could not do everything because there were things he had not yet known. And things that will later be developed by somebody else who's standing on his shoulders, and he did not know that there is such a mechanism or that there could be such a mechanism or that there ought to be such a mechanism. I think he might have figured that there ought to be but he didn't know what to do about it. He didn't know there could be a mechanism and therefore his choice was to publish or not publish. Not to do it by release to ARD vs. not to release by ARD. He had no such choice. He didn't know about it and so he published it. Very reluctantly, I might add. Very, very reluctantly. He's one of the most reluctant authors in history.

Edmond Halley: A Double Great Man

The world damn near lost the Newtonian integration but for Edmond Halley who himself should be remembered as a double great man. A great man in his own right for his own discoveries, his own achievements—much lesser than Newton's to be sure—but a great man nonetheless. If he were not being compared with Newton, he'd look like a giant. Only next to Newton does he look small. Compared with other people, he's a giant and he's another great man in another sense, he's a double great man, because he took a still greater man, Newton, and preserved him from oblivion and preserved the world from the loss of the Newton. And were it not for Halley, we would have lost Newton's achievement. Were it only up to the Linuses, we would have lost Newton.

Linus: The Generic Type of Primary Murderer

A Linus is a primary murderer. He's a murderer of primary property sources. I don't mean he would have killed Newton, the primordial property of Newton which is his body, his life. He would have killed his achievement. That's a concept of murder not used in flatland. That's a bigger concept of murder, ladies and gentlemen, than even the one that is considered in flatland. The destruction of Newton's achievement, had it happened, would have been a far greater loss to mankind than the death of most people or many people. I would say the loss of Newton's primary property would have been a greater tragedy to mankind than the combined deaths due to all the wars in history except if those deaths included an Archimedes or a Newton sometime, measured in terms of property. And yet it came damn close to happening and we don't know how many times it did happen to other people of Newton's caliber whom we don't know existed for the very reason that they also had a Linus or more than one Linus. Linus is not the only one, he's just the generic type. When a person is like Linus, it would have been better for him not to have been remembered at all.

What Newton Didn't Publish Is Enormously Larger, Probably

When a person such as a Newton, a Newton caliber person is lost to the world, the world loses the achievement, and the potential Newton caliber person has lost his life's work and he lived in vain. Except for his own private satisfaction, which Newton referred to, that he would only do things which were for his own private satisfaction or leave to come out after him, which in fact he did not leave much to come out after him after he stopped writing. And he wouldn't have even written this much but for Halley. You might say that even though he did publish what he did publish, what he didn't publish is enormously larger, probably. You know, he spent forty years at the mint and in the parliament and administrative tasks as president of the Royal Society as a purely nonsensical office to hold, as an administrator of a scientific body that votes on things. What more preposterous thing can you find scientists doing? [*AJG chuckles.*] Can you imagine voting on the law of gravitation?

Most Scientists Believe Their Achievements Belong to Mankind

In general, when I said before that a scientist's work is done for himself but it's also done for mankind, there is an apparent contradiction, but it's only an apparent one. There's no real contradiction. The reason there is no real contradiction is because the scientist, from his own point of view, if he is not altruistically deluded, he is doing it for his own ego. If he is altruistically deluded, he will say he's doing it for mankind and think he owes it to the world which is the delusion the scientist holds, and scientists are not immune from delusions either. They are brainwashed to say you owe this to mankind, and I daresay most scientists believe that. And they have readily consented to become enslaved by their inferiors, by their intellectual and moral inferiors.

The United States Approaching Its Termination as a Culture

Which is the main topic of the decay of civilization pursuant to the American Revolution, which is what happened after the American Revolution with our present society which led to the high production and low durability, which is covered basically, but not only, in V-111. How the United States with its tremendous, potent beginning, fizzled out and in the end of its second century, is approaching its termination as a culture. I'm sure that that will not be the main theme of the noise next year in the bicentennial year, how close we are to the end, and yet that is the truth. I'm talking about the political culture. I'm not talking about the primary development that here is potent and alive and capable of being built upon. That can continue.

For the First Time in History, Mankind Can Build a Superior Civilization Before It Dies

That's why for the first time in history, mankind in general and you in particular, since you represent the small component of mankind that knows about this theory at this time, you have the unique—to this date in history—the absolutely unique opportunity to be unlucky enough to be living, through no choice of your own, at the terminal stage of a dying culture. Since you could not choose the date of your birth nor could I. You could not help that you came into the United States' history not at its birth, but at its death, and that you're at the end of the civilization. That is not your fault nor mine. But you have the first and only and unique, thus far, opportunity to say I won't accept that. Because for the first time in the entire history of man a civilization is dying and a superior one can be built before it dies. Now that's the question. You can just sit on your you-know-what, or do something about it and build something. That depends on how strongly you react not just to the theory of primary property, but to the much more painful subvolitional derivative called the theory of psychology. Because you're not going to be coddled and told that you're the victim of your environment.

You Are the Master of Your Own Fate

You are the master of your own fate, within the limitations of the laws of the universe, the laws of nature. Within those limitations, and you cannot violate natural law, you have volitional choices and you don't have to accept this dead man's philosophy that we have to perish because that's written on the wall that we are now decaying. With political structure there is no possible return. There's absolutely no possible return and I'll put my entire reputation on that. There is never going to be a return from this political collapse. It's just a one-way street to oblivion. We just have one element of luck to hope for that the catastrophe won't happen so soon that it's too late.

But even if it does you have nothing more dignified to do with what's left of the time than to not agree to it and spend your time, invest your time, to do something more constructive. And should it fail, at least you did the most dignified and most progressive and most self-satisfying thing, namely that which is productive and useful and therefore adding to your ego major. And nobody can do more than add to his ego major because that's his ultimate maximum potential. Whatever his size of

41

ego major is, he can add to it. You can also abdicate that choice and be the, quote, victim of your circumstances, unquote.

The Motivation and Moral Behavior of Innovators

Mankind is not a slaveholder. Innovation is not done for mankind in the sense that you owe it to mankind. Innovation is done for mankind by the innovator in a quite different sense. He owes it first of all to his own self-esteem to do the best he knows how, and he owes it to his ancestors, intellectual ancestors, his upstream shoulder accounts, to pay for what he has used and use it only with consent. If it has been released to ARD that's fine. If it has not been then you seek NRD. And if it has been in the political or flatland public domain then you treat it as ARD because, essentially, it was published without the restriction. You put on self-imposed restrictions that you won't swindle the innovator and you won't use it to injure other people.

How Can the Innovator Benefit Himself?

Now, the innovator, when he says he does it for himself but believes he's doing it for mankind is usually under a delusion when he's been brainwashed to think that this is altruistic. There is no true contradiction because I explained that. If he does it for himself, what does that mean? How can he benefit himself? At the present time man's durability is quite superficial. He lives very a short time. With our present life expectancy we are around for less than a century, at best. This can be improved but it takes time and this theory is needed to precede the improvement. Right now we're around for a short time. Even if we lived a thousand years or two thousand years, that's still a short time in the cosmic time scale, that's just nothing. So basically, measured in the cosmic time scale he's still around for a short time. That can be prolonged perhaps to greater length, probably.

Well, in the present world, certainly, and even in the future world it's not likely anybody will live for the duration of the cosmos. What does it mean he's doing it for himself? Well let's say he did live for the duration of the cosmos, which is at the moment rather unlikely. Well then of course he's doing it for himself because he expects to get the benefits of the achievement. He'll be around. But what if that isn't the case, which is the present situation. Now how is he doing it for himself? What is the mechanism of doing it for yourself?

What Motivates the Major Achievers?

Okay. The mechanism of it is this. Everyone has an ego but not everybody has a positive ego. A major achiever certainly has one. We will talk about other people also and that's, of course, a major problem. Ego minor is a major problem. That's the universal disease. Right now I'm not on that point. I'm on the point of those who have the ego majors on a large scale. Major achievers. Okay. What does it mean to do something for yourself when you know you will die? Let's say a person is doing something at a certain age of his life and he knows he has a normal life expectancy from the moment he's doing it, of so many years. It certainly isn't many because the total number of years he's going to live altogether isn't many. So from wherever he is at the moment he's thinking about it, it will be not too many years before he's dead. How can he be doing it for himself?

Most People Have Trivial Objectives

Most people don't think about that. You know, the line of William Jennings Bryan: "He doesn't think about the things he doesn't think about." Most people don't think about these things. They only think about what is here and now, which is why most people have trivial objectives. Trivial time scale. That's why the World Series is popular. That's why people have short-term concepts. This is the basis of short term, that life is so transient and so flimsy and so nondurable and short. Most people have no long-term goals, no long-term concepts. Can't even cultivate it after they're stimulated. Long-term relationship to his own lifetime, the whole of his life.

A Rare Person Who Can Think 100 Years after His Death

That's a rare person who can think in terms of the whole of his life as a unit. Treat it as an investment unit, as a single package deal. A very rare person can do it. Those, however, are the most successful people in flatland. Hardly anyone thinks in terms of what happens a hundred years after his death. It would have to be a very strong family tie for someone to care even what will happen fifty years after his death when perhaps his youngest child might be getting old. That'd be a very strong family tie in love and affection and personal allegiance to his family derivatives to think in terms of fifty years beyond your life. "Well, let the

43

kids worry about it, that's their problem. When I'm gone, who cares?" Or, "As far as I'm concerned, the day after I die the universe could blow up for all I care."

Somebody who claimed he was a major cosmological innovator once told me that statement and I knew immediately that was the red flag, the warning, this guy is not an innovator of any kind. This guy's a fraud. You know, he had read a lot of other people's works and pretended that they were his own. He got people deluded, including me for a short while. But this was the clear warning signal. This guy's a phony. He made that proclamation to me one day, "The day after I die universe could blow up for all I care." That man couldn't innovate his way out of a thimble. [*Some light laughter in the audience*.] And I'll bank on that.

What Kind of Person Cares about What Mankind Will Look Like in 2,000 Years

What kind of a person can think in terms of not only beyond his own life, the significance of anything or what does he care about, what will happen in the world after he's dead? What kind of a man can think of this other than a person who thinks in terms of family derivatives which might take him another generation or two into the future that he might care about? What family man is so affectionate towards his family offspring that he cares about what the universe will look like or even the earth will look like or mankind will look like in two thousand years? How can he imagine what his own descendants will be doing or what they will look like or who they will be two thousand years from now, when that would represent something like sixty more generations? How many people care about their sixtieth-generation derivatives? How many of you know your sixtieth-generation ancestors? So the point I'm driving at is, family ties aren't strong enough for that. No family could be that strong.

Genealogical Ancestry vs. Ideological Ancestry

The only thing that could be that strong is the creative achievement of the human mind. Archimedes is two thousand years ago and as far as I know, I'm not biologically descended from him, but I care about him far more than anybody in my ancestry who was his contemporary because I'm his descendant more. I'm Archimedes' descendant personally, because that's my ideological ancestry, not my genealogical ancestry. I care about Archimedes as my intellectual ancestor and I empathize with every problem that man ever suffered, including his murder and I feel for him as though he were my contemporary. That's an emotion which is rational because I identify with that man's problems.

Receiving Value from and to Archimedes

And that man is more significant, that man of two thousand years ago is more significant to me today and has done more for me personally that I can directly identify with than the vast bulk of all contemporary living mankind. Out of the 3.8 billion people now presumably sharing this planet with me I think that Archimedes has provided me with a greater value than all 3.8 billion combined, less maybe a hundred people close to me. That's a powerful statement, I presume you recognize? Now there, with all that remoteness dilution, that man is that powerful. That to me is the potency of an idea.

Would you like to know what I further think about Archimedes? I think about Archimedes two thousand years back, that though he had nobody adequate to comprehend his importance and his greatness and his significance and his heroic stature, and I mean heroic in a non-swashbuckling and non-military sense, but I mean heroic in terms of grandeur of the power of his intellect. That he can shape a civilization two thousand years later, where political civilizations have collapsed. Several cycles of political civilizations have collapsed in the interim. Religions have come and gone, ethnic groups have come and gone, but Archimedes stands as a towering giant of two thousand years ago.

When I contemplate that, I also say to you not only has he provided me value and also other scientists who can appreciate him. My predecessors, also giants such as Newton and Galileo, also held Archimedes as their firm pillar of strength before them. But not only have we received value from Archimedes, but let me point out that Archimedes, the dead great giant, is receiving value from me, now. He is getting market value from me, now. I am his customer. I am paying him gratitude. I am paying him primary acknowledgement. I'm even paying him monetary royalties, which in a century from now, if this will continue and does not die on the vine with me, will be so gigantic that Archimedes' natural estate could—I won't say will—could, in principle if this continues properly, it could dwarf the financial wealth of the American Telephone Company, or the Metropolitan Life Insurance Company, and these other

45

financial giants. Archimedes is receiving value from me now.

He's also receiving value from every other person who has done any productive creative work who has not failed to acknowledge him in some manner. He also received value long before I was born from Galileo for having resurrected his achievement. Not resurrected the spirit or the soul of Archimedes, which there is no such thing known to observational world as spirits and souls. We cannot resurrect Archimedes the man, the biological organism. You cannot resurrect even Archimedes in a spiritual form. And I don't think anybody will conduct a successful seance to convince me that we can talk to him. I know there are crackpots who would claim it, but we'll drop it here.

Galileo resurrected the recognition of his achievement. What did he resurrect? Something real? Yes. The knowledge. The ability to create a better world, a stronger culture, a higher cosmology and a vast technology which ensued therefrom. And in the Renaissance, among others but especially Galileo started that, Newton continued it in the modern world as Archimedian as well as Newtonian. So Archimedes, yes, has provided us value which is fully acknowledged and gratefully accepted and paid for. But because it is paid for—because it is paid for!—it is no longer mooched. You don't just dip in the trough and help yourself and say, "This belongs to everybody. This is public property."

It is not public property, it is Archimedes' property! Now, two thousand years later. It will remain Archimedes' property two million years later, or two billion or whatever duration we can make it into. If we can make it to two million, we'll make it to the duration of the cosmos. As a matter of fact, I'll go farther than that. If we can make it another thousand years, we'll make it to the duration of the cosmos. Man's critical era is now. That will always be Archimedes' property. That's the innovation on this point. And therefore, not only have I received value from Archimedes, not only has Galileo and Newton received value from Archimedes, we've also paid him a value. He's dead but his work is alive.

Most People Cannot Understand an Ego as Large as That of Archimedes

That's what he lived for! That's his ego! When he lived, he had an ego you know. He was not immune to ego. He had an ego and he lived for the fact that his work should have a significance greater than his death, greater than his lifespan. And the fact that he knows that he didn't live in vain and what he has done will outlive him. Now that's an ego so large that most people can't cope with the understanding of it, and so not understanding it they say, "Well, he did it for mankind," and they call that altruism. That's because they, in their ignorant attitude, do not know he did it for himself.

The Highest Form of Egoism There Is

The highest form of ego masquerades as a false altruism. As Ayn Rand has shown, much phoniness which is called altruism is in fact the way of our present world of claimed selflessness and public service and devotion to public duty and mankind and society and the community and the commune or whatever, and this she exposed beautifully as a sham and a fraud. There's another point which she didn't cover, and that is that there is a true operation called humanitarianism, doing something for mankind, which is the highest form of egoism there is. When you do something that you have so much self-esteem about that you respect yourself to the point that you want and expect and fully intend to succeed, that what you have done will outlast your lifespan and will be used by your intellectual descendants a hundred years from now, a thousand years from now, and possibly indefinitely. Those are the real big ones. The ones that last indefinitely. And that ego has been well-served in its own lifetime. That self-esteem has been well-earned in his lifetime, if the person has the capability to make the innovation, he also has the capability of having the self-esteem.

Einstein and Newton: Examples of the Highest Form of Egoism

I again call your attention to that story not for repetition's sake, but for emphasis' sake and to highlight it that the man who was publicly known for his modesty, Einstein, when he said, "I don't need the eclipse expedition to show me that it's right. It's the rest of the world that has to be convinced." That's a high ego even though he is gentle and not ostentatious about displaying it. And when Isaac Newton says, "If I have seen farther than others it is because I have stood on the shoulders of giants," that's both a high ego and gratitude to his antecedents, beautifully combined in the same sentence and therefore integrated even within itself. That's the flowstream, incidentally. That's where my expression the farther account and the shoulder account came from, that sentence. "If I have seen farther than others, it is because I have stood

47

on the shoulders of giants." He is acknowledging the giants without whom he could not have done what he did but he pretends no false modesty and saying, "Well, what I have done isn't really very much. I'm just a little dwarf." He says he's taller than the giants whose shoulders he's standing on. Do I make the point clearly? [*Some "Yeses" in the audience.*] This is the highest form of egoism. That's the powerhouse of civilization.

The Non-Diseased Version of the Theory of Psychology

Those who attack that are serving not only to immunize themselves from survival and cause their own destruction, they are causing the destruction of their whole species. Fortunately, it does not depend on the masses. If it did, we'd be lost. It depends on the strongest egoism of all. Now this is the non-diseased version of the theory of psychology. The egoism that is so strong that while you live a short life you have enough self-esteem that what you have done, you expect, fully expect, will outlast you indefinitely. And with no theological mysticism about spirits and souls and reincarnation. Something will last longer than you, and that's your ideas and the products that are derivable therefrom. That's the powerful, positive nature of the psychology. This will certainly not be terminated here; I expect to discuss more of it.

The Dwarf Machine: The Obstacle to Civilization

But there's also the thing that is the obstacle to civilization. The dwarf machine. The ego minor. That which keeps most people children when they shouldn't be children. The characteristic of mankind that is abhorrent. When what is tolerable in a child becomes intolerable in an adult and makes most mankind contemptible. And you may wonder, what kind of a person am I that I make such contemptuous remarks about my fellow species members? It's because most of them are contemptible.

Misanthropes as the Only Humanitarians

And I have long ago come to the conclusion that misanthropes are the only humanitarians. Now I'm going to make a clear-cut statement. You know what a misanthrope is? Technically, it's a hater of mankind, but I wish to subtract from that. I mean that's a technical meaning of it in this dictionary meaning. Let me just check that. I want to subtract from it the meaning of hatred in the emotional sense. It's more contempt, rather. The way I use it, contempt rather than hatred. Hatred implies a desire to counter-coerce. That's absent. But let me just read you the dictionary definition to see how it comes out.

Misanthrope. One who hates or distrusts mankind.

Well that's pretty close to what I just said. I would remove, as I say, the word *hate* from my own usage of the word. Distrust is closer. But I would prefer to use the word *contempt*. Who holds mankind in contempt. Not all mankind. Those of mankind who have not grown up and have injured their fellow species members. And I claim that all progress comes from people who are basically misanthropic in the way I define it. Not hating mankind, not harming mankind, not injuring their fellowmen, but holding their fellowmen in sharp distrust or contempt. Unless they have earned their way out of that.

The Personalities of Newton and Paine

Now I don't hold in contempt Newton. There's not everything I like about Newton. I think, reading his biography, he was probably a somewhat coldish man, not a warm, affectionate type. At least that's the way I interpret his biography. I don't think I would appreciate that. I like warmth, and personal—it's a difference in latitude. The Nordic types are not always affectionate types, and the Mediterranean and southern types get to be; it's a warmer temperature that comes out of their personality. [*AJG chuckles with the audience.*] I probably would have liked the personality of Thomas Paine much better than Isaac Newton even though they were both Englishmen. Somehow, Thomas Paine must have had some kind of warmer blood. [*Chuckles from the audience.*] He seems like a very compassionate and warm and kind man, and I like that. And also a dynamic man with a fiery temperament which goes with warmth. Well, anyhow. But personal liking or not, Isaac Newton I certainly do not hold in contempt. He has earned my respect.

Galambos Empathizes with Newton's Going to the Mint

How about the fact that he went into the mint? Well, that's not exactly what I consider his best thought. [*AJG chuckles along with some in the audience.*] On the other hand, I fully empathize with him. If I didn't have something better to do I would too. Except I don't have the same

49

attitude towards the state and the bureaucrats that he had. That's one of my farther accounts on him. That's one of the things I developed which he didn't know. I don't like statism.

Also, I many times feel a similar emotion as Newton must have felt when he was motivated to stop disclosing and become a hermit from science and entered into the political arena. I don't wish to enter the political arena but many times I feel when I have been plundered or otherwise annoyed, as far as I'm concerned you don't deserve what I have done. As far as I'm concerned, if I only had people who plunder me I would have long ago, if that's the only kind of people I'd ever met, I long ago would have stopped disclosing anything. Long, long ago. I'd do something else. As Newton said, I would do it for my private satisfaction, maybe write it up. Leave it for thousands of years from now. My ego would be satisfied if it doesn't get destroyed, even if it's in a time capsule. And I sure wouldn't care to donate this to unworthy ego minors and thieves and plunderers and assorted scoundrels. But for every thief there have been a few people who have come my way. No, not one for one, actually there are more plunderers than good people. But for every so many thieves, I've had one or two, here and there, quality people. That's I say, "I got a market which is nonzero." I just hope it's a durable market.

The Zero-to-One Transition

I don't know if I made the zero-to-one transition yet, by the way. But I know I have some people who have the potential for it and I hope they don't drop the ball. And the fact that they exist, and I have affection for some of them, and respect for all of them, the few who are in that category, all of whom have taken my courses and only some of the ones who have taken my courses are in this category. And the fact that there are some such and I hope that one or more of them makes the zero-to-one transition in reality, hopefully in my own lifetime.

You see, Archimedes finally had the zero-to-one transition but it took him two thousand years. Galileo was his one and Newton made it two, and now he's got more than two.

As I say, it's nicer if you get it in your own lifetime and you know it. I don't know it yet. That would be pleasant if I ever found that out. But I got some candidates for it. And as long as that's the case, as long as there's some qualified, strong and decent people who have come my way, I'll put up with the crap too. In shear bulk and magnitude that out-weighs the others. Literally. If you weigh them by the ton. [*Some chuckles from the audience.*] Or count them by the nose. But I can take a ratio of even ten-to-one, but not ten-to-zero. [*AJG takes a drink.*]

The Highest Form of Humanitarianism

Many times I fully empathize with Isaac Newton when he went to the mint. And but for certain things and people I would say to the whole of mankind, "Drown. Die. You deserve it." Yet, ladies and gentlemen, that's the highest form of humanitarianism. When you don't expect mankind to get what you have done free for nothing as a charity dona-tion, but you expect them to earn it. You may have my ideas, ladies and gentlemen, on my terms, but not as a donation. If you use it on any terms other than I offer it, you're a thief. I'm not looking at anybody. I'm just saying if you use it on other terms, you're a thief. That's why this is not an easy course to listen to.

But if you're a strong person, you will not take that as a personal affront because I did not mean that as a personal insult to any person unless the shoe fits. And only you know whether the shoe fits or not. Your own actions are your own judge, not I. I merely am your witness. You are your own judge, and that's the way it is. The terms under which innovation is offered is a pure and strong one. No innovator has previously made terms to mankind. They just say, "Okay, here it is. Here's what I have done. Good luck." And then the airplane is used for war. Mass is transformable into high amounts of energy and used for atomic bombs, when the man who developed the concept was a pacifist. That is not only immoral, it is not only criminal, it is destructive of the very survival of the species.

Teach Children What Property Is

I would like to say that the ego minor and the ego major, as the names are intended to imply, are at opposite ends of a spectral distri-bution of producers and destroyers. When an ego minor is what it starts out to be, a child, the destruction is limited to the effect that a child can perform. Now some children can be quite destructive. If you give an in-fant a box of matches and put him in a room full of dynamite he can do a lot of harm. Without any malice. It's just out of lack of adequate intel-ligence.

51

Have the child grow a bit and start to think a bit and give him guns to play with, play guns, when he grows up he might like real guns. Teach him not to respect property. If you have a house full of children, tell your children, "Now children, don't fight. The toys are all yours. You can share. They belong to all of you. Be fair to your brothers and sisters and you share alike. They belong to all of you." Now that's a nice, harmless parental guidance. That's a nice, loving way to teach your children that they should honor each other as brothers and sisters and love each other and share each other's property. And they grow up to think that property sharing is a way of life and you have built a Communist in your family.

The toys should be their own. Each has his own, and if they wish to exchange either temporarily by trading or permanently by exchanging the use of the toys or the ownership of the toys, that could teach them what property is as two-year-old or three-year-old or four-year-old children.

Santa Claus: A Harmful Fable

It's actually quite natural for children to like property. You have to actually brainwash them not to like it. But it's done all the time, with false delusions under the guise of sweetness and brotherly love. You know, like Santa Claus. He's the bad guy in this theory. I can imagine what some of you people must be thinking. "My goodness, this fellow is really something. He makes a criminal out of Santa Claus." [*Audience laughter*.] As a matter of fact, that's true, except that Santa Claus is not really a criminal because he doesn't exist. It is the people who make this mythical nonexistent creature into a hero because he does what he's supposed to do. That's the crime. To propagandize this delusion and make a belief out of a fable which is a harmful fable.

Ego Minor Naturally Begins in Childhood

But the ego minor naturally begins in childhood because that's where it's automatic. Everyone is born with an ego minor. Nobody is born with an ego major. That's the nature of the definition. You have to accumulate positive achievements and you have nothing to draw upon. There's no bank account to draw yourself credits from when you're born.

The Theory of Psychology Rests on Practical Aspects
of Financial Market Exchanges

I'll discuss the balance sheet discussion of this later. There's a very direct analogy with financial balance sheets, and profit and loss statements. That's why I say this theory of psychology rests very much on the practical aspects of human market exchanges, which is financial. And yet, financial concepts are subordinate and quite correctly secondary to primary achievement, but they have to be integrated properly in the subvolitional domain as they already have been in the volitional domain.

The Ego Minor Has an Enormous Number
of Characteristics

The ego minor, which is not particularly harmful in most children and ultimately should be outgrown, becomes dangerous in adults. Of course, I'll discuss many of the characteristics and many of the varieties of ego minor and I obviously cannot cover every possible case because it's as many as there are people and the number of events that they experience, because ego minor is basically a characteristic, not of a person but of an action. And a person has many actions in his life and he can be ego minor all the time or sometimes or never. Never is almost utopian. But rarely, so rarely, that it's negligible. There's a whole spectral range there as to how frequently and how dangerously and how often and how large the ego minor behaviors occur.

But basically, the ego minor concept is as varied as the number of actions of all the people in the world. So obviously I can only discuss certain categories of them. I'll cover large categories, which cover many occurrences and many different characteristics. I've named a few already. Megalomania, which can be an enormous disaster in a Hitler, a smaller variety of it is dreams of glory, on a smaller scale. Where someone believes he'll be the World Series pitching champ or batting champ, which is trivial by comparison, but it's still a dream of glory for most people, because most people don't even play baseball that well. They just dream about it and talk about it. Most people are big talkers and non-doers and they can't duplicate with their achievement what they talk with their mouth. The real proof of the pudding is, can you apply what you're talking about? What have you got to show for it?

That's a V-30 concept. It's also V-201 concept. It's also a psychological concept. What have you got to show for what you have done? What is your proof of what you have done? Where's your corroboration of your achievement?

As I say, the ego minor has an enormous number of characteristics. Chintziness is another abhorrent one. As a matter of fact, all ego minor characteristics in adults are abhorrent, some more so than others. The chintziness variety is getting very prevalent. The whole society is succumbing to it through this environmental and energy double-hoax.

I believe that I will continue from here in the next session. [*Applause from the audience for twenty-nine seconds.*]

SESSION 7

PART A

Good evening, ladies and gentlemen.

I was talking about the time scales in the previous session and I mentioned that actually, from a political standpoint, one could add a fifth scale between personal and species. Political societies, in general, last longer than one lifetime, but far less than a period that could be compatible with the duration of the species, and the species has so far experienced the rise and fall of many, many cycles of civilization.

There Is Only One Reason for the Rise and Fall of Civilization

There are so-called historians who have explained the cyclic nature of history, and I have explained that they are full of poppycock. There is no reason for the rise and fall of civilization other than the nondurability of producers to keep what they had done and that is the only reason, and I have failed to notice that is the reason amongst those who claim that they can explain the rise and fall of civilization. They have political interpretations. Naturally, they are not compatible with reality.

I have even explained how come it is that the highest production ever attained in a civilization has produced an unbelievably short life cycle for this country, and that it has even followed in a less dramatic way, that earlier civilizations had shorter life cycles the higher the culture. And the reason for that is because the coercion and the production are working at counter purposes and the greater the coercion, the greater the envy of property and the greater, thereby, the internal mechanism which favors the masses who are less competent, less productive. And therefore, all political societies ultimately rule with the consent of the people even though that has not been formally acknowledged, except since the American Revolution. Even the worst tyrant cannot rule without the concurrence of the masses. But the concurrence of the masses could also lead to an overthrow and violence.

The United States Is in Its Terminal Stage

Now of course it does do that, this does happen, but it usually happens during the decaying—not during, it's always, not usually—it's always during the decaying phase of a historical cycle. First the civilization rises and then it falls. And always the violence is during the fall. And that is when there is rebellion and there is assassination, there is terror, there is overthrow. It is unpleasant for all of us to experience this, these events happening in this country now. It is clearly exclusively because the country is in its terminal stages.

Of course, nobody likes to say that because that's not nice. It's not pleasant to hear it. It is not pleasant for me to say it. I am no more happy about it than you are. As a matter of fact, except for the fact that this led to a solution for the first time in history, it's my unhappiness that I'm living in this period.

18th Century:
The Age of Enlightenment

It would have been easier to work on physics in the 18th century when things were quiet, and when occasionally there was some kind of minor violence it quickly stabilized and when there was some violence it was only on a short term. As a matter of fact, the wars of the 18th century even in Europe were, for Europe, quite minor compared with wars they had in other centuries. That really was the Age of Enlightenment, the Age of Reason as Paine called it, because that was the first century after Newton. That was the first century when these ideas were beginning to take hold in the Western world.

And what violence there was, was not a continued, prolonged, protracted one but minor things relatively speaking, and it is always for a person who wants to achieve a more pleasant time to live. For a person who does nothing it's also more pleasant because everyone has enough property in himself, in his primordial property, not to want that molested, except for various psycho-neurotics who enjoy swashbuckling and dueling and like to be in battles because that is the only way they can prove themselves to be something. Except for such mental perverts, I don't think anybody likes violence.

Political Structure Is Not Anywhere Near as Durable as the Species Time Scale

Anyhow, the political structure is not included in my time scales although if one wanted to include all aspects that would be an intermediate one between species and personal but much closer to personal. In other words, it's not anywhere near as durable as the species time scale. The species time scale itself can endure as long as the species can. Supposing we were to exterminate ourselves in the near-term future? That means any time from now until the next century or more. What is the duration of that? Recorded species time scale is less than ten thousand years. Eight would be on the high side, six would be more realistic. Recorded period of the species time scale.

The Species Time Scale in Terms of Unrecorded Prehistory

The biological species which has most of its period unrecorded, would go further back into prehistory, *prehistory* being defined as that period of man when there was no recording of what he did. That would go back to an arbitrary time which would be determined by just how you define when this particular species emerged from other earlier species by evolution.

If you took the biblical time scale it's even more preposterous because that's only six thousand years altogether without any period of evolution because everything happened all at once in six days and then God was so tired he had to rest. [*AJG chuckles with the audience.*] Not much of a God I would say. In any event, that six thousand years, the recorded history, is at least that long.

If you take the species time scale itself in terms of unrecorded prehistory, it could be anywhere from a hundred thousand years to at the very high side a million, which would be the time when man was sufficiently differentiated from other anthropoids and various other related species from which he evolved that you can say this is a separate, distinct biological specimen. So even if you took the highest end of that, one million years, even that's low duration contrasted with how long the dinosaurs lasted, which would generalize what I said before. The lower the culture the longer has been the duration of the species or the lifetime of that cycle.

The Dinosaurs Lasted Much Longer than Man Will Have Lasted

The dinosaurs lasted longer than man will have lasted if he termi-nates himself in the near future because, at the very most, it would be one million years and that is a very small part of the period of time that the dinosaurs lasted. They went on for hundreds of millions of years and maybe that's why we got enough fossil fuel for a while, despite all the noise. The fossil fuel, you know, really comes out of their hides as well as the old forests and the old, varied vegetation. Anyhow, it is remark-able how many dinosaurs have to die to propel one jet to England. I was thinking of that on my recent trip. [*Audience laughs with AJG.*] I'll bet most people don't give that a thought. How many dinosaurs had to live for getting one jet across? [*AJG takes a drink.*]

What Are We Going to Do after We Run Out of Oil and Coal?

By the way, this is a minor digression. I was talking about things I thought of as a kid. That's another thing I was thinking of as a child. You see, I didn't need the eco-nuts. That's my shorthand for ecological nuts. I didn't need the eco-nuts to tell me about the dangers of our running out of fuel. I knew that before I was ten years old and I worried about that too, not just the inventor of the wheel, which I mentioned in an earlier session. Who invented the wheel? That was one of the other worries I had as a little kid, instead of playing baseball and throwing spitballs in class and things like that which my classmates were very good at. And I was thinking about, "What are we going to do after we run out of oil and coal?" This was right after I heard in geography class that we have about two or three hundred years' supply. "What are we gonna do after that?" Even as a sub-ten-year-old kid I was worrying about the future of mankind. Of course, I knew that I wouldn't be alive then but it still mattered to me, but I didn't worry about it when these people started screaming about it.

Poor Evaluations of How Long Coal and Oil Will Last

By the way, to show how poorly they have made their evaluations, the time period stated for how long it will take us to run out of fuel—in oil and coal, for example—was about two or three hundred years when

I went to school. That's still the estimate for oil. How come we used all that oil? We've used more oil since I was a kid than all the past civilizations in the world combined. And that was figured, that two or three hundred years' supply, at the rate that it was being used in the *1930s!* which was vastly lower than today. You had fewer automobiles, hardly any commercial airplanes, just a few private airplanes and aviation had just begun to start up and most ocean liners were on coal and not oil and they use enormous amounts of oil. For just a one-week cruise they will use thousands and thousands of tons—*tons!*—not gallons, of fuel and they were on coal for the most part. Just a few oil ships existed in those days. In those days, the locomotives of the railroads were on coal and not on oil. In those days, houses burned coal and the oil burners came in later.

I remember the oil burners came in during World War II in New York and other eastern states because of John L. Lewis and his remarkable contribution to the war effort in creating a coal mining strike during the war. And so that's when the conversion to oil occurred on a massive scale in New York City. So the big oil consumption hadn't even begun yet and they were figuring, at the low consumption rate, two or three hundred years.

Now, we have used more oil since I was a kid than they had figured for the whole two or three hundred years at that rate. We still haven't run out of oil and there's not even the closest hint of it. As a matter of fact, they have found more reserves that have not even tapped yet than has all the oil that has ever been taken out. And that is not an answer to my sub-ten-year-old worry, "What are we going to do when we run out?" Because that just means it'll be later. Instead of two or three hundred years, it might be five hundred years from now. Although I admit, at the present rate of consumption, which is vastly higher, it could be less than three hundred years now.

Galambos' Knowledge of Nuclear Energy and Physics as a Child

But then of course, that was before nuclear energy. Before I was ten, I didn't know anything about nuclear energy. As a matter of fact, it hadn't been developed yet. When I was ten, that's two years before the atom was split, and I was sixteen before I wrote an essay on it. And that was four years after it was split [*AJG chuckles*] and five years before the atomic bomb came to thrust upon the world's cognition that such things exist. So since nuclear energy—I hadn't even known what physics was yet when I worried about it, and then when I found out what physics is, that consoled me. Physics consoles me on everything, as a matter of fact. If it weren't for physics, I think this planet would be a wall-to-wall mental case. And absolutely certain to perish, this civilization.

Coercion Is What Destroys Civilization

The cause of the destruction of civilization is the lack of a civilization that's based on the simple rationality of physics. It's based upon the irrationality of coercion. The dinosaurs lived by coercion too. What do you suppose they ate? Each other, the carnivores. There were also herbivorous ones. Well, they were the victims and they were simple vegetarians and they were the fair game for the carnivores. They ate each other. [*AJG chuckles.*] And that of course, the same is true for animals, in general. But they also ate other things. They would eat little animals too. They didn't discriminate against other animals. [*Chuckles in the audience.*] Anyhow. Coercion is what destroys civilization and it's as simple as that. How many Toynbee-types have figured that one out yet? He's one of my favorite non-historians who are masquerading as experts at history. Anyhow.

Time Scales: The Key to Understanding Human Motivation

The time scale discussion, of course, is both basic and also much of the psychological explanation of behavior to understand the psychology of human beings, which means to understand what determines and influences motivation of people. What I have in V-201 is the key to it. It's the time scales. People do not think in terms of the same time scale. What is important to them, their ego, is always without exception what they want to pursue which they consider good. In other words, there's the key to the whole thing. Everyone lives to pursue happiness, period. There are no exceptions. That includes long-termers and short-termers, that includes rational people and irrational people, that includes moral people and immoral people, that covers everybody in all cases. That covers people who are nice or obnoxious, that covers people who are docile and those who are violent, that covers bank robbers, that covers Hitlers, it covers also Einstein and Newton and the world's greatest. It covers Jesus Christ. It covers everybody.

The Pursuit of Happiness Would Affect the God Concept Too

And if god were a volitional being, which I don't think god is, then it would cover god, for those of you to whom that is a significant input. That if there is a volitional God, then the pursuit of happiness covers God. I don't want to go into a theological dissertation here, but even God would pursue happiness but he would pursue what he wants! He, she, it or whatever. It could be a third sex, you know. You haven't heard of yet. [*Audience chuckles.*] If God is that infallible and omnipotent he could invent a third sex. We haven't exhausted all possibilities of the universe. It could also have non-sex. It could also be nonvolitional, which I am certain it is. Anyhow. *He* is just a figure of speech, leftover. It's part of man's past heritage. So if it slips out, I don't mean *he*. And that's not a patronization of women's liberation, because you're full of baloney too if you believe in women's liberation. It's not a question of liberation. God isn't a he or a she. It could be an *it*, if it's volitional. God isn't even that, I don't think. Man has really missed the boat on this subject. They need God for a crutch so they can support themselves through their own fallibilities. But just to mention the subject of god just for a moment. If there were a volitional God in any form then the pursuit of happiness would affect the God concept too, because God would will the universe and everything in it according to what God wants. That makes the pursuit of happiness automatic. It goes with volition; you seek what you want.

Everyone Pursues Happiness; You Pursue What You Have Evaluated Is Desirable to You

Now, getting back to the time scales. Everyone pursues happiness and that means you pursue what you have evaluated is desirable to you. You choose those things that you think will improve your circumstances. You reject those things that you think will eliminate the things that you have a desire or want for. You do the best you can which is not always very good, but that's the purpose of the choice. Nobody willingly chooses things they don't want. Unless—and this so you don't give me any exceptions—unless it's a lesser evil. Now there you do. If you are told, "Do you want a broken arm or a broken leg?" That's rather hard to choose but if you were to be asked, "Would you like a broken arm or a broken fingernail?" That would be a little bit less difficult to make a decision on because I think there isn't anyone here who wouldn't

61

choose the broken fingernail. That does not make the broken fingernail desirable, except that you can live with it, and it'll heal in a day or two and it'll stop hurting soon and you will not be totally disabled unless you're a concert pianist or you need that particular finger for whatever you do. A dentist might be disabled with one damaged finger for the duration of the time that his finger is incapacitated. But in general, a broken fingernail, that would be a less difficult choice than a broken arm and a broken leg.

But the point is, you're still choosing what you would want and that is interpreted: if you're offered two things that you don't want either one, or three things or four things or ten things, you will choose the one that you think is in your proprietary interest which means which will cause you either the most property gained or the least property lost. Always you would choose a property gain if that's the alternative to a property loss. That's always.

The False Alternative:
Characteristic of a Coercive Society

But if you have no such choice and your choices are only property losses, you will take the one that is least damaging. That's true when you pay your income tax. Those people who pay their income tax, which is most people, they pay it either out of the fact that they're deluded to thinking this is good for the world and good for the country and it's their patriotic duty and how else could you do it? Or, they pay it because of a much more pressing reason. If you don't pay it, you consider the alternative and that is pursuing happiness because not that it makes you happy to lose the money or to have these constant, Big Brother controls where you're watched as to where you picked up every penny of income and what you did with it to the last penny and you've got to account for it to people who have no proprietary interest in your lifetime and your achievement. That is not what you want. But you consider the alternative.

The alternative is being arrested, prosecuted and imprisoned. Another alternative is if you refuse that, is to be shot while you're resisting arrest. Another alternative is that you flee the country before they catch you. Another alternative is suicide. Now these are all alternatives. Not one of them is the kind that people want. Not one of them is something where you gain property. All of them are cases wherein you lose property, in each and every case. And then you choose the one that you think will have the least damaging effect upon you. Unfortunately, that is the characteristic of a coercive society, or flatland. Because in a flat society, especially in the terminal stages of its decay, these are the bulk of your external alternatives.

Only within your own household can you have better alternatives. You can determine whether you put sugar in your coffee or you drink it without sugar. You can determine whether you eat spinach or you eat potatoes or you don't eat either one. You can determine whether you eat fish or you don't eat fish. You can determine which room you choose for a bedroom and which one you choose for a den. You can determine within your own household things and perhaps inside your business minor things, not the big things, the minor things you can choose. What color stationery you're going to have or what style typeface you're going to use. You can't choose whether you pay corporation income taxes or not. There you have no choice in the matter. There you have zero choice in the matter altogether. And you don't have any choice as to whether, if you have a corporation, whether you have to get a corporate charter approved by the state and have it maintained without the revocation because if your corporate charter is revoked there goes your business. All right.

The Most Unpleasant Part of Living in the Late Stage of a Civilization

Now, the fact of the matter is that in the development of a civilization there are all kinds of people from beginning to end, from the beginning of a culture to the end of a culture. But the kinds of people that, you might say, the probability distribution of the quality of the people alters from an early stage of a civilization to a late stage. The most unpleasant part of living in the late stage of a civilization as we have is not that ultimately you know that civilization will end because the land will still be here. The birds may or may not be around, probably longer than we are. Certainly, the ants will still be around. They will inherit the earth if we abdicate it. Even if nuclear energy should be used to destroy mankind the ants will survive it and they'll have another crack at the deal and as I have already said, the main purpose of the Institute, in a very blunt language, is that the ants shall not inherit the earth because I'm not planning to abdicate it to them. I am not interested. They can make out their own civilization. When they go to the moon independently of us, that's fine. Or if they want to learn to do commercial transactions with us, that's also fine. But I'm not planning to abdicate the earth to the ants or to anybody else. If we have to abdicate the earth, I'd choose the cats but I don't think they're qualified, intellectually. [*AJG chuckles.*]

Measuring the Distribution Change in Quality of People at the Terminal Stages of a Civilization

I'd like to point out that during the period of a civilization, there is an actual distribution change in the quality of people and there are many ways to measure that. Morality, of course. It deteriorates. Rationality, that deteriorates. Not that ever there has been an overabundance supply of either one of these, but it decreases. It gets worse because there's nothing as contagious as looting, which comes from envy of others, and looting can be primary or secondary or both. Envy of others. Never is this more critical than at the terminal stages of a civilization. That's why the rapid decline after the hump has been reached. The terminal stages of civilization do not last as long as the build-up phase. The build-up phase takes longer and there's a decay in the quality of people, both rationality and morality.

Rationality and Morality Influenced by Time Scale Attitudes

Both of these are influenced by the time scales in which they do their thinking. They don't know that they have a time scale, that concept doesn't arise in their mindlets, but what does happen is, there is a, not a conscious discussion, "Well, how long do I consider is important to me?" I don't think that thought has crossed the minds of as many as 1% of the people who have ever lived. "How long a time matters to me?" This conscious, explicit, articulated question does not arise. But what does arise is, what kind of thoughts do run in their minds? Such as, what's important is today's dinner, the next paycheck, how quickly the days will flow by because they're running out of money and they need it to pay their bills which they incurred earlier. And what kind of recreation they seek, what kind of things they do when they are not working.

Leisure Time: One of the Troubles of Civilization

And by the way, they got too much damn leisure time. That's one of the troubles of civilization. They're too damn many people who are doing nothing. Partly because there's welfare and they don't have to work, it's just as profitable for them not to work as to work, and partly because of this great pity for the poor overworked person. Forty-hour week!

Oh, my goodness! How hard the poor people have to work. We have to reduce this to thirty-five or thirty or twenty-five. And then we have to pay them more, the poor little dears, what they're making. They should make more money in less time. As I have said, the insane irrational attitudes of what the politicians and unions have always been the ultimate goal: infinite pay for zero work. And both ends of that are impossible.

There Are Only Three Ways to Acquire Property

You cannot be paid out of nothing. You can only be paid out of what you do, or else you have taken the money out of someone else's achievement. There's nothing to pay out of. Value comes from achievement, and either the one who is paid performs the achievement or he does not. If he gets paid, and it comes out of someone else's achievement, that's called theft—if it's without his consent. And it's mooching if it's with. That reduces all possible alternatives to three. There are only three ways to acquire property. Through your own efforts or through someone else's efforts. That's two. And if it's with someone else's efforts it can be broken down into two sub-categories: with his consent or without his consent. There is no fourth possibility.

There Is Only One Way for a Person to Earn a Living

Theft is immoral. Mooching is, in the short run irrational, and the long run it's both irrational and immoral. Mooching is not immoral only in the short run. It has long term consequences which become immoral, I've discussed that elsewhere. That leaves one. There is only one way for a person to earn a living and have the things that he gets, and that's through his own accomplishment. It must come out of his own work and everything else is wrong. Now that's as simple as that. Unfortunately, most people neither know that nor care. What is worse, they cannot be taught. Therefore, it's hopeless, almost, but for one thing. The one thing is the recognition which comes from historical inputs, which is step one of the scientific method.

The Comprehension and Acceptance of the Majority

We've gotten to know plenty of things in the past without majority comprehension, without majority acceptance, and later, the majority not only accepted it but couldn't live without it. Example: television set. What did the majority of people do to get it? Nothing. The most positive thing they did is nothing. Most of them acted as an obstacle to the innovation by their very existence. The inertia against anything that's useful and new. Their very existence was an obstacle. Yet now that they have it, they would consider they have been deprived of one of their most important rights. Unless the state gives them the witchcraft that, "Oh, it's unpatriotic to have a television set on when you're not looking at it, and to have two sets in the same home." Then they will, as a brainwashed group of infantile and completely deluded creatures say, "Yes, father." The same paternal attitude of a Santa Claus—be good and you will be rewarded—now comes to the benefit of the state. For the state, they will do it. Provided it is presented to them, "It's in the national interest."

Supposing it were the various companies that manufacture television sets that said, "Well, we've had enough of this union agitation and state control over our industries. We decided we're not going to make television sets anymore and we're going to shut down our factories and the hell with it." Then they would think that they had been deprived of their most fundamental right that they've had since pre-biblical times. [*Some chuckles in the audience.*] It came with Genesis, and that they have been deprived of a God-given right and that those people who they always knew to be enemies of society, those evil greedy capitalist manufacturers who make these sets, they ought to be punished. "Take their plants away, seize them!" They'd favor immediate nationalization of the television industry. How many of you think this is far-fetched? Or is this the way it would be? [*Murmurs of agreement in the audience.*] Okay.

Galambos Never Expected to Do Something Such as This Course

Now let's talk about time scales more. As I say this is, you might say, the fundamental analysis, really, of the whole subject of psychology. I had it in V-201, which is, I never expected originally—say five, six years ago or earlier—to ever do something such as this course. Certain circumstances came into my life that made me so conscious of this entire nature of defects in human beings that were easily generalizable that I ultimately thought it was my own profit, my only profit actually, in having lived in a time that I have to deal with people that I don't wish to deal with, and all kinds of anti-contractual behavior which are exceedingly, sensitively annoying to me. I would much prefer to just be living off by myself and see hardly anybody except those I wish to invite. I really don't have the personality for running companies.

I don't have the personality for dealing with people who are anti-contractual, who are dishonest, who violate and default all the time. And I would like to see only a few people in my life, and can easily be happy with, say, fewer than a hundred people all of my life of my selection. I'm not that lucky. I live in a turbulent society, which everything is thrusting itself on my consciousness. I'm a physicist. What the hell do I have to do with economics, politics, psychology? This was an alien thought to me, years ago. I wanted to go to the moon. You don't need to deal with the whole world for that. I would think up scientific concepts. What interested me was relativity, and the cosmos and the structure of the cosmos. I was interested in the stars. I was interested in galaxies.

Felix Ehrenhaft:
A Completely Unaccepted Scientist of the 20th Century

I am interested in electromagnetic theory. I knew one great, but completely unaccepted scientist of the 20th century: Felix Ehrenhaft. I had the privilege, and great value to me, of knowing him and being very close to him. He has never been accepted by physicists in general. I personally do not know if he's right, but I know he's got things that he has done which are not explained by anybody else and I have preserved it. I have a lecture on this in the physics course, and I have preserved them. I would have liked to have spent years of my life on that and come up with a conclusion as to whether I think it is completely right, or that he has right experimental observations but not the right interpretation, or whether the observations aren't right. By the way, I'm convinced the observations are mainly right because I've done some of them, not many, but some of them. Every one I've repeated and is corroborated by me personally, and they're not explained by what is called classical physics. This is what I enjoy.

A Painful Trauma for Galambos to Go to School as a Child

Here I have to deal with people I don't like to, and I end up with a course I never thought I'd give. I didn't even know I would do something such as V-50 or V-201, twenty years earlier. But I did know even forty years ago that I didn't like even my fellow children in school. It was a painful trauma for me to go to school. Not for the school's sake. The school was fine. I didn't mind that at all. I minded the children. Even then, because they were mainly rowdies, they threw spitballs, they weren't interested in learning. They were there to make noise and yimmer and yammer and scuffle with each other, and after class run out and play ball games, and I went home. My Father would ask me, "Why don't you ever have any friends?" Why don't I have any friends? "Well, there aren't any! [*AJG chuckles.*] What am I suppose to do, manufacture them?" [*Some light audience laughter and AJG chuckles.*]

The Meaning of Friendship

So as I say, the psychology course actually is a product of the fact that that problem is still on, except that I have a few friends now. It depends on how you define *friend*, and if you define it very stringently it's still only a few. Although I've had access to thousands of people who like my ideas, but friendship is deeper than that. It has to be a two-way contractual matter. Two-way, not unilateral. You can't always give and get nothing in return. That's not friendship. Most friendships are basically mutual mooch operations where each uses the other one as a crutch, and one of them might be stronger than the other, which means that one guy gets a good crutch when he needs it and the other one gets nothing when he needs it, and then the friendship breaks up. [*AJG drinks.*] These are the defects in people I have never enjoyed.

Galambos' Expectations of Mankind

You may say, "Well, aren't you expecting too much of mankind?" Probably. Probably. Which is why I don't enjoy the dealings with people. That doesn't mean all people. There are people who I have come to enjoy knowing. And by the way, they all came through my theory based on how they reacted to it. And I don't mean empty platitudes and they say, "Oh, that's great." That doesn't do it. I could throw you a ton of paper like that. I can match up letters from the same person where one says this is greatest thing that has ever happened to him, this has been the most important thing in his whole life. He says this has saved his marriage, saved his life, saved his business. I've had all kinds of letters, and then two, three years later *the very same person* denounces me as though I was the greatest enemy of mankind. I have such paired-up letters. What do you suppose I think of mankind as a whole when it produces garbage like that?

No Two People Alike in Terms of Their Time Scale Attitude

When have you ever heard me turn on someone who befriended me? I have even paid royalties to people after they had plundered me, because I had value from them before they plundered me. And therefore, I do not refuse to pay for what I got even after they have plundered me. What's so difficult about that? Would you like to know the answer? In two words: time scale. Because I don't think in a short time scale. That's the answer. Time scale, that's the answer. And you say, "You mean you want everybody to be a long-termer?" I can't say that I'd want it, I don't think that's possible. It's not important whether I want it or not. I haven't figured out yet whether that would be desirable or undesirable. I tend to think it'd be undesirable. But I know that, whether it's desirable or not, it is not possible.

And you may say, "Wait a minute, you just said something is not possible. What law of nature does that violate?" It violates the distribution law, that when you have such a great variety in nature as you have with people, no two are alike. Not even two snowflakes are alike, they're all hexagonal because of the crystalline structure of water and ice. I don't think any two snowflakes have ever been found that were exactly identical under a microscope. No two fingerprints are alike. No two halitoses are alike. No two thoughtprints are alike, and here we're coming to a much greater variety.

Thoughtprints: The Most Fundamental Characteristic of Man

The thoughtprints. That's the most fundamental characteristic of man as a species. The fact that there are thoughts which exceed the more primitive thoughts that must prevail in the minds of animals such as the lower mammals, the reptiles, the fishes, the birds. They have thoughts too. It's clear from the fact that they can make a choice. A fly can determine whether he moves this way or that way and whose ear is he going to annoy or which garbage can he's headed for. To make that choice, I don't think it's a very penetrating thought although it must have something going through what passes for its brain; it must have some kind of an image. That's a thought, in a most primitive way. A species such as man which has produced as great a variety of things as man has, including production and destruction—and don't forget the production must precede destruction. There's nothing to destroy until it's produced first.

And when man has produced as great a variety of things as he has, the thoughtprints must be very much varied and very much distributed. When you have such a great distribution as this you will not duplicate two, except under the most improbable-to-assume circumstances. The probability of two identical human beings in terms of thoughtprints would be probably less than of a kettle of water freezing after you put it on a stove.

For those of you who haven't had my physics course, that is not impossible. That is improbable by the laws of statistical thermodynamics. That is not impossible, as you may think, it is just improbable. And the probability of that happening, I tell people in my physics class, don't hold your breath and don't spend your time in front of the stove watching kettles of water in the hope that you will be the first one to see it because the probability of that happening to any person is not zero, but so little you'll waste your life. You'd get a better rate of return on your life if you spend your life at the crap table. Which is not a recommendation, by the way, that you spend your life there. [*Light laughter in the audience.*] I'm saying your probability of it having a value to your life would be greater there.

The Probability of High-Quality People

On the other hand, the thoughtprints are so varied you won't find two alike people, and because of this you cannot assume that the quality of their personality will be equal and since I'm talking about a rather idealistic quality, you will not find any form of probability that even two people will be alike on a given standard of quality, let alone that the whole of mankind would be that way. For it to happen, however improbable that this could happen, it is still not zero but it's close to zero, very close to zero. But even if it did happen, what would be the probability it would be sustained for more than a second?

So therefore, that's why it's not possible. I mean, if you want a detailed explanation. There is no chance whatsoever that everybody will have high quality and because high quality requires greater personal self-control, greater discipline, self-discipline. I haven't given a lecture on discipline yet. I'm going to. I'm not decided at that the moment whether I'm going to put in this course or not. It might be put in the open-end course. I might put it in both.

Self-Discipline: The Solution for High Quality

Discipline is an exceedingly important topic. It may or may not be put into this course. Self-discipline is the solution for high quality. Only a person who can make a distinction between the greater significance of a long-term goal and a short-term goal and then choose the not easy, but hard alternative, to not do something which would be in his short-term interest because it would conflict with something that would be in his long-term interest. You may say, "How would that be illustrated?" The answer is very easy. Any form of investment will fall in that category, whether it is a financial investment, a temporal investment, an investment of your time. You only have a limited amount of time to live, that's clear. It's very short.

The Difference between Spending and Investing

You can only invest your time in so many things. You can spend it or invest it. What is the difference? What's the difference by the way, in general, between spending and investing? This comes from my investment courses. The difference is that when you spend something you get something in exchange for it. When you spend money or time and you get something back in exchange for it, which is a transaction, then the durability of that thing you got back is short. For example, if you buy a banana, that's correctly called spending money. You could call it an investment because the nourishment would keep you going on for a few hours, then you're hungry again. In that sense, even that's an investment.

In that case, every expenditure is an investment except that the short-term nature of that investment, if you eat a banana and you get hungry again in a few hours and you want something else or another banana, therefore it won't last very long. Such a short-term value that you get back is usually called an expenditure and so you call your food

71

purchases, expenditures. Although they are investments in your survival, you can't survive without them so in that sense, but you have to have a continuing supply of food.

And any particular amount of food that you eat at a given time—one thing we know we cannot do, however desirable it would be, for example, because eating is really a waste of time, however enjoyable it could be. If I could do it, and there's no natural mechanism I can fulfill this under but if I could, I would choose it. I would determine how much food I need between now and let's say the next ten years, or even the rest of my life because I don't know how long that is, and I'd say okay, I'll eat it all now as quickly as I can and then I'll use it up for the next ten years and I do it in a compressed way and don't waste my time. Then I would spend the rest of the time not wasting my time with either eating or sleeping. I'd like to invest it now, store it up and have continuous expenditure of effort. I know most people wouldn't like that. It would interfere with their lifestyle. It would not interfere with mine, however [AJG chuckles] and, after all, that's what I'm talking about.

Things That Have Short-Term Value

Now in the same way, in general, when something you buy has a short-term return and value to you and the value of it is used up soon. Like you buy a ballpoint pen. That will last until the ink runs dry, then you throw it away and buy another one. Unless you have a refillable kind in which case you need a refill, which is also an expenditure. When you buy a pad of paper, it has a value to you until you use up the last paper. That could have a permanent value. You might put down something permanently that has value and file it. The filed paper would be an investment but it's not so much the paper, it's what you put on it. The paper is simply the vehicle that carries it on the pen with which you write it.

Things That Have a Longer-Term Value

The things that have a longer-term value would be things which will produce value for you, not in a day or a week or a month or even a year, but something which will produce value for you in ten years and twenty years. And what's more, there should be a return which means it should produce a greater value than you put into it. If you put in $1,000 into an investment and ten years later it's worth $10,000, that is something worth withholding the $1,000 for this year. So, that $1,000 could also have been spent on something of short-term value which would be enjoyable, but you would not have it long. You know, like you take a vacation trip—$1,000. These days you can't go very far on that but anyhow, whatever it is, $1,000, $2,000, $5,000. That's spent and you have pleasure for it, and you may not even begrudge it and it's worth it. But after that all you have, in general, is a memory, which has not necessarily a zero value. And that's why I say such a thing would have a residual value.

Buying a Dress for Every Occasion:
A Shorter-Term Value

You could also do something which is less valuable. You could buy, say, something of shorter-term value. There are some women I know who buy a dress for every occasion. They go to a party, they can't possibly be seen in the same dress that they have already been seen in before and they have to have a brand-new dress for every party. Fine. Well, it's their money and they can do with it as they damn please. But what is the value of such a thing?

It's a one-shot occurrence because that dress once worn to that party cannot be seen again, by her own standards and her own rules, and for the next time she goes someplace she has to have a new dress again. That same money is not available for something like buying an investment in, let's say, a stock or something else which would have a greater value in ten years.

There Is No Such Thing as a Thousand-Year Investment

Now of course when we talk about a thousand-year investment—which just slipped out because that's my normal way of thinking. [*AJG chuckles.*] Well, yes, that is something people don't think about. Now what's a thousand-year investment? Name me a thousand-year investment now, in flatland. [*Long pause.*] Quiet, huh? [*Audience laughter.*] There isn't any. You're absolutely right. It's not your low imagination. There isn't any. There isn't any! There never has been. There has never been a thousand-year investment, ever, in the entire history of man. Hitler tried to make a thousand-year Reich. That would be hardly an investment, but even that lasted only twelve years and the more coercive, the worse it is. There is no such thing as a thousand-year investment.

73

Time Investment Is Even More Important
than Money Investment

You say, "Well, who wants it?" I don't want to give you 201 now, you've had that. So, who wants it? I want it. "Well, you're an oddball." Fine. When oddballs like me put out things like that then you have a choice of two things: find that it has value, or not. That's your choice. If it has value, you will find that you will have a greater pleasure in the short run than in the long run if you make an investment in it. Now of course, I was talking in terms of personal time scale investment. I said ten years or twenty years, but that's what investment is. The difference between an investment and an expenditure is the duration of the time during which some value comes in on what you have put out; comes in to the one who put out the money, or the time. The time is the same. Time investment is even more important than money investment because time is harder to come by than money. And that's another point. That's another point. How many people are sensitive to that?

Payment on the Basis of Time

Let me talk about that. It just occurs to me. The basis of payment of people in this country is time payment. Almost everybody is paid by the hour, either directly or indirectly. Either directly hourly wages and a weekly salary, which is spelled two ways; monthly salaries, annual salaries. Even presidents of corporations are on annual salary. It maybe larger than the janitor, at least I hope it is. In some cases it isn't. Most janitors make more money on union wages than I made at FEI in the first couple years. But then again, they weren't building something of durable significance. On the other hand, most people in flatland work for time payments; payments by the time that they expend.

This is also explained to include professional people, as I explain in V-201 which I will not go through again but most doctors, dentists, lawyers, consulting engineers. Regular engineers usually are paid on an annual salary, employees. Consulting engineers usually charge by the hour. These are all wrong, and even a professional man such as a doctor or lawyer says, "No, I don't charge by the hour." Oh yes you do. You charge by the ritual you perform, except that you determine how much income you want and you divide the number of rituals you're capable of performing a year and determine what the ritual fee shall be, but it's based upon how much you expect to earn by the year, which is the same as by the hour except it's a longer unit of time. It is not based upon whether you succeed or not and unless it is, that's still a payment by time.

The Whole of Man's Mental Outlook Depends on Time Scale Attitude

That's why many people don't react well to 201 even though they pretend they do because it's easier to hear it than to live it. That's why the yield is low. It's not zero. If it were zero, it would be hopeless. But it's big enough to have a shot at the possibility of having a zero-to-one transition within the next decade or two. There's a shot at that. "You sure are a pessimist." No, I'm a realist and I'll be coming to that. That's my next topic. When I get to it, it'll blend into this smoothly. I'm a realist. And then you may say, "Wait a minute. You sound like some kind of an idealistic visionary whose got his head in the clouds." No, I'm that too. I'm an idealist, but I'm not in the clouds. I'll discuss realism and idealism both but that depends on the time scales which is why I want to discuss the time scales first because the idealism and realism will— as a matter of fact, almost everything really depends on time scales. The whole of man's mental outlook. What kind of a person he is. What kind of a personality he is. What kinds of things motivate him. What he wants out of life, the pursuit of happiness itself.

One's Own Pursuit of Happiness Is Dependent on Time Scale Attitude

The pursuit of happiness is simply the selection of goods and the rejection of bads, and everything that he seeks and puts in a value scale of preference, you can't have everything for the same money for the same time. You cannot have everything. With the given finite amount of time available to you, you cannot do everything you'd like to try to do. With the amount of money you have—and it doesn't matter whether you earn a very small income or whether you have a millionaire's income or a multimillionaire's income, your amount of money that you have available to you is finite and whatever it is you have to make choices as to what you wish to spend it for and what you wish to invest it upon. And an investment, of course, is an expenditure with a longer time scale. That's all the difference there is. It's a time scale concept too. Therefore, what you seek and what motivates you to do what you do, which is always some kind of a desire to increase your happiness, all of these are influenced by what kind of an attitude you have as to what's important to you in terms of how long a duration are you expecting to get value from it and which duration matters more to you: today, tomorrow, the end of the week, a month later, two months later, ten months later, a thousand years later, and so on. I mean, there's something in between, I just jumped ahead. It doesn't go from a few months to a thousand years. [*AJG chuckles.*]

A Person's Quality Depends upon His Time Scale Attitude

And you can see because of the distribution of people and their respective qualities, the quality depends upon the time scale, exclusively. Every other rendition of quality, any other criterion of quality, every other characteristic of a human being is ultimately dependent upon his time scale of operation. If he has basically a short-term view, he will be not a moral person, or if he's moral he is simply *honestly* moral, not *integrity* moral. Do you understand the distinction? He will be honest because of the surrounding pressures on him, whether they are coercive pressures, fearing capture by state authorities in case he commits a crime or fearing the opinion of his neighbors and his friends, which are really acquaintances because he doesn't know the difference.

The Majority of Human Beings Have Never Had One Friend

Shallow people cannot have any friends, incidentally. I haven't gotten to that one in this course. Every time I say something, I think of something, I extend the course another three sessions. [*AJG chuckles with the audience.*] The quality of friends, too. That's another topic. Most people do not have any friends at all. Never have had any, by my, I'm sorry—I have to say what friend means. *Friend* means what I define it to mean. It's a long-term concept where there is a mutual interaction which is mutually profitable and where neither side is mooching from the other. By that standard, and there's a more elaborate set of criteria I could furnish, I would say the majority of human beings have never had one friend. And those people who say, "I have lots of friends!" This applies to the strongest. Those people who say, "Everybody is my friend," or, "I like everybody." Will Rogers made such an utterance, "I never met a man I didn't like." Somebody has told me, my wife has told me and she got it someplace, that that's a misquotation. That what he really said was, "I never met a man I couldn't like." That's a little better, but not much. [*AJG chuckles.*] Either way, it has that flavor to it.

Everybody Cannot Be a Friend

Everybody can be a friend. That's ridiculous. Try to make a friend out of Hitler. And try to make a friend out of a person less damaging than Hitler but whose personal character is as bad. Now his character was very bad but he was very potent because he had a mesmerizing personality and he had the ability to get people to kowtow to him. Most people don't have that capability. That's a talent he had attached to very bad character. Now if you had the bad character, but you don't have his personality of getting people to bow and scrape the floor before you, then you're just obnoxious and untrustworthy. But you're not a Hitler, but you have the same character you could be a Hitler if you had the talent. He had bad character and the talent to make people follow him.

The Distinction between a Friend and an Acquaintance

Now, I would like to point out most people cannot have a friend because they don't have what it takes to be a friend. You have to be a friend before you can have a friend. You have to be able to pay value as well as receive value and most people don't have that ability. They don't have the time scale for it. And by my definition, nothing in the short term can mean friendship. That's just acquaintance. Just because you have acquaintances that don't spit in your face when they pass you and give you the normal amenities of life and grin and slap your back and, "Hi friend! Nice seeing you! Nice day, ain't it? Take it easy now, when you leave." [*Some chuckles in the audience.*] That's not a friend. It's an animated turnip. [*AJG takes a drink.*] Well, that's an exaggeration. Turnips don't really have thoughts, but that's really a form of exaggeration. I haven't talked about that either.

Exaggeration Is a Form of Frustration

That's another thing I have on my agenda to discuss exaggeration, the role of exaggeration in psychology. The role of exaggeration is a form of frustration. When you call a person a turnip, well actually, no person is a turnip. The most stupid person in the world is smarter than a turnip. The turnip has no thoughts at all as far as I can determine. But when you get so frustrated meeting people who have a zero reaction to anything that has any sense, then you call him a turnip. Now obviously that's an exaggeration, it's a form of frustration. All right. That could be generalized to other forms of exaggeration. Exaggeration is an ulcer relieving frustration release.

A Friendship Is Always a Primary Relationship

In any event, to get back to the business of friendship. Friendships are such that most people cannot have one because their time scale is so short that they cannot know the person well enough to know what his true characteristics are. Now that can be true with a person you have known for fifteen years or twenty years or thirty years. You still don't know them well because all your contacts are superficial and your encounters are on the most superficial level. You never get any deep penetrating interactions when you have nothing to do but exchange gossip or play games such as home games like cards or outdoor games such as badminton. These are not friendships. These are passing relationships and they might pass for long periods of time. What is there any real deep relationship which could be called a friendship? A friendship depends upon a primary concept. I don't even acknowledge such a thing as a secondary friendship. A secondary friendship is an acquaintanceship. It's a misnomer. A friendship is always a primary relationship.

Most People Are Not Deep Enough to Have a Friend

Most people are not deep enough to have a friend. And this is never more true than amongst the people who think that everybody is their friend or that they have hundreds of friends or thousands of friends. This is the Dale Carnegie salesman philosophy. A salesman, to him everybody is his friend but he's a prospect and therefore as long as there's something in it for him, he'll pretend that the sky is green if that'll make a sale. Or he'll convert to a Baptist if it'll get him a big enough commission. [*Some audience laughter.*] He'd sell the gold out of his dead grandmother's teeth to land a sale. That's an expression my wife has used. [*AJG takes a drink.*] These are real cold end concepts. Cold end and short term go together here, of course. Cold end refers to consumption, hot end refers to production.

Hot End, Long-Term Attitude, Rationality and Morality All Go Together

Production gets longer term the more important the product, the more durable the product. Intellectual products, discoveries of nature, major inventions are so durable that they are the longest-term products of all, therefore, that produces the longest time scale and the hottest end of man's culture of the production scale. That's why, by the way, it's so important for anybody who gets something out of this and my other courses to try to upgrade himself and the means of upgrading refer to the upstream aspects of the ideological program. The hot end is the upstream end; the arrow points downstream. The beginning of it is the upstream end, to go upstream, and the farther upstream you can get the stronger you are.

Also, it makes your life more difficult. You say, "Well, then it's hard." Of course it's hard. That's why it's scarce. That's the reason that there are more cold-enders. That's the reason there are more short-termers. That's the reason that there are more people of low morality than high morality. That's the reason why there are more people of low rationality than high rationality. These all go together.

Morality Comes from Rationality

First of all, nobody can be moral unless he is rational. That has already been covered in V-201 that morality and rationality, though two separate concepts, one is upstream of the other. Everyone who is rational is not moral. But everyone who is moral has to be rational first, because morality is a rational concept and therefore it's a derivative of rationality. Some of the people who are rational are also moral, and freedom and durable culture depends on both. So essentially, it's a sub classification of those people who are rational who can also be moral and when they're both, they can be right. Rightness is necessary to be a producer of a major, durable consequence.

Flatland Is Full of Irrational and Immoral Producers

You may say, "Well aren't there producers who are not rational or moral?" Of course. Flatland is full of them. That's why their companies are not durable. That's why their products are not durable. That's why there are no entrepreneurs of two thousand years ago or one thousand years ago whom you know about today. Do you know any? Name me the richest businessman in the Roman Empire. Any century of it. We've been through this.

Volitional Science: A New Branch of Physics

But you can see the observational corroboration in that decay. Now generalize what that significance is and put together the many different observational facts and explain it with a single, interwoven network of ideas which ultimately heads upstream and shows only a few things that it depends upon. It depends on the laws of nature and the application of these to volitional affairs, which is a new branch of physics to apply what was previously the laws applying to inanimate nature to the animate component of nature, which is a smaller but to us, exceedingly important part because we are part of that component of the universe, and as far as we're concerned the universe wouldn't have any significance if we didn't exist. The universe would still be here. It wouldn't have any significance to us without us, which is the only reason I'm doing this, which is how I got into this trap.

Making the World Safe for Physicists

As I have many times said, this whole thing that I've been doing ever since I founded FEI is a detour on making the world safe for physicists. That was all I wanted in the first place, just to be left alone. And since I can't find a way to do that in the natural course of events, I decided I had to undertake it. I guess it is still the most important thing to do, as I once thought going to the moon was. But I think I would have had personal enjoyment from the other one better. I wouldn't have had to know so many people.

A True Friendship Is a Long-Term Concept

Now, getting back to the time scale and the friendship thing. A true friendship is a long-term concept. The least friend that you can have is one that's good for a lifetime. A friend that's good for this week only is a passing casual acquaintance. If he lasts a year and then he betrays you or attacks you or even just disappears and forgets you, even that, that's a more passive way of doing it. That's not a friend and he really never was and it was a mistake to call him a friend. That was an error in overestimation.

FEI: A Market University of the Long Term

I have learned all of these things in the university I have founded which is the University of Hard Knocks. [*AJG chuckles.*] That is also FEI's alternate—it's not its formal name, but it describes it. It's a market university. It does not feed off subsidies, endowments, and state support and therefore it's in the true market. And what's more, it's not in the market of the short term, it's in the market of the long term. These ideas, with or without this particular school, will endure with or without your concurrence, I might add. It doesn't depend upon whether you like this course or not or any of my courses. I hope you do, but hope is cheap. Results are hard to attain.

Realism and Idealism Go with Pessimism and Optimism

This is not cynicism, incidentally. Although it sounds like it. Do you know what cynicism is? Cynicism is, in fact, a disbelief in the ability of anything good to happen. I don't disbelieve that; I only believe it is difficult. I don't believe it's impossible. A cynic is a permanent pessimist. I'm only a short-term pessimist. I am a long-term optimist because I'm generating the long term and it depends on me and not upon haphazard fate or chance. It depends on some other people. I'm not the only one it depends upon. But it depends upon what I've done. That's a major component of it, and that, by the way, is an introduction to the discussion of realism and idealism that I said before and pessimism and optimism, the two go together.

The Shortest-Term Concept of Friendship There Can Be

But the concept of friendship itself is a concept which has to be at least long enough term to be meaningful, that it should last the rest of your life. That's the shortest-term concept of friendship that there can truly be. You see, I can understand that not everybody's in the species time scale, you're not talking about friends into eternity. Although I look upon Thomas Paine as my friend though I never met him. Because I have all of the ingredients that a friendship has, a true friendship, available from Paine. He has provided me the values I would have liked in the best of friends. And I, though I've never met him, have provided him payment of like kind—the values he would have gotten from the best friend he would have ever had. And although we never met and cannot

81

meet, it's physically impossible, a concept of friendship in species time scale is definitely possible. On the other hand, I don't expect everybody to practice such long-term concepts of friendship where two people have never even met and they're friends. But within one lifetime where you do, in general, meet people or know of people.

Friendship between People Who Have Never Met, Living in Different Times

By the way, that's another thing. You don't even have to know a person to have a friend. Two people can affect each other's lives and never meet and still be friends. Not just in different times, but at the same time. Someone can befriend you from afar by something he does that affects you, and if you don't take it as a mooch acceptance or a favor but do him something of like value in reciprocation, that makes it a friendship. But the minimum time that a friendship has to endure to be a true friendship is a lifetime. The lifetime of the persons involved. Because anything shorter than that is a misevaluation which leads to disappointment and bitter feelings, which of course is always the case. That clearly means there was no friendship. If a friendship can terminate it means it never existed. It means there was a defect in the original identification of what is a friend, or whether that person is a friend or not. Now since most people don't have any such concepts as these, they cannot even have a friend. Never is this more true than a person whose encounter with people is always on the shallow level of cold-end relationships. Those people are not having any friends at all.

Galambos' One Friend in Childhood: His Father

By the way, I did have a friend in childhood. The person who asked me why don't I have any friends: my Father. He was the friend I had. I had a very happy childhood. Most people who say they had no friends in childhood usually come out as bitter, frustrated persons who hate the world and want to take out their hatred and their bitterness against the world. They could turn either to crime or to some form of personal bitterness, melancholy, cynicism. I don't have any of these. I have no bitterness against the world. I have bitterness against individual people, and there I let that get subordinated to trying to avoid seeing them again, that's all.

Ending a Bad Relationship

I don't carry out vendettas against such people. I just would like to say, "Okay, well, we've had a bad relationship. Why don't you get lost, and I'll get lost as far as you're concerned. Let's forget about each other." It's the bitter people, the people with a vendetta who continue the association when it's no longer desired, who hound you with their presence. I have no desire to be near a person who obviously resents my presence. That's humiliating. I don't go after people, "Please like me," or, "Please deal with me." The hell with you. I got the way I live with the terms I have, take it or leave it. Most people can't stand that. They can't stand especially the openness with which it's stated. The fact that I don't conceal it with Dale Carnegie acts, putting on fronts of all smiles and handshakes and backslapping and bootlicking. Or pretending I like someone I don't like.

People Want to Remake Others in Their Own Image

Most people can't take human beings as they are. They would like to see them as *they* are. I'll take that back. I will rephrase it. I will restate it. *All* people cannot accept people as they are, they want to remake them in their own image. I am not an exception. There is a question of, however, how you remake them in your own image. Coercively, or otherwise? Hitler did it coercively. He failed. Most people don't have enough dynamism to try anyhow.

An Example of a Mutual Admiration Society: The Hippies

A slob would like to see other slobs so they find solace in each other and they form mutual admiration societies. That's what the hippy movement is, among other things. They're all societal, self-appointed rejects. They have rejected society and they have formed the society of their own. They fit each other's standards which is low and casual, and they call it permissive, but in fact it is slovenly, is the correct term. Their manners are slovenly. Their appearance is slovenly. Their morality is slovenly. Their goals are slovenly, and they are short-termers since everything is for the now. Otherwise, they're great. They remake each other by imitating each other. They haven't got the imagination to set an example. They think they're individualists. They are in fact the largest, most conformist group of people in a group today.

83

Galambos' Father: Leading by Example

My Father made me in his image to as great an extent as he could, with zero coercion. He was the least coercive parent a man could have, and yet he tolerated no insolence. He got some from me sometimes, I'm sorry to say, but that's part of the ego minor of children. He got quite a bit of insolence from me as a child and I deserved the reprimands and the occasional slaps in the face I got which I richly deserved. I don't resent him for that. Quite the contrary. I'm glad he shaped me up before it was too late. He was the most kind and compassionate man in the world. He coerced me never. He did not forbid me to smoke. He never forbid me to smoke but it was because of him I don't, and he has spared me the misery of a lifetime of a filthy, dirty, dangerous, unhealthy habit. Also expensive. He smoked himself, which is a mistake he made in his childhood. He had the willpower to stop but most of the time he didn't give a damn. He was so frustrated as a person, he frequently smoked because it gave him pleasure. He knew it was dangerous and he told me I should not make the same mistake by telling me what was bad about it and how he regretted it.

He did not say, "I forbid you to smoke." He did not say to me, "If I ever catch you smoking you'll be punished thus and such." Whereas other parents whom I have known have forbidden their children to smoke. I have several such relatives where the children were forbidden to smoke by their parent and that's why they smoked. It was a challenge to supersede the command and to do it behind the back and then they say, "Oh, I'm just doing this as a lark." And then it became a lifetime addiction.

The reason I don't smoke is because I have respect for the one to whom the opinion of not smoking mattered, and he said he made a mistake and he hopes I won't. And I didn't. That is making me in the image he would like to have me in, but that's because I respected him. How many people command such respect? How many people have what it takes to know what value there is in following the good ideas that are available to them? Which though, in the short term, may represent you giving up something. In this case, not smoking. You give up being one of the gang of other people with whom you are associating. They will ridicule you. "How come you don't do this?" I got a little bit of that, although not much because I just turned my back on such people and had nothing to do with them. "Here, try it." "No, thank you." "Well why not?

Don't you have any courage?" If that's what it takes to have courage, Ha! That's some concept of courage. To abdicate your mind.

"The Authority of Respect"

As a matter of fact, an early colleague of mine, Alvin Lowi, it just occurs to me, once made a rather interesting play on words which has merit in the context I'm just talking about. There is the prevalent usage of the expression in flatland and coercive society: respect for authority. You've heard about it. That's what the state demands. Respect for authority. He said the correct concept is the authority of respect. It's backwards. My Father had the authority of respect. I respected him so what he said to me mattered.

The Story of Galambos' Crew Haircut and His Father's Authority of Respect

For example, I remember when I was twenty-four years old, I was going to the University of Minnesota, I was not living at home. This was two years after I got out of the army. I'd gone back to school, got my undergraduate degree in New York and went to graduate school in Minnesota. And for the first time, I was neither at home nor in the army. It was the first time I was ever away from home other than the army. And it was then the fad for young men to wear crew haircuts. Which is the exact opposite of today's slovenliness. It was a fad. I got a crew haircut. As a matter of fact, that's when I met my wife. That's the period of time my hair was cut that way, that's when I met my wife. A few weeks after this happened, I went home, it was after the semester. My Father took a look at me and all he said was, "I did not bring you to America so that you should look like a Prussian Junker." [*AJG chuckles with the audience.*] Do you know what a Prussian Junker is? The Prussian ruling aristocracy. They are the landed aristocracy from whom the officer class came for the German Wehrmacht, the German army. The very powerful army class that made Germany the most potent military machine in the world. And they had this type of hairdo, originally. The original reason, incidentally, was quite intelligent. It was to keep yourself clean when you're in the field in the army where it's easy to catch lice and with short hair the lice don't have an easy time of it. But my Father didn't like the discipline of Prussianism, he didn't like the Prussian concept. He said, "I didn't bring you to America to look like a Prussian Junker."

Please note, he didn't say, "You are forbidden to do that." Or, "I won't allow you in my house if you look like that." Which he would have had every right to say, by the way. It's his house. He could have gone that far and still be moral. It's his house. He just told me what he thought of it. But unlike cutting off long hair, you can't grow back lost hair very fast. It takes a while for it to grow out. Frankly, I don't know how some of these long-haired characters grow their hair that long because I should think it would take twenty years to get it down to as far as they got it, because I know how slowly my hair grew out, it seemed like an eternity. [*Audience laughter.*] I lost all interest in crew haircuts that day. [*More laughter.*] Because I respect my Father's opinion.

You say, "Well, you're a slave." Oh, no, I'm not. When you are the great man that he was, I'd respect you too. When you're one-hundredth of what he was, I would respect you. He didn't think it was right. That means a hell of a lot to me. I don't know of any significant occasion my Father was wrong on anything. Just think about the significance of that remark. You say, "Well, that's filial devotion." The hell it is! That's absolutely false. That's an observation of twenty-seven and a half years that I knew him, which is unfortunately all I had, knowing him. And I couldn't alter the problem immediately, but it grew out. I never had one again. The other extreme is even more repugnant. Fortunately for him, he was spared the seeing of that.

Other People Who Have Authority of Respect

I brought it up not for the haircut. I brought it up for the authority of respect, which is Lowi's well-coined expression. "The authority of respect." That's the truest authority there is. Other people who have authority of respect with me, for example, are Isaac Newton, Thomas Paine, Galileo, Archimedes, and many others that you've heard about. You say, "Well, that's the same list as your gratitude list." Of course. "Do you mean the two lists are the same? The authority list and the gratitude list?" Of course! How else could it be? I am grateful to them, they have provided me a positive value and that balance of their value to me is so positive it would be not only ungrateful, but insanity—irrationality on a permanent continuing basis, which is insanity—for me not to look upon them as authorities. "Oh, you believe in authority too, huh? Like all people in dogmatism!" There are people who have earned that authority. That doesn't mean I take it on blind faith.

An Example of Earning the Authority of Respect: Thomas Paine

There are plenty of things that Thomas Paine did in the political writing field that I consider completely wrong. I did not know this when I first read Thomas Paine. There was a time, I might say, when I was twenty years old, when I read Thomas Paine, I was so absolutely enthralled with what he wrote, I said, "I do not disagree with one word that Thomas Paine has ever written." And I had read his complete works when I was twenty years old when I first ran across him. I mean, I don't take things in half measure. I didn't know who Thomas Paine was, one day I found out who he was. *The Age of Reason* is the first thing I read, I thought it was fantastic. He said exactly what I already believed thanks to my Father, but he said it so beautifully and he said it so positively and so firmly and so correctly.

I read the other things and I said, "There isn't a single word that Thomas Paine has ever written that I would disagree with." That included what I now totally disagree with: *Agrarian Justice*, *Rights of Man*, the discussion of taxation and things like that. Well, what happened? What happened is that in the meantime I learned more about political philosophy, and I found out that the whole concept is wrong, and that the entire basis of the American republic is based on wrong principles, namely majority rule, representative government and a coercive state. The whole thing is wrong, and Paine was part of that. Well, the fact that he's Thomas Paine does not alter the fact that he's wrong. The fact that I like Thomas Paine, that I respect Thomas Paine—I even love Thomas Paine in the sense of deep affection which I describe in V-215.

Love for Thomas Paine

There is sexual love and there's nonsexual love. You can love somebody without sexual attitude. You can love a brother, a father, a friend. You can love a man you've never met. I love Thomas Paine. In the truest sense of the word. I have the deepest respect for him. I think the man was one of the finest human beings who ever lived. If he had never done anything right, if everything he wrote were wrong in its terminal effect, his character was so kind and so decent and so warm, I would still love the man. Because he was a fine human being who has been unjustly reviled by people who weren't fit to shine his shoes. He was also right on many things.

He was right on the principal part of the American Revolution: man needs no king to worship or to bow down before; he needs no ruler. I generalized that from no king to no ruler, or no rulership, which includes parliaments and representative government. The only rule he needs is self-rule. I don't want to go into that; you've heard this. I hope you took it seriously.

Thomas Paine Respected the Same Authority as Galambos

The fact of the matter is, what constitutes a linkage between two people? Two people who never met and cannot meet? Because he did things that are right and he did things that are fine, and kind. He was moral, even if his conclusions were all wrong in terms of all of his political and even religious philosophies were wrong, which they are not. They're not all wrong. Quite the contrary. I will still make this statement about Thomas Paine. His basic cosmological concepts are all right. I have not yet found one that I would disagree with. It's the technological applications, like he said that man should not be under the authority of a king. Well, that is absolutely right. I can generalize that: no authority under any other person either. He, however, himself respected authority. Paine. He respected the same authority I do. He respected Copernicus and Newton because he wrote about them. He wrote about Copernican astronomy and Newtonian gravitation favorably, therefore he accepted their authority. That's the authority of respect. Because you respect the work and that is a correct concept of both authority and followership. Ideological followership.

The Concepts of Friendship, Gratitude and Authority of Respect Are Integrated

Of course Newton is the leader of man's intellectual achievements. He had his shoulder accounts, but he's the one who integrated what we've got. It would be total ingratitude as well as irrationality to the point of continuing disaster not to accept Isaac Newton and his work. This is also a friendship. It's also the gratitude list. It's also the authority list. And these are people who don't hurt you and these are people who did not threaten to punish you for failing to believe, and you can accept the right things and still not be a slave for the wrong things.

Thomas Paine Would Concur Political Philosophy Has Failed

If Thomas Paine made a mistake on political affairs, I'm not enslaved to those thoughts, and I'm capable of adding shoulder accounts which my concepts are basically the continuation of the American Revolution. I am probably, I cannot prove this because there's no observational corroboration possible, but I believe that it is highly probable that if Thomas Paine were alive today and had access to what I teach in V-50, V-201, V-76, that he would concur that political philosophy has failed. I don't think he'd be a little man. I cannot believe he would have such an ego minor.

Galambos Does Not Claim Infallibility of His Theory

I knew Ehrenhaft and he believed he was right, but I also know that if he had evidence, experimental evidence, proving that what he did was not right, and you had experimental observations that prove his observations to be faulty, that he would have accepted that. That's in the book that he wrote, which I helped him put into English. And that's in my statement here. I don't claim infallibility. I just claim that what I have done is right as I see it, and I've given plenty of evidence for it and everything fits and there have been no failures of corroboration. But I have said rather sharp, out-on-a-limb statements. I said for example, in comparing my work with let's say Jesus Christ's philosophy, I said if my work were around for two thousand years and generally used, and there were still wars after two thousand years, I will admit now that the theory is wrong. Fine. If two thousand years after my ideas have been circulating there are still wars—I don't mean two thousand years later and nobody read it!

And let's say my book were buried and you people are all dead and nobody followed you and it was completely lost, and then somebody found one ancient copy of the book ten thousand years from now and did something with it, then it will come ten thousand years from now. I don't mean that if it were lost for two thousand years. I mean, if the ideas were in general knowledge and general circulation for two thousand years, then I will tell you it cannot be right. But of course, there is nothing to bring something out into the field of clean rightness more than if it's still around thousands of years later. That is why, to anyone who is a major producer, the concept of the long term cannot be ignored.

It Matters to Your Ego That You Will Be Remembered

It's not a matter of just self-vainglory. That you have such a vainglorious attitude about yourself that it matters to your ego that you will be remembered. Of course that's natural that you'd want that, if you have done something significant. It would be completely anti-expectation, anti the nature of man, the pursuit of happiness, to say, "Well I'm doing this altruistically. The world can have this but it doesn't matter whether I get the credit or not." There are people who make statements like that, and you know what I have to say about that? And this can be included in my predictions. That's called the third step, extrapolation, not crystal ball gazing. Anybody who talks that way: "Here I am, donating this to the world and it doesn't matter who gets the credit, as long as the world's got it." Anybody who says that, his work is of very small significance, if any at all. Because he doesn't have the proper attitude and self-esteem to have the ability to do something better. There's one possibility. [*AJG takes a drink.*]

False Modesty

That's one possibility, there's one other. He knows its value, it is good, and he recognizes it and it does matter to him but he says this because it will make him more popular. Or because he is fearful of the ridicule he will get for having any public display of self-opinion and self-respect, because man has been cruel to his fellow man and every great man has been condemned to various forms of injury. And the greater the accomplishment, the greater has been the harassment at the hands of his fellow human beings. There are men who have done some value, but have thought it in poor taste and feared, or at least if not feared, at least would have preferred not to have the consequences of their neighbors and their associates and their acquaintances thinking poorly of them and so they will pretend a false modesty which, to the extent that it is offered, is not a true statement of their real convictions. That is the other possibility. If he really believes that it doesn't matter, then he has done nothing, or so little that it matters not in the long term.

Long-Term Success Depends on Long-Term Time Scale Attitude

Now, whether a person has a long-term success or not depends on whether he had a long-term attitude or not. Now on the basis of that, clearly, most people have not got that ability and therefore there is no concomitant time scale with them. The two go together. You say, "Which causes which? Which comes first, the chicken or the egg?" Does a long-term view produce competence, or does competence produce a long-term view? They assist each other. Where the nucleus or the egg is, I'm not certain. I'm not sure it's a unique answer, but they feed on each other. The more competent a person is, the more long term he gets, and vice versa. The more long term he is, the more competent he becomes.

The Counterfeit of a Long-Term Time Scale Attitude

Now I want to warn you here about a counterfeit of that. There are people who talk long term, pretend long term, especially after my courses. I don't think there's too much of that elsewhere. But after they hear—"Oh yeah, oh it's long term. That's a good team to get on!" [*Some audience laughter*.] Some out of genuine, sincere desire say, "Oh yeah, oh boy, this is really exciting!" The person who previously figured that when he dies it's over and he doesn't care and nobody else cares, and all of a sudden he says, "Well, maybe my life can have significance after all." And he gets all excited, the carrot on the stick is in front of him, "Maybe in a thousand years people will know who I am. Gee, hot-ziggity, I never thought I could do that!" And so he gets all excited, "I'm a long-termer, hip-hip-hip-hooray!"

One of the Biggest Psychological Internal Disasters People Have

He doesn't have the ability to match it and when the wet herring of reality slaps him in the face [*audience laughter*] he turns hostile to my theory and, of course, to me, personally. Because he has demonstrably failed at doing anything and just the excitement of being important in the long term—he's got that bug—but not the bug of competence! He didn't catch that one! I mean, if the competence doesn't go with it, the long-term desire is as flat as a blow-out tire! And that person has neither long term, nor gratitude, nor rationality, and worse than that he can't even expect to develop it. So you might as well stop kidding yourself. And that's one of the biggest psychological internal disasters people have. They cannot face up to accepting what is.

Dreams of Glory Not a Substitute for Achievement

Now a person can improve himself, but not beyond what nature provided him. You've got to realize both of two things which seem contradictory, but actually they're complementary. You must head as high as you can go and not take small goals, but they must be realistic. They must fit your talents, your aptitudes, your lifespan, your capabilities. You must not dream about things that you cannot attain. Dreams of glory will not be a substitute for achievement, and I'll have much to say on dreams of glory. You say, "Well that's a childish concept." Of course. All ego minor is. You should not expect to attain what is not attainable, but you should head higher than you will attain so you have the goal higher. And you won't fail because you set your goals too low, but don't set them too high either.

How Do You Know What Your Goals Should Be?

Where is the right place to do it? Figure out what you really can do and add some more to it, but not so much more that it's unattainable and then if you fall short of what you expected that means you still have some incentive left at all times. As I pointed out, how do you know what these goals should be? Well, what's within your present knowledge? Extend what you know to what you don't know. Ask yourself: what is the range of things that I can imagine that I don't know but I think I can find out? What is the range of things that I haven't done that I think I can do in the real world with my present base that I'm standing upon? That means financial base, emotional base, even political base.

Flatland Culture Restricts All of Our Output

We're living in a flatland culture which restricts all of our output. My output is reduced enormously by politics and the political state in flatland, so is yours. And so therefore, we have to live within the real world, which at the moment is a bad one, a flat one, and we have to accept that limitation not because we enjoy it but because it's there. And someday it won't be there and it's because of what I'm talking about, that someday that flat base, that flatland base, will be improved with a spaceland base, but right now that's one of the limitations. The flatland base prevented me from being the first one on the moon, not my lack of knowledge. I had no opportunity in the short time that I was doing this to develop the capital for it, but I had developed the means to acquire it and it would have taken about ten, fifteen years. In the meantime, they got there through money that was not invested, but taxed, and I don't use that technique. All right. By the way, we wouldn't need twenty-four billion dollars to reach the moon, either. [*AJG chuckles.*] It would have been less than one. As a matter of fact, a small percentage of one billion, on a rational basis. Never mind.

You Must Not Disintegrate When Accomplishment Is Not Attained

That's another thing. When you don't achieve something you would have wanted to, you must not disintegrate. That's another point, bringing up the wanting to go to the moon. Can you think of anything more frustrating for a person who at age nine wanted to go to the moon and it was no childish joke and it wasn't even a dream of glory, although my Father for a long time worried that it was. Nine-year-olds don't normally want to go to the moon and he thought that I'm going to waste my whole life on stupid little dreams. It turned out it wasn't a stupid little dream. That's why I'm in California. Because the rocket industry's capital is here in the Western world. And I had every intention of doing it. On the other hand, as I say, because of the fact that to develop honestly and morally the capital through investment means is a very long and difficult thing and you have to know something about the investment world to accomplish it. In the meantime, the state, which is not limited by this problem, did it with an expenditure of an enormous amounts of tax money.

Well, just how do you suppose I reacted to seeing what I could have done better for a scientific and commercial purpose without taking the taxpayers' money and making a profit at it and make it a long-term base for future expansion of the species, and not merely leave six piles of junk on the moon and go on to some other political adventure after that. What do you suppose my reaction to that was?

Galambos' Reaction to the First Moon Landing

Do you suppose I said, "Well, if I can't do it, I won't even watch," and not even watch the landing on the moon? I was glued to the television the whole time. I saw the whole television broadcast of Apollo 11, all of it, even though it went into my sleep and everything. And I don't even say that it was incompetently done. As a matter of fact, it was competently done. It was successfully done. I respect the successful accomplishment. I don't respect who did it, why it was done, who paid for it, and the political consequences of it, and the fact that it was dropped after a few such adventures and they didn't know any of the reasons why they were there. And they used relatively minor people to behave as state robots to do it. None of these do I respect. Nonetheless, the accomplishment is there and I didn't boycott it or denounce it and say that I could have done it better.

As a matter of fact, I could have. That doesn't detract from this accomplishment. I could have done it without expense to you as a taxpayer, and profit to those who invested in it. That doesn't mean that I am frustrated. And I didn't say, "Well, I failed in my life's ambition." No, I didn't fail and I never did it. The time scale wasn't available to me. And I did not go into my cocoon and retire and say, "Well, if that's the way it is, I won't do anything." As a matter of fact, I was doing something towards that at the very time it happened. FEI was formed for that purpose.

Why Galambos Founded FEI

FEI was formed for the purpose that I realized I couldn't raise the kind of capital that this result required. It needed, as a precondition, an earlier result. I needed to have a mechanism whereby it is possible for science to dominate the production world; science and knowledge and morality to dominate the hot end, that is. To dominate the production world and not political misbehavior. Even though it looks like I went off on a tangent, I didn't because I still intended to return to astronautics at a later time, probably after my death. But man is still going to the stars and to the galaxies, but it must be done proprietarily. If it's going to be done by the political method it won't succeed. The technology can succeed because Newtonian physics works for bureaucrats too.

Political Mechanism Is Not Compatible with Physics

Newtonian physics is not wrong just because a bureaucrat's paying for it. I'm sorry. He's not paying for it. The bureaucrat is controlling it, the mechanism. The physics of it will work but man will destroy himself both economically and militarily in the process. The knowledge is too dangerous to handle for such people and they'll of course wreck the economy of any country that will try to go to the stars. If it took as much to go to the moon, what do you think it would cost to go to the stars? It's already wrecked the economy to the extent that the politicians had to disband the entire moon operation because there was so much political pressure on: "The money could be used for urban renewal and for welfare and for cancer research." That was the reason. They couldn't stand the political guff they were getting from their constituents.

Proprietary Mechanism: The Path to Man's Survival

This could not happen with a proprietary mechanism, which would be both efficient and therefore inexpensive relative to the results obtained and also would not injure people, but reward those who invested and not those who did not invest. So this is still intended for that except that I had to have an earlier product accomplished first: man's survival. If man doesn't survive, he doesn't have to go to the moon because it's all in vain. And so I did not die of frustration, nor am I bitter. I'm sad, but I'm not bitter. I'm unhappy over the fact that it wasn't done proprietarily, in general, and by me, in particular. But I've not committed either physical or moral suicide. I have not committed psychological suicide. I work all the harder now. That's something which is important and will supersede in importance even that. But just imagine how frustrating that is to see someone do it in the wrong way for the wrong reason by the wrong people and it's financed for the wrong reason for the wrong goals.

Stability Needed to Accept Even Bad Results and Survive

But you have to develop, above all and before everything, stability. Next year, I'm giving a course on stability. I thought of giving it this year but I'd already scheduled this one, and that's quite good that I did because this course is a good introduction to stability. This course is emotional stability. I've already given courses on financial stability. The two are integrated. They're integrable and they're integrated now. A person has to have the stability to accept even bad results and survive. How do you suppose I reacted to dealing with people I don't like to deal with? I've mentioned that too. I might have faced some other frustration.

Do you know about the fact that I had in college only one subject I really had any difficulty getting through without flunking, just one, and that was public speaking. The closest I ever came to flunking a course.

95

The teacher didn't like my way of speaking. I can't tell you his name, I don't remember it, fortunately for him. [*Light audience laughter and AJG laughs.*] I just thought I'd pass that on. Did I overcome that frustration? [*Some murmurs in the audience and AJG takes a drink.*] Do you suppose anybody who teaches public speaking would have anything to say for fifteen hours and keep people listening? [*Audience laughter.*] It's not how you say it, it's what you say. I still know I'm a lousy speaker, but I have something to say, and that's what counts.

I'll discuss the time scales more after the intermission plus I'll bring up the discussion of idealism and realism. And of course, stability weaves into all this too. Stability is a large part of this course. [*Audience begins to clap.*] The reduction of the ego minor is the stability mechanism. [*Applause from the audience for fourteen seconds.*]

PART B

Before I go on with the main lecture, I have a message that was given to my wife at the end of the previous session by my senior lecturer, about my lecture, and I thought that you might like to get some feedback about what he has written. He said to my wife:

> There must be something seriously wrong with this Galambos fellow's ideas. V-201 has been in the market now for over ten years and we still don't know who invented the wheel. [*Loud audience laughter.*]

I hope that will be the worst historical problem I will ever face. [*More laughter.*]

Additional Points from the Previous Half-Session

I have another thing to add before I continue the lecture. This comes directly from my wife and she has pointed out to me that there is a third possibility which I fully concur with. It's just, I wasn't looking at it from this point of view. In reference to someone making the statement that, "I don't care who gets the credit as long as the world has the benefit of my work." And I pointed out that either this person has not much of significance, or else he is pretending false modesty for either social circumstances or market consequences, or that's his native personality to not like to invoke upon himself critical remarks and when someone has a high opinion of himself in today's pseudo-altruistic culture, it is automatically going to be the case that someone will be disliked just because he does have a high opinion of himself. That's an automatic response and some people will not bear with it.

There is a third possibility, she points out, and this is actually the most fundamental. Well no, it's not as fundamental as the first one which is that it isn't right or he didn't do much. The third possibility is: it is not his work, and he is getting credit for something which isn't his own. A person who has no proprietary interest in the creation of something can afford to be generous and let it be cross pollinated, and many people do that all the time. They use somebody else's work on which they are coasting by, getting money, selling somebody else's product and then somebody else copies it or claims credit for it, they don't get bent out of shape. They're in it for something less important than primary; for the money or for short-term prestige. There's also a variation of that, also pointed out by my wife, and that is that it may be his work partly, but not fully.

It may have been a group effort and he was part of the group or the nominal head of the group or a component of the group, and it got so cross pollinated and so intermeshed one with the other, that they have absorbed some of this team spirit philosophy, and they look upon it as teamwork. This is part of the brainwashing of scientists. They're not immune from brainwashing too. An illustration of this that she knows about better than I do—she read a biography of Jonas Salk. It appears from such, the biography she read, that his work is actually due to the combined efforts of a number of people and he got the credit, so he can afford to be so gracious. It isn't all his. Okay.

The Time Scale Attitude Difference of Innovators and Entrepreneurs

I was discussing the importance of time scales which by no means is over. It will run through all of this course implicitly, and explicitly at many times it will come and rise to the surface. The significance of the time scales. When I introduced those time scales into V-201, I was essentially bringing the key problem to the surface that has prevented freedom from ever coming about. The key problem is the difference in time scale attitude of the two most significant groups of people in the history of the world. They are the innovators and the entrepreneurs.

97

Von Mises Not Looking upon Discovery as Property

This is pointed out, I might add, in Ludwig von Mises's writings too, the significance of three groups of people. This is in my earlier course's discussion of von Mises in 201, that von Mises also has pointed out before me that the three most important elements of production are: the investor—that's the capitalist—the inventor, and the entrepreneur. And I have in V-201 said that I have reduced these three to one: the capitalist. The capitalist includes all three of the others. They are different kinds of capitalists. The inventor is a primary capitalist; he accumulates knowledge.

Moreover, I also commented that von Mises said *inventor*, and I say *innovator*. And I don't think the distinction is superficial or accidental. First of all, von Mises certainly is a master of literature, both in German and in English, and I don't think it was a literary error on his part. I think he meant inventor. I don't think he meant innovator. I think he meant inventor, because the other kind of innovator is the cosmological innovator or cosmovator. And he, I think not accidentally, but intentionally, omitted that because he does not look upon discovery as property. That is not part of his basic concept. That is part of mine, however. To look upon the discoveries about nature, the laws of nature, the theories about nature are property.

A Natural Phenomenon Is Not Property; Laws of Nature Are Property

As a matter of fact, that by itself out of context seemingly is sheer lunacy and I'm perfectly aware of what this sounds like to a person who has not heard my course on primary property, to say the laws of nature are private property. It sounds like somebody is a raving lunatic. But of course, if you look upon it from the standpoint that ideas are property, and a law of nature is a human idea. It's a volitional idea which happens to fit the reality of observational corroboration about nature, that makes a law of nature property. And the reason that most people would seem to rebel when they hear it is because they are not capable, semantically, to distinguish between a law of nature and a phenomenon of nature.

For example, the average person when he hears this will rebel and have his hackles rise, because he'll say, "Well, how can anybody own gravitation?" Well, nobody said that gravitation can be owned and that isn't what I said. The law of gravitation is owned by Newton with a farther account by Einstein and the two of those, separately, are the owners of the law of gravitation. The one for the first one, which is the first approximation, and the second for the second one, for the second approximation, which couldn't have come about without the Newtonian one. These are the owners of the law of gravitation in the two forms it has become known so far. The law of gravitation is ownable because that's an idea of a human being. The gravitation is the phenomenon that it describes and that is in nature. It's inherent in nature and cannot be owned and does not fit my definition of property.

Gravitation Is Not Property

My definition of property is razor-sharp enough that it makes this distinction in it, in its structure. The definition of property as you well know in *Thrust for Freedom—No. 2* where it was originally printed. It was originally in Course 100 but it was printed here, originally:

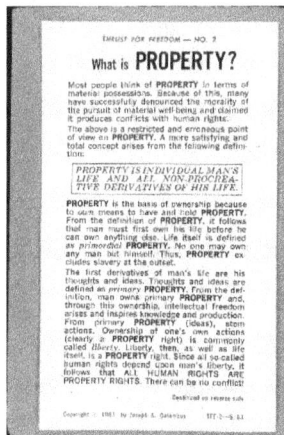

THRUST FOR FREEDOM — NO. 3

What is PROPERTY?

Most people think of PROPERTY in terms of material possessions. Because of this, many have successfully denounced the morality of the pursuit of material well-being and claimed it produces conflicts with human rights.

The above is a restricted and erroneous point of view on PROPERTY. A more satisfying and total concept arises from the following definition:

PROPERTY IS INDIVIDUAL MAN'S LIFE AND ALL NON-PROCREATIVE DERIVATIVES OF HIS LIFE.

PROPERTY is the basis of ownership because to own means to have and hold PROPERTY. From the definition of PROPERTY, it follows that man must first own his life before he can own anything else. Life itself is defined as primordial PROPERTY. No one may own any man but himself. Thus, PROPERTY excludes slavery at the outset.

The first derivatives of man's life. Thoughts and ideas are defined as primary PROPERTY. from the definition, man owns primary PROPERTY and, through this ownership, intellectual freedom arises and inspires knowledge and production. From primary PROPERTY (ideas), stem actions. Ownership of one's own actions (clearly a PROPERTY right) is commonly called liberty. Liberty, then, as well as life itself, is a PROPERTY right. Since all so-called human rights depend upon man's liberty, it follows that ALL HUMAN RIGHTS ARE PROPERTY RIGHTS. There can be no conflict!

Continued on reverse side.

Copyright 1961 by Joseph A. Galambos   TFF-3—6.61

Property is individual man's—later improved to volitional being, to generalize it—property is an individual volitional being's life and all non-procreative derivatives of his life. All right. Gravitation is neither any volitional being's life, so it doesn't fit that part, nor is it a derivative of that life. Therefore, it does not fit that. Therefore, gravitation is not property. Simple. And nobody claims it is. But the *law* of gravitation is a derivative of a person's life. That was developed by Isaac Newton. Gravitation existed before Newton formulated the generalization that explains how you can determine both the magnitude and the direction in which gravitational force acts and he also developed a theory by

99

which you could use it to compute motion in the celestial world of the orbital behavior of various objects in a gravitational field. And all of these are his property. He thought of this. Therefore, the knowledge of how to do this is his property.

Primary Capitalists

And so, since von Mises did not say innovator, and since von Mises certainly cannot be accused of not being articulate and he cannot be accused of failing in the precise use of words, he clearly intentionally meant inventor, and there's a difference in our concept there. I include innovator because to me primary property of a cosmological type is also property, whereas to von Mises only the direct applications, the technological applications which are called invention, are property. However, with the broadening of this, both *discoverer*—which is cosmological innovator—and *inventor*—which is technological innovator—both are primary capitalists. The entrepreneur is clearly a primary capitalist. He's investing his technical knowledge of how to organize the factors of production including, but not limited to, the ideas that enter into the product to make a profitable business out of it. That requires his knowledge, which is skill, which is capability, which is talent, all of which are primary property.

The Progressive Elements of Civilization

Furthermore, the investor, which is also called the capitalist, the men who put up the money to invest in a business, that's clearly a capitalist, but that's a secondary capitalist. The others are also capitalists except they're primary capitalists. So if you include both kinds of primary capitalism, innovation and entrepreneurialship, plus the secondary capitalism of putting up the P_2 to operate a business, then that single concept stands, both including the other three. Plus the cosmological innovator that von Mises does not include, probably deliberately. Plus, it made it simpler to reduce it to one concept instead of three. And those are the progressive elements of civilization as von Mises, before me, said.

What I have added is, besides the simplification and the addition of cosmological innovator—which is not the slightest bit immaterial. That's the most important part, he left out the most important one.

That's like describing an animal and forgetting the head. In addition to all this, this represents a mechanism of identifying the source of progress. It is these people upon whom all progress depends. The others are the beneficiaries.

The Problem of All Civilization to This Date

Now, the problem of all civilization to this date—why freedom could not exist—is that all freedom depends on property. Any concept of freedom that does not anchor itself upon the protection of property is a political, religious farce. There is no possibility of even identifying freedom in the absence of a discussion of property and its full protection. Everyone wants freedom as an intuitive, fuzzy goal. What he means by it is he wants to be happy, and a world where he's happy he would call free. And he's usually happy when he gets what he wants, whether he worked for it or not, and whether he likes to work or not. That kind of freedom is, in fact, the worst tyranny, where everybody gets something for nothing because that produces the enslavement of the major producers. But it makes people who would like to be loafers, it puts them in a position where they don't get something for nothing and they call that terrible, hence they are not free.

Today's Misconceptions of the Political Majority

See if that doesn't fit today's misconceptions of the political majority. Well, does it fit? Isn't that what people want, really? Not freedom. Irresponsibility. They want to get without doing. Whether it's outright thievery or simply parasitism. A parasite can only get what he gets without working for it by either mooching or stealing and if he doesn't want to steal and he's too embarrassed to say he's begging then he demands it as a political right, so he appoints the state to steal for him, and he's got the political clout because they ask the majority opinion and they count noses, and those who do little outnumber those who do much. Simple. *N'est-ce pas*?

That explains more about politics than all the politicians who have ever given their programs, combined. They can talk from now till hell freezes over which is not easily expectable, for two reasons, because the temperature is too high, and secondly, it doesn't exist. [*Audience laughter.*] Except here. [*AJG takes a drink.*]

Faith:
A Substitute for Thinking and Corroboration

Religious hell was invented to pacify men who are living in the real hell which is the world they have to take it and expect for something better by going to heaven as a substitute. Tremendous invention of mysticism. That's called incentive generation by mystical standards. [*Light audience laughter* as *AJG drinks*.] Also based on the Santa Claus principle: be a good boy, and you'll get rewarded. That means be obedient to your intellectual, mystical masters, and believe as we tell you to believe and you will enter the kingdom of heaven. Do not, and you will roast in hell.

If you don't think religion has done much to reduce man's psychological capabilities, in case you think I'm dwelling on it too much, oh no, I'm not. I'm very gentle on the subject. You ought to take a course on religion from me sometime. Because religion has conditioned man to be a nonthinker. To accept on authority without knowledge and to believe for reasons only because he has been told to believe, or else. That has reduced man's psychological potential and has put him into a world of dreams rather than reality, where acceptance of corroboration is considered unnecessary as long as you have faith, which is a substitute for thinking and corroboration.

Getting Back to the Sources of Progress

Well, anyhow, getting back to the sources of progress as von Mises stated it and as I have stated it with Occam's Razor simplification with the addition of the cosmological innovator. There is only one class of people involved in progress and that is those who invest for the long term, that's the capitalist. That's the one who accumulates property and puts it to work to create more property. However, there are basically two categories of time scales involved in the accumulation of property, called investment. They are the personal time scale and the species time scale. It so happens that of the three types of capitalists, there is the primary capitalist of innovation; there's the primary capitalist of entrepreneurialship; and there's the secondary capitalist of investment of P_2.

The Distinction between Investing and Speculating

The P_2 capitalist is always a relatively short-termer. He always wants his return soon. Now there's a difference in the world of investment, as explained in V-30, between the long-term investor and the short-term investor. The short-term investor wants to buy today and sell tomorrow. Some want to even do it on the same day; it's called day traders. They're not really investors, this is in V-30 explained; this is not investing. This is speculating. Calling it investing doesn't make it so. It's like Abraham Lincoln's well-established story about when he got tired of somebody who was annoying him he said to him, "Tell me, if you call a cow's tail a leg, how many legs does the cow have?" And the other party said, "Five." And Abraham Lincoln said, "No, four. Calling a cow's tail a leg doesn't make it a leg." Well, you can call day trading *investing* or even *in-and-out speculating*, trading. Now you can call that investing but it doesn't make it so.

Investing is a longer term concept and there are people who are such short-termers that they watch the ticker tape and they see something that's in a rising trend and they buy something at fourteen and three-eighths and they're willing to sell it at fifteen and a half because they can make up their commission and make a few bucks on it and that's the way they try to make a go of it. I assure you it is nerve racking. I assure you it is not impossible to succeed at this, but it takes certain skills that most people don't have and certain temperaments that most people don't have and it takes all of the time. At least all the time the market's open and quite a bit more to think about it. It is not the easiest way to make a living even though it is possible, but not the easiest. That is not investing, it's speculating.

The Concept of the Dividend Is a Short-Term Concept

An investor, in general however, does expect some return on his money and most company managements know that. And for that reason, they invented the concept of the dividend: "Here's something to show you that we appreciated your money and we are giving you something back to show you that we're still functioning and we're doing well and we want you to keep up the good spirits and liking us." And that's what a dividend is. It's a morale booster in P_2 to keep them investing, so they pay out dividends. And there are people who invest for the dividends, which I point out in V-30 is not the strongest form of investing. It's not impossible to do this, even with success, except that it's not the strongest form of investing. This is the income approach. A dividend is only important in the short run. It is the long-term growth of the company, and I of course make a distinction between speculative run up of a stock's price and true growth of a company's productive output and the profit it makes thereon.

People's Emotional Stability Connected to Financial Stability

I don't want to go into V-30 here but that is of course not in any way inseparable for it. So although I'm not going to repeat the course here, I draw upon it from time to time to show that this has much to do with psychology, because people's emotional stability is unavoidably connected to their financial stability.

Speculating Based on Pessimism

In any event, the reason that investors put money into whatever they invest in is clearly because they expect to have property in greater value back later. That's the whole concept of investing. They don't always succeed but they never invest saying, "Hot ziggety, this is a real lose racket, I lose my shirt on it!" [*Audience laughter.*] Nobody does that intentionally. If they have a real loser, they are speculators who short sell which means that they sell the stock instead of buy it and expect it to go down and make a profit by buying it back later and there's an entire mechanism for doing this. You can borrow the stock, sell the stock you don't own and then buy it back to repay the one who's loaned it to you. In the meantime, they also hold some money from you as collateral, so you don't abscond with the stock that you sold which wasn't yours but which was loaned to you. And there's a whole mechanism for this. And that's speculating based on pessimism, by the way. I have never been a short seller. I just do not have the attitude for it because you have to be a born professional long-term pessimist to have that kind of—never mind. I'm coming to pessimism, anyhow. And optimism.

Optimistically Oriented People Should Never Try Short Selling

The entire concept of short selling is a pessimistic one. People who are optimistically oriented should never even try it. It is an absolute losing game. You have to have people who are cynical and possibly hostile to the world who would make good short sellers. Or born speculators, people who minimize their emotions to the point where they take a dispassionate and totally detached attitude where they don't have any attitude towards the survival of companies or civilization or the market stability and to whom the only issue at hand is making money. People to whom achievement is important cannot make good short sellers. I have long, long ago detached myself from any intention to short sell, not because you can't make money at it if you know how, but frankly, I don't have the proper attitude for it and I wouldn't want to take the risk. I'll take risk on things that are rational risks. That's not a rational risk for someone with my temperament.

No Upper Limit to What Short Sellers Can Lose

Now in general, except for short sellers who expect a stock to go down and then if it doesn't go down, boy have they lost. To anybody who bought the stock it's a gain, to them it's a loss, and it's unlimited. If you buy a stock for $30 a share you always have the consolation of knowing the maximum you can lose is $30 a share, if you didn't buy it on margin. That's all that you can lose. If it went to zero and disappeared from sight forever you lost $30 per share, plus the commission in buying it. But if you short sell it at $30 a share and it goes to $1,000, you have lost $970 a share. Of course it's not likely it will do that but there's no upper limit to what you can lose. So it's a rather dangerous sport. [AJG chuckles.] It's probable that even football is less dangerous, so that's pretty dangerous too.

The Principle of Investing

In any event, the intention in investing, true investing, is that you want to take some of your property and you want it back later at a higher value. Not necessarily higher dollars, but higher real values, purchasing power, not inflated value. In these days of inflation, one has to distinguish between value and price. You always have to, in principle, but here it's especially important. You expect a greater value later than what, you might say, withheld from circulation for a different purpose. You always have alternate choices as to where you invest your money or your time, and you have a decision to make. Are you willing to wait? Is the time of waiting, that you're not getting some earlier and more desired value, worth it? And if you can wait, then you should expect the reward for such waiting. That's the principle of investing. As I say, nobody invests to lose money but you can lose money. But the intention is to make either money or some other value. There are things more important than money. Achievement, of course, which should be measured by money. And that's the principle of investing.

P_2 Investing Is a Short-Term Endeavor

But since P_2 investing is not done, in general, by major achievers, but just anyone who has the wherewithal to do this and the willingness to do it and the imagination to do it, which in itself is a small percentage of all the people because most people would rather spend it as they make it and even overspend what they make and go into debt and then spend their whole life as derelicts trying to work their way through one crisis after another and every day represents to them another bill, another debt. And so those who have enough, you might say, longer-term view that they wish to take the position that they're willing to wait a bit for a reward which is bigger than what they can get now for what they have, those are the investors. But here, it is almost inconceivable in flatland to find anyone who would be willing to invest for something longer than he's going to live. The only exception to that would be perhaps an old man who invests for his children or grandchildren, but then he expects somebody close to him to reap the benefits. [*AJG drinks.*] But nobody in general would be willing to invest when the return is a thousand years from now. And so P_2 investing is, per force, naturally and innately, a short-term endeavor.

The Total and True Entrepreneur

The primary capitalist, which represents the entrepreneur, has also been, in general, a relatively short-term capitalist. He invests his skills and knowledge with a view to making a profit. That's the true profit, the entrepreneurial profit. That's the P_1E's profit, except that there haven't been any P_1Es. There have been businessmen and promoters thus far. There has never been a P_1E before this in the sense that I have developed it. That's the total and true entrepreneur. After V-212 I have tended to drop the word *entrepreneur* for those who think they are entrepreneurs, and use it only for P_1Es in the full sense of V-201. On that basis the world has damn few at the moment, not zero, but damn few. And before the last two decades, there were none.

IBM at the Top of the Ladder as an Investment Vehicle for Flatland Purposes

Okay, on that basis all entrepreneurial effect and achievement so far has been in a time scale which varies from very short term like this year—the profits of this year and all that matter—to up to ten years, is better than average. Twenty years is superior. A person who thinks for the rest of his life in business is at the top of the ladder. And there are a few exceptions who will even build something for the next generation, extend it, say one generation. And the IBM-type businesses have been in the top category of quality businesses, which is why they receive the mention that they do and the examples and illustrations that they do in V-30. Not because I like their ideological posture. I explained there I do not. They are rather damn fools on politics.

Thomas Watson who built up IBM, the late Thomas Watson, having read a biography of IBM, I found the man was unbelievably deluded on almost everything political, whether it's Franklin Roosevelt or Harry Truman or Adolf Hitler. The man was an absolute ass on politics. Yet he knew how to run a company and for that he is to be commended. Everyone is commended for what he does right, not what he does wrong. So in that sense, he did know how to build and create a successful and powerful business. Of course, he indulged in the usual P_1 thefts that every other business is guilty of, but judging him by flatland standards, not by spaceland standards and by flatland's non-integrity, he was a great businessman. For P_2 investing in flatland I mentioned IBM is at the top of the ladder as an investment vehicle for flatland purposes and within flatland time scales and with flatland moral concepts, which are a pretty low standard.

Entrepreneurs Have Always Been Short-Termers in the Past

Okay, on that basis, we have gotten to recognize that the other kind of capitalist, the entrepreneur, is also basically a short-termer and always has been in the past. *Always* has been in the past. With no failure. That's why when I asked you to name entrepreneurs of a thousand years ago you don't find any because they're not important enough to be remembered that long. There were people in business and who took risks then. The Roman Empire was basically a PSC country too. Not as far developed as today because their technology was lower. It was pre-Newtonian technology.

But the Roman Empire recognized what a merchant's profit is. They didn't have much in the line of factories because they had no Industrial Revolution, but they had individual craftsmen and people who could do things like build houses, much better than our American houses are built, I'll say. Somebody had to build those houses. Some of them are still standing! They built aqueducts, they built roads. Well, people had to know how to do that. But the significance of their P_2 enterprises has disappeared. Well, that's because the long-term time scale has never been recognized by entrepreneurs, businessmen, merchants, bankers, they've never done anything in this line. And therefore, businesses remained short-term in quality and scope.

The Innovator:
The Third, Final and Highest Class of Capitalist

That leaves the third class of capitalist who has never been looked upon as a capitalist or as a property holder at all, until myself. That is the innovator. Now, the innovator is the third, final and highest class of capitalist there is. The innovator can be in any time scale from personal on up to the top of the species time scale, which blends into the cosmic time scale, depending on the quality and the magnitude of his innovation. If he's a mousetrap inventor he's in the personal time scale, bordering on the trivial time scale. It's a short-term and relatively unimportant invention, but even that is on the short end of the personal time scale. Big inventions like the wheel, that's in the species time scale from the very beginning.

Mammoth Technological Innovations Here for All Time

The wheel is as important to mankind as a major law of nature. That's not a law of nature, but it's sure as hell is a major application and that's here for all time. There's a future for the wheel. Too bad there was none for the inventor of the wheel. There is a future for flying. I consider the invention of man's ability to fly as of the same magnitude as the wheel itself. These are technological innovations but they are so mammoth that to say they're long term is an understatement. As long as there's any form of volitional being to come after us in our thoughtsteps—that's better than footsteps—there will be flying too, and there will be astronautics.

Astronautics Is a Misnomer

Right now it's too ambitious to call it astronautics at the moment, it's just local planetonautics. Then it becomes planetonautics, then it becomes astronautics, then it becomes galactonautics and then it becomes cosmonautics. The Russians, in their usual boastful exuberance, have somewhat jumped the gun in calling their parrots cosmonauts. We have jumped the gun in calling them astronauts. They haven't made it quite that far yet.

All Cosmological Discoveries Are in the Species Time Scale

Anyhow, the development of these technological major innovations is most emphatically the species time scale. All cosmological discoveries, without exception, are in the species time scale. The work, the product, the achievement lasts far longer than the person involved who developed it. That means that a person to pursue happiness satisfactorily, to do any of these things that I just described, has to have a personal outlook that fits the time scale his work is immersed in or he couldn't do what he has done.

The Disparity in Time Scales and Mismatch of Goals of Entrepreneurs and Innovators

There is the problem of history; why there has never been freedom. It couldn't even be identified until now. Because there has never been a mesh, an interlocking, of the proprietary interests of the innovator, especially the cosmological innovator, to whom can be joined the major technological innovators such as wheel innovator, flying innovator and such things. The major innovator—the cosmological innovator is automatically major—the major innovator's time scale is naturally species. There is no other way to do these things. Those who make a practical application which can be put into the commercial world, into the market for sale to people who are not interested in what happens in a thousand or ten thousand years or even a hundred years, but they eat now, they wear clothes now, they need vehicles now, they need the necessities and they like the luxuries now, or they cannot survive now—these people are supplied the things they want now by people who make for now, and later. And those are the entrepreneurial people who should be in the species time scale but because of their own personal shortcomings, and the lack of imagination and not being in the innovation field, they haven't got the imagination to know what to do with the innovation at a full scale, which is why innovators have had to wait centuries before their work was even noticed or applied or successfully used.

And by that time they are good and dead and, "Well, we'll just help ourselves. There's nobody to pay. There's nobody to talk to." Even if they were moral, even if an entrepreneur were moral, and say, "Well yeah, I'd be willing to pay for it, I think it's only fair to pay somebody who invented this, but he died four hundred years ago. I can't pay him. So, line of least resistance, let's forget about it. I would be happy to do it if I could, but I can't so therefore, let's do something more practical."

So naturally, the disparity in time scales and the mismatch of goals because of the disparity and time scales of entrepreneurs and innovators has always favored the short scale for the entrepreneur and necessitated the long scale for the innovator. And the man who puts up the money, the capitalist, the so-called secondary capitalist, well, he has to be in the short scale since the money means only something to him in the now. And there has been a natural and easy linkage between investors and businessmen, and an almost impossible relationship between both of these two and the third. Which is why von Mises said *inventor* and not *innovator*. That is not an accident. I'm absolutely sure of that. And yet the omission is more important than the inclusion. [*AJG takes a drink.*]

The Masses Have a Market Function in the Ideological Program

In this course, psychology, I have to discuss both the hot and the cold end. The hot end because it matters. The cold end because they exist too, and reality requires non-overlooking what exists and there is an interaction and ultimately, as I pointed out in V-201, if that interaction did not exist there wouldn't be any market for innovation. There just aren't enough people in the world to use automobiles and packaged food and all the other goodies of life if the only customers were major innovators. They just don't make a big enough market pull. Their combined size and number wouldn't make it possible to put up one automobile plant, or a chewing gum factory or a lightbulb factory. Therefore, not because of personal merit, but because of market size, even the masses have a market function in the ideological program.

The fourth step of the ideological program, which is maintenance, is absolutely essential for the first step to exist. Because without the maintenance, which is the proprietary profit mechanism, to justify the very expensive cost of production in the second step—the first step expenses are relatively modest. The second step ones are exceedingly expensive.

The Ideological Importance of the Cold End

To discover the law of electromagnetic induction, it didn't cost much, either to Michael Faraday or to Joseph Henry, who are the two discoverers of this principle. On the other hand, to develop just one large electrical generator at Niagara Falls, or at Hoover Dam, or Boulder Dam or whichever one it's called these days, that generator that's installed there, where a steam turbine driving a generator, this is a very expensive operation. And to set up as an electric utility will be, on the low side, millions of dollars. Large electric utilities are capitalized at a billion dollars or more. Not million. Billion for the large ones. So obviously, to make such a large-scale production of high technology costs a lot of money, which means it requires a lot of sales, which requires a lot of people to buy it. You have to have mass production to get mass sales to justify the cost.

And so for the first step at the hot end of the ideological program to have a continuous sustained existence without mooching in the marketplace, which is the spaceland way of innovating, which will remove the disclosure barrier, even the cold end has an ideological importance because that's the source of the P_2 generation.

The Incompatible Time Scale of the First Step
of the Ideological Program

And of course it's very difficult for a hot-ender to deal with a cold-ender. I bare personal witness to that, I'll tell you. I've had other people say that I had a miserable childhood. No I didn't. I had a very pleasant childhood, I had a very happy childhood. I had a great friend, my Father. I didn't need any more. I consider my childhood very happy and unfrustrated. The frustration begins when I met other people.

I would like to point out that the nature of this interaction is a very intricate one. It's closely interwoven. And because only the three can get along but not the fourth in the ideological program:

111

The second step which is secondary production, the third step which is distribution, and the fourth step which is consumption—only those three have had any compatibility in history. The first step is totally incompatible with the other three. There has been a disclosure barrier, which is the cause of all coercion and is the source of the state itself. If that barrier weren't there, the state couldn't exist. You don't have to fight the state, it just couldn't exist.

And the reason why the three steps that do get along and the one that doesn't, why this condition prevails is that all three—the second, third and fourth steps—are basically agreed upon short-term time scales. They have mutually compatible goals which makes it possible to make contracts. The first one you can't make contracts with when the time scale orientation is different. Vastly different. You say, "All right, do it without contract."

Only Two Possible Volitional Interactions:
Coercion and Contract

Ladies and gentlemen, I hope you have understood V-50 well enough that there are two ways human beings, or volitional beings, in general, can interact and only two: coercion and contract. There is no third possibility. And where contract is not possible or won't work out, coercion is all that remains. How about accidental encounter? That's neither. Accidental encounter results in a ricocheting of two people off each other and there's no contract, they just meet and leave. If they meet and stick, it's contractual, or coercion. It's one or the other. And if there's going to be anything other than coercion, it must be contractual.

Impossible to Coerce Innovators to Adopt Short-Term
Time Scale Attitude

Yet there can't be a contract between the first step and the others in production and consumption and distribution, unless they have a time scale alignment that they can deal with one another. And you say, "Okay. Well clearly, since there are more people in the second, third and fourth steps than the first, the first step's got to shape up, and you got to shape him up and get him into line." That is called coercion, against the most able producers, which in fact will wipe out the very thing they do! That's clear? Therefore, that won't work. The only way you could do that is to kill them or to prevent them from working. That's what they do do, and this of course is historically corroborated. Bruno was killed. Semmelweis was pushed into early death either by suicide or accident which I've heard two versions of it, and he deliberately infected himself or the other one, it was an accident. Either way he was pushed into it. Or, it produces a condition where they cannot produce what they do. Either by outright theft, getting them discouraged to do something else like going to the mint, fighting the Smithsonian Institution, immobilizing them by getting them to fail to disclose such as Tesla, you name it. There's different ways to react, but the net result is it either totally eliminates or grossly reduces their output in innovation.

Impossible to Coerce Everyone to Adopt Long-Term
Time Scale Attitude

Okay, let's take the other possibility. Let's get everybody else shaped up on anything in terms of long term. That's also impossible. Because that also would take coercion by a very few people against all of the others, and that would be equally wrong and also impossible because you cannot force long term where there is none. You can get somebody to say, "Hip-hip-hooray for long term." I can even get that voluntarily from people who get excited about it. I told you that in the previous half-session. A lot of people get all excited about the long term, all of a sudden that means a great deal to them. But if their natural thinking does not include that, and it does not go with their attitude, their ability, their drive, and they're not willing to make the proper investment for it, that's not easy. It's easy to say and hard to do. You've got to live it. To live it you have to endure a large amount of decisions, all voluntary.

Qualities a Person Must Have to Be a Long-Termer

Pursuit of happiness involves taking choices which favor something which you get in the near term or later or still later or still later. And it requires a lot of self-discipline not to take something as a choice which favors you immediately, but say, "I'll wait with the expectation of some reward till later, because I'd rather wait and get still better results later." If that person hasn't got the temperament or the attitude for that, saying you're long term doesn't make it long term. It's the fifth leg of the cow that the tail is not. And that means one of the aspects or qualities that a person must have to be able to be a long-termer, he must have the self-oriented internal discipline to know what a value judgment is which favors the long term instead of the short term, which means that he's willing to wait.

Patience Is Necessary

That's why patience is necessary, which I've always stressed in courses such as V-30 and other courses, V-201. It's easy to be impatient, that takes no skill whatsoever. "When do you want freedom, Galambos?" Right now. Yesterday. A hundred years ago. If I can't have it that soon I'll wait till Thursday next week. [*Audience laughter and AJG chuckles.*] This is preposterous. That's one of the things that makes people get disenchanted with my courses. "I've taken this course. I put my money out for V-50, I even put my money out for V-201, I sat through the whole damn thing and we still got taxes, and we still got one clod after another running for office and getting elected. And we still have wars and we still have communism and we still have terrorists and we still have inflation. Look, obviously it can't work."

And we haven't even discovered the inventor of the wheel, as Dr. Snelson so ably pointed out. [*Some light laughter in the audience.*] I don't know where he was when I said in 201 that it might take a thousand years to find it out, or more. Weren't you there when I said that? [*Snelson responds, "I was absent that day," followed by audience laughter.*] I was absent that day. [*AJG laughs with the audience.*] As a matter of fact, he heard the whole course five times.

Impatience: A Natural Reaction of the Shortest of Termers

Anyhow, the situation is, it's easy to be impatient. That takes no talent whatsoever. You can be a fly and be impatient. You can be a cat meowing for its food, it can't wait till the can is open. I can't get my stupid cats to understand, "Look, while the can is being opened, keep your nose out. [*Audience laughter.*] Move away from the can. Your nose could get cut by the can and move away and while I'm opening it up, I'll be more efficient at opening it up and I'll open it up sooner so I can give you the food a half a second earlier." But no, they cluster around the can and won't move away. They can't wait to get at it, and they actually delay the receipt of the food because they're so damn impatient.

I use this illustration because you know I love cats and here are these charming and lovely and wonderful animals who happen to be stupid but exceedingly nice. [*Audience laughter.*] They have many desirable features, brains not being one of them. And yet, I use this illustration to indicate to you, you don't need to be smart to be impatient. You don't have to know anything at all! That's a natural reaction of the shortest of termers.

Patience:
The Combined Total of Character and Courage

To become patient requires both the knowledge of what's in it for you to wait and that there will be a greater reward if you are patient and wait, and then to have the discipline—self-discipline, not by Prussian army command—but by your telling yourself, "I am mature enough to know enough that there is a reward in it for me and I will pursue happiness better and abler by waiting for a bigger return than by demanding something immediately which will prevent me from getting the bigger thing." Now that takes thinking. It takes some character and courage, and the combined total is called patience which is cultivated and not innate.

Once You Learn Patience, It's Learned

I don't know anybody who is naturally patient. I am naturally impatient too. It was a hell of frightful delay for me to get out of the boondoggle, which is what I describe in V-30, the mechanism which I sought financial independence with. That's what caused most of my mistakes. I was in such a hurry, I couldn't wait for it to happen fast enough at the rate I was able to withhold money from my income and not spend it but invest it. It wasn't growing fast enough, so I said, okay, I took bigger risks. So I bought some stupid things at a stupid time under stupid

115

conditions. I even margined it which I shouldn't have because I was an amateur and I lost my shirt on this occasion or that one or the other one. And it delayed it. And that's how, the hard way, one learns patience. But hard or easy, once you learn it, it's learned.

Most People Have to Learn the Hard Way

There are people who can't learn. There are people who can experience the same thing I did ten times over and they still won't learn. As a matter of fact, for anybody really intelligent, hearing what's in V-30 would prevent it for yourself from ever occurring. And for the price of the tuition, which is actually nominal compared with what you could lose on one transaction, any one transaction could cost you more if it's a boo-boo than the full tuition. If you could get just enough out of the course to avoid one lousy transaction, you've already gained. But most people won't learn until they get it the hard way. The course still has a value because if they didn't hear the course and had made the mistake, they wouldn't know what they made a mistake in and they would blame it on the broker, the tip sheet they read or on something other than what it really is.

And what it really is, I can tell you in advance without knowing what the mistake is. It's you. You made a mistake. [*AJG chuckles.*] And you haven't had the courage to face up to it. "I made a mistake." Find out what is the mistake and don't do it again, and if you hear V-30 and then you make the mistake then you can find which category of mistake it is for the various types of mistakes that were covered. And then you might be able to generalize, maybe. The fourteenth time. To get it on a twelfth time you're better. To get it on a tenth time it's still better. To get it on a second time it's still better. To get it on the first bounce, that's tremendous.

The Whole Point of the Learning Process

But you see, the learning process includes, but is not limited to, finding out in what you do do, most of what you do tends not to be right because on a haphazard basis, the random thought and the random behavior will not be right by the standards of rightness if it's not thought well out in advance. So if you just take some course of action by random selection, the chances are it's wrong. But then when you have been burned by it, and you remember what caused it and try to find the blame instead of trying to find an excuse as to who to blame, then you have learned something. If you refuse to accept this, you have not learned something. And there is the whole point of the learning process. You have to be able to filter out the right results from the wrong results and that comes from corroboration. Did you get the result from your behavior that you anticipated? When you started to do something you had a goal. You had some reason to do it. You had some expectation. Did it materialize in full, in part, or not at all? Or did it go so bad that you ruined everything?

The Difference between Lucky and Right

Obviously, if you realized your expectation in full, then you were right, or else you were very lucky. You can tell the difference between those two, however, by doing it again and again and again. The luck won't hold out. The random probability of being lucky ten times in a row is small. But if it always works, then it's not luck. Then you're right. If it never works, you're wrong. If it works partly, that means there may have been some merit to what you did but some part of it didn't work, and therefore you have some improvement to make, and therefore the result gives you the feedback of what's your next step, and that's the learning process. That's the scientific method. The two are the same, by the way.

The Difference between Ordinary Use of the Scientific Method and the Scientist's Use of It

The difference between ordinary use of the scientific method by the average person and the scientist's use of it is that the ordinary person uses it for daily routine tasks. That's how the cook learns how to cook, that's how the plumber learns how to plumb, that's the way the average person learns to do the various things that are to be done, by applying the scientific method to some extent, successfully. And you get better and increase your skills by saying, "Did it work?" Did you get what you expected to get out of it? The difference between that and the scientist is he applies it to usually abstract things that most people never give a hoot about and wouldn't concern themselves with and which, "Well who cares about that? I mean, who's interested in going to the moon?" was the prevalent thing in the world I heard as a kid. "Who would want to go to the moon?" Me. And I was all alone.

117

"Why would you want to go to the moon? What's there that's better on Earth?" Or, "That's ridiculous. You know that's impossible." I said, "No, I don't know that's impossible." And that made me unpopular with adults as well as children. [*AJG chuckles.*] Also, I've been considered crazy. As a matter of fact, most people felt sorry for my parents for having a nut for a child. [*Audience laughs as AJG takes a drink.*] So I had an early exposure to this type of thing. Which is why I can take some of the crap I've been taking the last couple of years. [*AJG chuckles.*] To tell you, if I were running for political office, I couldn't take it. I'd have to be much more popular.

All Achievement Comes from Long-Term Time Scale

I'd like to point out to you that the knowledge that a scientist has, the knowledge that a layman has, comes from the same source. They both get it from the scientific method. The only difference is the scientist does it by applying it to abstract things, and usually to things which involve much greater generalizations which would never concern other people which, to a large extent, is a difference in the time scale attitude. Therefore, this discussion of time scales is more than minor. It is very major to this theory, and when we can go about talking about motivation of people it will become very clear before this course gets finished that all achievement comes from long-term time scale, which you've heard in V-201 but not as far as the reason for it.

Short Term Is the Source of Ego Minor Behavior

Now, other things than that, the short-term time scale produces not just criminality—which I said it does. All criminal behavior is short term in nature. It also produces incompetence, which is not necessarily criminal. The short term is the source of ego minor behavior. That's true in children, *ab ovo*. They haven't got a long term in mind; they don't know what long term is.

Children Learn by Generalizing

A little kid comes into the world, looks around, he doesn't even know what a person is. He has to find out he's a person. He has to know how to communicate first. He has to pick that up gradually, slowly, learn how to talk, find out what he or she is. A child. What's a child? A child, well, that's something that later on will become a grown up like the parent.

"Oh, you're gonna be like that someday?" [*AJG chuckles*.] And he has to connect up the two that he or she, something that is just barely off the ground and just barely begun to exist, is just barely beginning to even recognize its existence, that is self-cognition. This kid has to learn to generalize that this little character that doesn't look at all like the bigger one that his parents are, that they are actually replicas of each other except at a different time, and that the parent was once a child too. That's a very difficult thing to imagine. That's called the generation gap today. [*AJG chuckles*.] It's very difficult to figure out for a little kid that once upon a time those ogres that are his parents were once little children and as irresponsible and as happy-go-lucky and as non-thinking as they are.

The Child's Formative Years Are Exceedingly Important

Then they grow bigger and they start thinking, up to a point, and what point it is depends mainly upon the parents. Secondly upon the rest of their environment, including their so-called school. The parents are the most responsible, for two reasons. One is by the law of logarithmic stimulation, they are the closest and they have the highest proprietary interest. Secondly, because the rest of the environment of the child to a large extent depends on the parents' selection as to where they live and what they do and who they wish to associate with. For this reason, the child's formative years are exceedingly important and in the beginning they have no conception of generalization and later they get a little bit and the faster the generalization concept develops, the sooner they're able to do something other than just consume and exist and be total parasites.

All Infants Are Parasites

Because that of course, whether you like it or not, is the case. All infants are parasites. I know the maternal types, "Oh my, you call my child a parasite?" Yes. Flatly and clearly. Produces nothing and consumes a lot. [*Audience laughter*.] All children at birth start out as parasites. That's what a parasite is. Someone who lives off other persons' effort. [*AJG drinks*.]

A Human Being Takes Two Decades to Grow Up, Biologically

Human beings are actually among the slowest operating creatures in history, in the history of the world that is, not the history of man. Do you realize that it takes two decades for a person to grow up? We're real left behind. A cat's grown up at age one. You see, cats don't live as long. A cat can live ten or fifteen years. Some have lived even longer. I have heard of a twenty-year-old cat and I just read about one that's twenty-nine. Cats in general, have a normal life span of about ten to fifteen years. It takes less than one year for a cat to be full grown. A female cat of one year of age I believe can also give birth and can pro-create. All right. One year, that's on the high side.

My present office cat, Sam, one of them, he is the original, he's the head cat in the office. [*Light audience laughter.*] I once brought him in to the open-end course, some of you saw him. He's a very handsome black cat. Very obnoxious, though. Very gently obnoxious, very tame and very sweet, but he never shows any affection at all. When I say obnoxious, I'm referring to a lack of affection, not to wildness. Anyhow, Sam, I remember he was as big as he is now when he was nine months old.

All right. It takes a cat less than one year to grow up in the sense of biological growth. Intellectual growth hasn't increased either since then. [*AJG chuckles.*] So psychologically they were grown up, as far as they were going. One out of ten. That's a low end of the normal lifespan for a cat, that's 10% of his life. If he lives to be fifteen years, then one year represents only 6⅔%. And if you actually take nine months as the time it takes to grow up, then it's even a smaller percentage of the time that the cat lives that it takes to grow up.

A human being takes two decades to grow up, biologically. Psychologically, it depends. Some grow up sooner. Most grow up later, or not at all. So let's take two decades, and the normal life expectancy these days is approximately seventy years. It varies a bit with latitudes, races, climate, sex, and so on, but let's say seventy years is an average lifespan for a human being these days. All right. Twenty years out of seventy is over 28% of one's life is wasted growing up, and many people don't live to be seventy. That's, let's say, the average life expectancy these days, give or take a few years.

The Inefficiency of the Human Life Cycle

There are many people who don't live anywhere near that long. Spinoza died when he was forty-five, to name a major person who lived a short time. He died when he was forty-five. So did Semmelweis. Galois, one of the world's great mathematicians, died before he was twenty-one, and that means he spent his whole life growing up and he did what he did in a very short time between, let's say, age eighteen and age twenty-one. As a matter of fact, everything that we know him for and everything that he will be remembered for and much of 20th century mathematics depends on what he did the last night he lived. In the last thirteen hours of his life he wrote sixty pages, which is his life's work and he's one of the major mathematicians of all time. And yet he spent all these years growing up to get to the point where he could do that one thing.

So you can see the inefficiency of the human life cycle because of our biological backwardness in terms of how long it takes to grow up and what's left of life after that. Now for example, if it took us twenty years to grow up and we lived a thousand years that would be perfectly fine and dandy, but when you only live forty, fifty, sixty, seventy years that's a hell of a big investment for such a small return. That means that you better be efficient about it or you are going to waste all of it. [*AJG chuckles.*]

Ego Minor: Totally a Short-Term Phenomenon

Now, the ego minor, which is a major topic here, and the short term are inseparably and inescapably intertwined. The ego minor is totally a short-term phenomenon. The ego minor is always caused by the inability to self-discipline oneself to say, "This is less important than something else." And whatever form the ego minor comes out in, whether it's megalomania, dreams of glory, chintziness, outright crime, criminal action, these are just some that I have mentioned—it has many manifestations—whatever form it comes out under, or whatever combination of forms it comes out under, it always boils down to a single thing. The person who is behaving with an ego minor, at any given moment, for any given reason, for any given act, is unable to put the longer view ahead of the shorter view and he's not rationally thinking it, he's emoting it. The emotion governs his behavior.

121

The Characteristic of Emotion

Now I have already said I am not one of those who wishes to down-grade emotion to an animal-like characteristic. I like the characteristic of emotion. I'm glad human beings have it. Man would be even duller than he is. Most people I find dull, as well as unpleasant. Only a few people have any excitement around them and any real pleasure in knowing them. All such people are not unemotional to me. In other words, anybody I have real affection for has some form of emotional characteristic, because otherwise the person is flat, insipid and color-less. If a person has no sense of humor, I find him not only dull but bor-ing and even suspicious. There's something wrong with him. I would like to point out, these are emotional characteristics.

I'm not downgrading emotion and saying only rational behavior is necessary, emotion must be suppressed. Then you get the wooden character that Ayn Rand writes about. Her characters who never smile, thinking that smiling is a show of weakness and a form of triviality. Who don't laugh, who don't have a sense of humor and who are stoat-and-bottle faced. They have an absolutely featureless face and they're more like automatons with a brain. Her characters are, mainly, very brainy automatons. Not real people. I mention that because some people have looked at her characters as models of perfection. They are not. Only the intellectual part is acceptable to me.

I think that it is necessary for people to show their emotions. What I would like to point out is not the emotional part being wrong or needed to be downgraded or reduced or absented from one's existence alto-gether, but to recognize that emotions have a function in life too; to have enjoyment and show enjoyment. There's no way to be happy more easily than to recognize the beauties of the having of a life and the doing something of it, and it's perfectly fine to enjoy it and show that enjoyment. If you have been injured it is perfectly fine and proper to be angry, upset, even frustrated. These are proper release mecha-nisms to prevent disease.

Human Biological Disease Very Closely Related to Psychology

I'll discuss human biological diseases later. They are very closely re-lated to psychology too. They are psychological connections, many times, psychological causes. I don't think any person is afflicted with a disease without psychological inputs. The disease can exist but a human being afflicted with a disease has psychological connections. You heard me correctly. I think all illnesses and even accidents have psychological inputs.

Sometimes they're the total causative agency. Other times they are merely a factor in the receptivity of the individual to be so afflicted. You may say, "That's going too far. As usual, you're exaggerating." No. Exaggeration is a form of frustration. When I'm giving a lecture that is not the frustration. It's the way it's received that's the frustration. I'm just giving it now. No, that's not an exaggeration. It is not an exaggeration to say that.

You may say, "You mean if a person gets a cold, that has a psychological input?" Absolutely. Absolutely 100%. As a matter of fact, you can eliminate colds from your life altogether if you have the right attitude. You say, "Well it's caused by virus." I know. A virus is the name of anything that a doctor cannot identify which is contagious. [*Some laughter in the audience.*] And an allergy is anything that a person may be afflicted with which you can't identify which is not contagious. [*AJG chuckles.*]

You may say I'm not being fair. That's not true, I'm being very fair. And you say, "Well, aren't there such things as viruses?" Of course. Electron microscopes are used to look at them. But the illness is not only caused by a virus, but whether one can survive the presence of the virus without getting sick or not. There is an interaction between the virus and the other organism and the attitude of that person towards his life will affect whether he gets sick or how sick he becomes or how long he stays sick.

You say, "A broken leg is psychological?" No, the broken bone is not psychological, but whether you break it or not might be. "You mean, if a person has a broken leg, he has a psychological defect?" He probably had his mind on something else when it happened. [*AJG chuckles.*] Unless it was externally induced by coercion or somebody broke it for him. Even in that case you must investigate everything. But in general, yes. I didn't say it's the sole causative agency in all cases, but it has psychological inputs and much of human medical diseases will be reduced when people have the right attitude towards life. I can't claim that I understand how to do this, totally, but I know that this is the way to go. I know for example—I'll discuss it some other time.

The Vanishing of the Ego Minor Comes from a Person's Ability to Think Better

Anyhow, the entire mechanism of a person's outlook on life and all aspects of life, the healthy parts, the unhealthy parts, is affected by whether he has short- or long-term attitude. The existence of the ego minor reduces with maturity to the point where it trends towards vanishing, which is of course total maturity, the vanishing of the ego minor. Then you're not afflicted with this disease. What causes the maturity to come? Where does the ego minor's shrinking come from? It comes from a person's ability to think better, which is why it's so rare that it vanishes.

The thinking better includes the ability to generalize unequal or dissimilar things, to form a connecting link between things that seemingly have no relationship to each other. When one generalizes, he understands more of what's around him. He also naturally develops a longer-term view because the source of the long-term view comes from the ability to generalize. Since that's the hardest of all intellectual feats, the long-term view is rare and the long-term view, therefore, to reduce the ego minor, which is necessary for it, does not come about easily. And for some people, not at all.

The Role of Generalization in Childhood Development

The reason why the generalization produces the increase of the point of view is what I gave you an illustration of before. For example, when a little child comes into the world and knows nothing about everything and doesn't even know what it is; it had to be explained. You are a person, a little one, but you'll become bigger later. He had to be told that he's a human being. What's a human being? Well, we're all human beings. Whoever tells it to him, mother, father, parent, teacher or whatever. What's a human being? That had to be explained, and then as a generalization. Well, there are several human beings around. Let's say it's just his father and mother. Well, they don't look alike. "You mean you're a human being and you're a human being?" Yes. "Well, you don't look the same to me." Well, we're both human beings. "Well, what's in common?" Well, a head, two arms, one nose, a mouth. "I mean, you still don't look alike." That's right. And later, the enormously important Freudian concept. Oh, they find out they're two different sexes and they have to know what a sex is. The mother is one sex and the father is a

124

different sex. "What is that supposed to mean?" Then the child had to find out which one it is. That's a great discovery to find out whether it's like the mother or like the father, and then it still can't identify because if it's a girl it doesn't look like the mother and if it's a boy it doesn't look like the father. So the generalization problem is a continuing one, so this is a permanent crisis for a child to have to face until he understands all this. "You mean I'm like you?" No, you're not like me, but you will be later. "What is later?" [*Light audience laughter.*] That's the concept of time. "What's time?" Well, how can the parent explain when the parent does know it?

The Various Definitions of Time

Even physicists find that time is the most difficult; the single most difficult entity in all of physics is to define is time. Nothing is that difficult as time is. Even in physics it's still open up for grabs for the supreme definition of time. We've gotten some distance along. We've got operational definitions to identify time in different ways in connection with the rotation of the earth, in connection with the vibration of atoms and the emission of light, and we have various definitions of time. We're not even sure it's the same time, did you know that? It's in our physics course. Did you know it otherwise? The concept that we're not even sure, in physics, we don't even know whether the atom time, the atomic time, is the same as the pendulum swinging time, or the earth's rotation time is the same time.

There is a major hypothesis in cosmology, which is little recognized and even less understood which has been around now for a generation and I'm aware of it because I came across it as a student. Most students don't even know it exists and most professors don't know it exists. There's a very interesting hypothesis put forth by E.A. Milne and an intellectual derivative of his, F.L. Arnot, which is a dual time scale cosmology, which would explain the redshift of the nebulae without an expanding universe, which is commonly held in the physics world that the redshift of the nebulae is explained by the fact that the universe is expanding. It may have another explanation. It may be that the time that the atom keeps, and the time that we keep, is not the same.

We're talking about two different physical entities which have been confused and thought to be the same. That is not proven. So when physicists have difficulty understanding what time is, and the problem has

125

been approached, it has been investigated, it has been identified operationally and it is well explored and much is known about it, but nowhere can any scientist ever claim that we have reached the end of knowledge about what time is. And upon that depends on the fate of the universe. When we talk about how many billions of years it will be before the universe runs down and the heat death of the universe. That may have a meaning and it may have no meaning whatsoever. It may be that we have the wrong concept of time. Everything depends on that.

What is Meant by the Word "Later"?

So, in physics we do not know what time is in full and just are exploring the subject and beginning to know something. Now what the hell do you expect a two-year-old infant to know about what it means, *later*. And how can its parent explain to a child what is meant by later when the parent has studied no physics? Which is almost the universal case. Or if he studied it, didn't understand much, which is among the people who have studied physics is close to universality amongst that small group. What is meant by later to the average person, anyhow, by the way?

Would you like to know one of the physics concepts of later? When the universe is more disorganized, it's later. When you die, it's later than when you're born. You're also more disorganized upon death. Your personal entropy has hit you a permanent blow [*AJG chuckles*] from which you will not recover. As a matter of fact, if the entropy increase could be controlled, you won't die. That, by the way, is the track to a longer-term life. To reduce the entropy increase within a given biological organism, and what is called aging is cellular deterioration. What in fact it is, in physics, it's the increase of entropy. Okay. So later is the more disorganized condition.

Developing an Operational Definition of Time

Now anyhow, I was just trying to explain to you some of the problems that are involved in getting a concept of time across, at all, whether it's short or long. Before you can have a decision as to whether you prefer a shorter, a medium, or a longer-term time scale, you first have to have a concept of time. It has to have some intuitive, if not precise, meaning for you. You have searched such words in the vocabulary which are intuitively connected with the word *time* but the connection is very vague, such as *wait*, or *patience*, which means to wait for a result later. Waiting implies time. What does that mean? It means that when you wait you are willing to receive a result not at the time you're thinking about it, but upon the expanding of your elapsed existence. As you continue to exist, that continuation implies time. It's not a rigid connection, it's a very vague connection, but it implies continuation of your existence. Waiting means to expect a result at a time in your existence which is not now, but later. But the word *later* itself implies time, there's circular reasoning there, but the circularity is not easily overcome.

Another way you can do it, and this is of course the technical way to do it in physics: later is when the earth has turned around again. We measure time that way. Every time the earth turns around once it's a unit of weighting, it's a unit of this time. You say well, "Wait a minute, that doesn't tell us what time is." Exactly. It does not. It tells you how to measure it, and that's all you can do. That's what an operational definition is in physics. It doesn't describe it in terms of other words. You will notice, I have not defined time. What is more I cannot define time and neither can anyone else in terms of other entities of physics, which are more fundamental and which you know before, or earlier, which implies time by the way. You get stuck on things like that: before, earlier, later, wait, patience, continuity. All of these imply time! Find a word to describe time that is earlier than time itself! The point is, you can't find a word; you get stuck.

There is no word that is connected with time that is earlier in experience and observation than time itself. So you cannot find any words to describe it in terms that it gives you a defining criteria that says everything contained in this defining criteria, which is called a definition, inside that definition is called time, everything outside is called something else. You cannot find any word that is closer to observation than the word *time* itself, so you cannot define it. It is the fundamental entity of all knowledge, together with mass and length, which are easier to measure though. And the only thing that can be done, this is one of the great, great intellectual philosophical achievements, is to know what to do when you don't have a definition possible. You make an operational definition which differs from a regular definition. A regular definition, you use words that you have previously known to define what is being defined.

127

Defining Time through Measurement Techniques

When you have no words to define something previously because there's nothing previous, you have to start someplace, then you can only define it by specifying how you measure it. So if you say time is that which takes place when, let's say, the earth turns around its axis and the in-between periods between the two times that the earth is in such a position that a given star is at a given position overhead on a given meridian, and the elapsed duration is called the unit of time. One day. That's a sidereal day. If it's the sun that's used, then it's a solar day. They're not the same, for about four minutes. The correct day is the sidereal day. *Sidereal* means pertaining to the stars. I might add, that we use solar days. It's easier for us to live with because the sun is so near and the law of logarithmic stimulation makes it that the year is broken up into solar days and there are 365 and approximately a quarter of those in one year. That's also the elapsed duration when the earth in its orbit is at a given point and when it moves away from there, when it comes back to that point, that's one year, and that's 365 and a quarter, not exactly, but approximately, solar days. But it's 366 and the same fraction sidereal days. There's one more sidereal day because while you go around the sun once you pick up one extra sidereal day because the sun was inside the whole time. If you think of the geometrical configuration, the topology of it, you'll find that you lose one day in measuring the sun. It's the same period of time, except the sidereal day is a little shorter. There's one more of them. All right. Now that's one way to measure time.

Technical Evaluations of What Time Is

There's also a way to measure time by the number of times that a given kind of light, a given kind of monochromatic light—*monochromatic* means single color—how many vibrations there are per second, and the time it takes for one vibration that takes place is another measurement of time. That's called the atomic time scale, and we're not sure it's the same. We're assuming it's the same.

Now these technical evaluations of what time is, it's so difficult even in science to do. As I say, the intuitive time which all of us experience, we intuitively know that tomorrow is later than today, but it's more of an acceptance than a knowledge. We know it's later when we die than when we're born. But you don't know that, by the way, when you're born because you don't know anything when you're born. You know absolutely nothing when you're born. You have a complete blank mind, and your behavior matches it. [*Some light laughter in the audience.*] So, later is an intuitive concept which is somehow related to the physical concept of time.

Generalization and the Concept of Time for Children

So as I say, just imagine the problem of a child in beginning to acclimatize itself to the existing surrounding universe that it has, and to try to identify with things and people around it. The obvious, earliest people it's going to see, under normal circumstances, are its parents. As I pointed out, it has to know what a parent is. It has to know what a person is. It has to know what a human being is. It has to know the generalization that it also is a human being, but smaller and earlier, and later it'll be bigger. It has to identify with a large number of things, all of which involves generalization and it picks up a concept of time in which it is immersed.

The reason children are natural short-termers: they can't imagine anything longer. It takes a while to experience life long enough to be able to project oneself and extrapolate to something beyond what is today. What is a school year, when it starts going to school? That's a terribly long time, until it's over. And to imagine another one. And another one. "What happens after that?" Well then you don't go to school. "Well then what?" And so on.

Generalization Capability:
A Function of Intelligence and Environmental Input

The generalization concept hones your intuitive attitude toward the entity called time and whether a short period of time or a long period of time matters more to you depends on several things, basically reduced to one. It basically depends upon your intelligence, your environmental inputs. The intelligence is mainly hereditary. The environmental inputs are acquired and the combination of these things. But that also depends upon the generalization capability, and that's a function of intelligence and also to some, I think smaller extent, upon the environmental input. I think that being exposed to much generalization will improve generalization but if you don't have what to takes in the first place, it's hopeless.

And for most people, I'm afraid, generalization remains on a very low plane and the generalization ends with such things as being able to cross the street without being hit by a car. Because you generalize that you've seen lots of cars before and lots of moving traffic before and you can select a time when it's safe to cross and you generalize from past experience when it's safe to cross. Or you can generalize when you have to take an egg out of a pot of boiling water so you can eat it, hard boiled or soft boiled as you please. You can generalize it worked four times before, the fifth time might work too. [*AJG chuckles.*] That's a generalization. From a few, limited experiences, you say it will always happen.

Now the generalization usually gets more difficult. For example, to generalize from the falling apple and the rotating moon that we have a universal law of gravitation is not within the scope of most people. Because practically everybody has seen the moon who isn't blind, and practically everybody has seen something fall. If it's not an apple, then it's something else. It took all these millennia for Newton to figure out the connection. Good evening.

[*Applause from the audience for twenty seconds.*]

SESSION 8

PART A

Good evening, ladies and gentlemen.

In the previous session, I was discussing the concept of generalization and the necessity to identify the exceedingly fine filter and criterion of human accomplishment, namely long- and short-term view and their distinction in terms of what the source of this is.

Integrity Doesn't Exist on a Short-Term Basis

Obviously, from many considerations already obtained in V-201 as well as in V-212, which carried it farther, a substantial connection was established in V-212 between long term and integrity, myself pointing out that integrity doesn't exist on a short-term basis. There is no such thing as a short-termer who has integrity.

The Concept of Non-Term Integrity

The best you can hope from a short-termer is either honesty or what passes for, you might say, an innate form of integrity which occasionally some people with little education, sometimes even little background in thinking have and that is called no-term integrity or non-term integrity, where it is possible for a person naturally to recoil at harming someone else without having any basic reason for it other than his own natural instincts.

Non-Integrity in the Animal World: Galambos' Cat, Sam

There are such things observable even in the animal world where there are animals of an exceedingly gentle nature. In the jungle they usually perish. For example, my cat Sam, one of my cats, the black one whom I have referred to as obnoxious. He's not obnoxious in personal conduct, he is only totally aloof, which is essentially a form of rebuffing attention from other people. He's shy to the point of pain. I would say he has, however, innate non-term integrity in the sense of non-thinking. He has never scratched anything or anybody that I know of. And yet, he's supposedly a cat, which is a carnivore, and yet, he's just about the gentlest creature I've ever seen. He would not survive in the jungle. It's lucky for him he grew up in a, you might say, custodial care. [*AJG chuckles.*] Whereas there are cats who are affectionate but still are capable of, let's say, showing outbursts of emotion, which I have started to discuss and I want to continue.

Intentional, Purposeful Integrity: The Only Productive Integrity

Let me now talk about the true integrity, not the non-term integrity which is basically simply a refraining from doing anything coercive by instinct rather than by thinking. The only kind of integrity we're concerned with in volition is intentional, purposeful integrity. The other kind which I mentioned, non-term integrity, is not really anything productive.

Those with Non-Term Integrity Have No Time Scale Attitude

Non-termers do not produce anything. They simply don't think in terms of any time scale. They just live their lives as it happens, day by day as it occurs. They don't injure anyone. Instinctively they just do not consider that proper and I don't think they have any elaborate reason for it. When a person thinks about himself—his goals, his future—usually the type of person I'm mentioning now has very little to do in life. In other words, he just gets by, makes a living, does not interfere with other people, but does not create anything that's memorable. When someone creates something that's memorable, it's done for a thought-out goal, a person who has a goal; these are not non-termers. Now, their goal has to be immersed in some time scale. In that time scale, if it's short, it cannot have integrity as a mechanism. It might have honesty as a mechanism but not integrity. So what's the difference? Well, you know what the difference is in V-201.

The Distinction Between Long-Term Integrity and Non-Term Integrity

Honesty is imposed upon a person, integrity is not. Integrity goes with the person's character. That's true for both non-term integrity and long-term integrity, the only two kind there are. In both cases there is a total refraining from interfering with property of other people and minding one's own property as what he has control over. In the case of the long-termer, he has a conscious goal as to what he wants to do which does not, in turn, conflict with other person's property. In the case of the non-termer, there is no thought-out purpose. It's just a, you might say, routine day-to-day existence, but without any form of action to injure anyone else. It's simply a peaceful, tranquil type of person. They're not very frequent, I might add, but there are some.

The Only Class of People That Can Have Total Integrity

When someone purposefully seeks a conscious, deliberate goal in life it has to be done, as I say, within the framework of some reference as to how soon he has to accomplish that goal. For almost the bulk of the human species, almost all persons who have any identifiable known goals or sought goals, they have that time scale well within their expectancy of life. Because nothing else matters beyond that. A few people might extend that to their time of their children, or a little bit beyond to their grandchildren. No one, in general, with a very minute class of people known as hot-enders, have any other concept of time scale. For that reason, integrity is a rare commodity, because you say, "You mean no one but a hot-ender can have conscious, deliberate integrity?" The answer to that is yes. That is the only class that can have total integrity. You say, "Well, other people are honest. There are plenty of people who are honest." That's right. But that's because of the pressure of the external circumstance of the world.

Coercive Pressure: State Justice

I pointed out in V-201, there are two kinds of pressure. One is coercive pressure, where you get punishment from a coercive agency called the state or some other thing that has state-like characteristics. It could be a church, it could be a private organization which operates coercively and therefore is in fact a state, even though it doesn't have the nominal status of one. Only a coercive apparatus can produce the pressure that is called state justice. That is a pressure. It's a lousy pressure. It's lousy for two reasons. One is, state justice is a contradiction. It isn't justice, but it is a form of confining behavior to a certain groove and you don't deviate from that, or else. The groove is what's within the permitted sanctions of the state. You do what is permitted and you are law obeying, law abiding. You are okay as long as you don't get the notice of the state. That does not necessarily make it moral, by the way, you understand that. Morality in the state concept is called legality, not morality. Legality and morality are seldom parallel, and when they are parallel it's for the wrong reason.

For example, there's a law against theft in state justice, and it's also immoral to steal. And you say, "This is parallel, isn't it?" Yes, but the reason is different. The state wants a monopoly of it, and it cannot undertake to allow competitors to get into the act of stealing because in the event that they had such permitted competitors, they might get more proficient at stealing and become the new state, and it's for that reason that it's not permitted. On the other hand, in the case of a moral person, he won't steal because it's wrong—defined as in V-50, Sessions 2 and 3—and he doesn't need any external pressure. That would be integrity if he can keep his life totally in order on the subject of coercion by behaving on his own internal inputs.

Market Pressure: A Stronger Form of Justice

There is however one other external input that will make a person honest aside from state justice and fear of the state, and that is fear of market pressure. That, in fact, is a much stronger form of justice. It's an omnipotent justice, which hasn't been given an adequate opportunity to show its true strength. It does not involve coercion. It requires market refraining from dealing with those who have violated property. At this time, of course, there is no general structure for this. There are only individual examples where a person says, "Well I know so-and-so has behaved in a despicable manner. He has injured thus-and-so and I don't wish to have anything to do with him." This is a strong and potent concept, but so long as there's safe harbor for that criminal to go elsewhere and deal in the marketplace with someone else, and others patronize criminals, there is no possibility for this to be a universal system.

Market Justice in the Natural Republic Will Render State Justice Obsolete

The natural republic which is described in V-50 and developed in great detail, including the mechanism of accomplishment of this in V-201, is the natural market mechanism for the market justice to become the prevalent form of justice, which will thereby obsolete and render unnecessary state justice. And the state will do what the Marx-ists-Leninists have said but will not accomplish and that is, the state will wither away. And it truly will wither away for reasons of market justice. Because there will be no need for it, and who needs coercion when you don't have any injustices without it?

Coercion Increases Injustice

As a matter of fact, the coercion simply increases the injustice and transfers the source and focal point of it, the injustice, from an individual to a single organization which controls everyone. Both of these forms of external environmental effects upon an individual will reduce a person's desire to steal, however, because it is against the law, in practically every country, to steal.

State Justice Works Only in the Short Term

That was even true in Nazi Germany. It was illegal for an individual German to steal, for his own account. So long as it was self-entrepreneured theft. [*AJG chuckles.*] That is not a correct usage of the word *entrepreneur* but that means that he self-conceived it for his own benefit. That was not permitted even in Nazi Germany. [*AJG drinks.*] That was not permitted in any monarchy. That's a monopoly reserved to the state, to confiscate property for public and state and therefore community and national and therefore benevolent purposes, by definition. It does, however, keep people from stealing as long as the state justice is very strong, in the sense that it accomplishes what it sets out to do and captures and punishes those who steal. As long as the state has this ability and has not yet earned the disapproval of the mass of subjects, this will work. In rising cultures there is little theft and much production for this reason, and the only source of major crime is the state and it's localized and concentrated, but it is limited in quantity so it does not destroy incentive totally.

A Characteristic of Most of Western Culture at This Phase of Degeneracy

History has demonstrated that this is the situation with the rising phase of a human culture. In the declining phase, the state loses its image, its potency to cope with criminals, and the number of private

135

criminals gets so diffuse and so general and broad, it's not able to put everybody in jail. They don't have enough jails for that. The jail keepers are no more honest than the jail inmates, and usually less. And so therefore, there's a diffusion of quality of such a culture and it just drains the vitality of production out. The state, meanwhile, being incompetent of coping with what they fancy justice to be, they become desperate and they become more violent to offset their impotency, but they cannot control the criminal so they control the producers instead. And the producers get the brunt of the wrath of the state as though they were the criminals, and the criminals get away with everything. I presume you recognize your country at this moment. This is characteristic of most of Western culture at this phase of degeneracy.

The Only Source of True Justice

The only source of true justice—source, I said—not distribution mechanism; the only source of true justice is integrity. Now, in the natural society, as I pointed out in V-201, it is nowhere necessary for all people to have integrity. If that were desired—I don't know if it's even desired. If that were expected, you could forget about it, it would be hopeless and impossible to attain. If everyone were to be expected to attain integrity, in the very strenuous way as defined in V-201. That's self-honesty. That's honesty not because you are afraid of who may catch you, whether it's the state and punish you, or the market that may notice you and ignore you to the point where you can't function. The latter is the stronger and, also, it's the moral version of justice, the market justice. But it's not possible to have market justice until a structure is established which makes this mechanism available, which clearly does not exist in any present or past culture.

The Hot End:
The Source of the Human Life Achievement Mechanism

This technology comes about only through the understanding of primary property and the development of the distinction between hot and cold end. Hot end is the innovative end of human life achievement, it's the productive end. The second step is also production, but that's visible production or secondary production. The third step is distribution. And the fourth step is consumption which maintains the purpose for having the production in the first place by providing the sales that produce the market feedback, compensation in terms of revenue which generates profit which generates the market feedback to the process and makes it a functioning and vital and surviving whole. But the source of this mechanism is the hot end.

The Hottest Part of the Production Cycle Never a Market Concept Until the Theory of Primary Property

The hottest part of the production cycle, which is step one, has never been a market concept until I developed the theory of primary property. Innovators have never been recognized as a market mechanism. They were looked upon sort of as a natural resource you dip into and help yourself as you wish and when you have used it up you throw away the carcass. That's exactly how innovators have been used, as though they were a natural resource, not to be treated as human beings and not to be treated with compensation, respect and gratitude, and then when you have gotten the value out of them you can throw away what's left of them like garbage. That is the flatland, usual concept. Yet, they have thrown away the source of man's progress, the source of everything that man has ever accomplished and it's there that the only true integrity lies, because you cannot discover laws of nature without integrity. Because then, if you don't have integrity, you won't have the wherewithal to determine the difference between what is and what you'd like it to be. And the scientist's function is not to restructure the universe, but to understand how it functions and harness his own conduct to fit the nature of the universe.

Primary Goals of Innovators Stems from a Natural Integrity

A natural integrity stems from the studying of nature. These people naturally have a long-term view or they couldn't function at all. Can you imagine any P_2 goal that would have gotten Bruno to do what he did? Can you imagine any P_2 goal that would have allowed Newton to put up with his band of vandals around him? Even as long as he did, as you know, he ultimately retired to the mint. But while he did it, he still did it for about two decades. Can you imagine any P_2 goal would have satisfied him for that? Can you imagine that Einstein would have done what he did for P_2? That's how he developed his contempt for P_2. It has to be a primary goal and the primary goal is a sense of self-esteem. Ego major, for knowing that you have succeeded, which by doing something

137

which is right, again defined in the proper sense, as in V-50, Sessions 2 and 3, respectively.

True Integrity Lies Only in the Hot End of Ideological Program

Well, on that basis [*AJG takes a drink*] true integrity lies only in the hot end of the ideological program. There is no other place you could find it on a production scale. That's necessarily tied to long-term objectives that are long transcending any human lifespan. No human being can live so long as the time it takes for Bruno to get recognized or Aristarchus to be recognized or Galileo to be valued. He isn't valued adequately today. None of these men are. They're just barely recognized by a few people. Even Newton, who has some recognition and has some books written about him and he's mentioned in many places, but the true scope of the magnitude of his achievement is not understood by even the majority of people who sport Ph.D.'s in physics and chemistry and mathematics.

The Majority of Scientists Have No Sensitivity and Very Little Gratitude

The vast majority of them have not the slightest conception of his monumental achievement and no sensitivity to it and certainly very little gratitude. Why? Because they live in short term. They are mainly the minor technicians of the present generation of scientists. I'm not talking about every one of them, I'm saying the majority of them, even within the physics and mathematics profession, the majority of such people have not adequate sensitivity. They only have a view of a very small part of physics or mathematics or chemistry or some other division of knowledge. Physics includes all of the others, of course. They only have a sensitivity to what little interests them and most of them are interested in some very minute specialization or sub-specialization and they might become great experts on that and write esoteric papers using thirteen-syllable words to impress their colleagues and have a mutual admiration society where they understand each other, or they pretend to understand each other and no one else knows what they're talking about. Occasionally, some of it is meaningful, and you have to filter out an occasional significant paper from a thousand pieces of junk which is published under the current academic dictum, publish or perish. You get promoted according to how many pages of publication, not what you did but what you have put forth. Well, these represent a low recognition of the monumental achievements, which are rare. On the other hand, that's the source of true integrity. These are always long-term accomplishments.

Never Any Monumental Achievement of Secondary Production in the Species Time Scale

It is for this reason that no one has ever produced any monumental accomplishment in the second step of the ideological program which is the secondary production, which I have repeatedly pointed out, and in this psychology course I'm going to dig deeper into the subvolitional domain. I started this at the end of the previous session, I'll continue it now. I'm just re-establishing the context right now. In the second step of the ideological program, there has not been ever any monumental achievement that has been in the species time scale. There has never been a business or secondary production enterprise that has had species time scale significance, not one. And there cannot be at the present time because those who are in business or think they are entrepreneurs, are not. They're not entrepreneurs, as I have defined it. They're entrepreneurs as they have defined it, which is a matter of definition to be sure, but using mine, naturally, because I'm not going to pollute my new vocabulary with earlier misconceptions. I refer to those people anywhere from cheap, two-bit promoters who live in the trivial or even subtrivial time scale.

The Subtrivial Time Scale Has Psychological Use, Not Volitional Use

I just added a fifth time scale, by the way. [*Laughter from the audience.*] Subtrivial, that was, I believe, one or two sessions ago. That has psychological use, it does not have volitional use. I do not believe it's necessary to add that into the theory of volition. But in psychology, where you're dealing with inner emotions and things that happen to you in the very now that determine what you will do one second later, that is a very necessary time scale, the subtrivial time scale, where the trivial time scale is long term. What happens by the end of the week is a long-term achievement compared to what's happening this minute [*AJG snaps his fingers*] the moment you're thinking it.

The Political Time Scale Has No Species Time Scale Significance

I also added a sixth time scale, which I will not make any use of in this course other than mentioning it, which I mentioned which is between the personal and species time scales, which is the political time scale, which hopefully will vanish. It's just the time scale within which civilizations have risen and fallen, which is more than personal and less than species. That has a political usage, but it does not have a species usage because if politics persists, the species won't. And if the species persists, politics won't. So therefore, it enters in the analysis of history. From a history point of view, it should be recognized there was a political time scale once which now happens to be still once. The *once* is still on [*AJG chuckles*] but from the futuristic point of view, there once was such a thing as a political time scale during which cultures grew up and perished. Well, that time scale will vanish and have no significance later.

Flatland Businesses Operate Only in the Trivial Time Scale

But in the development of products and product development, and distribution, as I say, there has been a range of people in business who varied from the short end of the trivial time scale, bordering on the sub-trivial. Mainly in the trivial time scale. I would say the vast bulk of businesses are in the trivial time scale. They are in business for the current buck. What they can make this year. "We'll worry about next year when we get to it. We'll not concern ourselves what happens in two years or five years or ten years. Right now, we're in difficulty. We gotta make a profit, we gotta make a quick profit." Or, "We gotta survive." It depends on which end of the prosperity scale they're on. If they're on the prosperity scale low end then everything counts on, "How do we stay afloat? How do we prevent ourselves from sinking beneath the water and drowning?" That means going out of business.

Or if they are in the trivial time scale but on the prosperous side, that they are well-known blowhards, which is the majority of what businessmen are—blowhards—who are promoting and marketing products of mediocre or lower quality but they want to make a quick buck and they don't care about their reputation in ten years or twenty years. These guys will make a profit as quickly as possible. It may not even be profit. It may be plunder called profit. On the best basis, it might be profit. That would be the higher and less frequent end.

Most people who are promoters are shabby swindlers who will say anything to make a sale. They don't care what they say is true or not so long as they get their products sold and marketed. Whether they have a reputation in ten years or not, who cares? If things get rough, they'll have a new product, they'll drop what they're doing and start something else up again. There are people who have a business and milk it for all they can get, then they're out of business. Then they start up again with some other product, either in the same place or if they're too widely notorious in a negative manner by that time, they might start up in a different community and start up something else or with a new product and a new operation and they start that up. These are the lowest quality of businessmen. These are closer to being crooks and not businessmen. [*AJG takes a drink.*]

Higher Quality People May Be Honest for Different Reasons

When there are higher quality people, they will care about whether they're honest, either for state justice or market justice reasons, or both. On the other hand, there might be some instinctive honesty, but not well-thought-out reasons for it, amongst them. They may have had decent upbringing, which is getting to be a more and more rare characteristic because as the decent people of the past generations are dying out the new parents have grown up in the present, rather promiscuous era, and they will make lousy parents for an even lousier next generation, who will in turn rebel against them as they richly deserve, and there might be a reaction to the present insanity.

Honesty Important for Businesses in the Personal Time Scale

On the other hand, the nature of business in the longer than subtrivial, or short-end of the trivial time scale basis, is where you do care whether you make a profit not just this year, but next year too and the year after. However, your main emphasis is still on the now except you're thinking, "Well how will it be next year?" You do consider next year. It does enter your thought on a secondary plane. You do think, "Well, am I still going to be able to do this next year?" You start planning ahead, "If this fails, what'll I do next year?" And then there are people who will have a continuous business doing something relatively simple, marketing something, making something. Some widget or squibbleduck or whatever. Or some series of things like hardware goods or software goods or clothing or food, and they market these or make these things and they do, in general, stay in business year after year and they might make it through even a whole lifetime. Those whose trivial time scale merges into their personal time scale.

These people tend to be basically honest, because they have to be honest in order to last that long. In order to have customers coming to them not just this year, but next year, the next year, the next year, the next year and a decade later, and have a growing clientele. The longer they last, the more honest, basically, they have to be. A direct connection. The way they treat their customers, the way they treat their vendors, the way they treat their associates, which are now called employees, will basically have a bearing upon how long they do endure. They also have to deal with the state, unions, and other such less than desirable organizations, and they have to survive their interferences. That's another problem. Their honesty isn't the criterion, but the good luck, and whether you come to their notice or something like that. It's a question of durability on a time scale better than trivial but still in the personal time scale.

The Highest Quality Businesses at This Time

As I pointed out many times before, including the last session or so, the longest time scale business and therefore the highest quality business that exists on this planet at this time is a type of business which is characterized by, say, an IBM type of business. This is not the only one, it's the most famous and, at the moment, I would say the longest duration of such company. That's not true. There are longer histories. DuPont is older. It is no longer as much of a growth company but it certainly has lasted longer and has had very significant contribution to make the world culture. General Electric is older than IBM but less old than DuPont.

These companies have some major quality to them. They all have deteriorated in the last decade noticeably, visibly, both in management quality, attitude towards the future and even their profits are decaying. It's not catastrophic yet. It has not reached the catastrophe proportions. IBM is still a growing company but it is no longer as spectacular nor as predictable as it once was. These are the longest duration companies. Other companies, just to name a few, are a number of insurance companies that have lasted over a century or more. Some have lasted two or three centuries if they go back to Great Britain where they were earlier than in America. Also you will find [*AJG drinks*] some banks have lasted over a hundred years or more.

Companies like this are actually at the high end of the personal time scale and in very, very few cases they are in the very, very, very high end of the personal time scale, where the personal time scale is not one lifetime, but the lifetime maybe of the grandchildren. Possibly as far as the great-grandchildren. I start thinking that the personal time scale and the species time scale started blending together after a few hundred years, like maybe three to five centuries later. Well that means that the highest quality businesses that have ever existed have lasted to the high end of the personal time scale but have not yet entered deeply into the species time scale.

That is just, as you might say, an osculating encounter—*osculating* is kissing—encounter with it. It was not a very profound or deep kiss. It was not a real deep-rooted attachment. It just grazed it and it didn't like the contact and it perished. [*AJG chuckles*.] I don't know of any business that has lasted for a thousand years. Do you? You've had plenty of weeks to think about it now.

Everything Has a Reason

There is a reason for all this. Everything has a reason. Once you abandon mysticism as your savior and look upon reason as your guide you will find that everything has a reason. If you don't know it, seek it and ye shall find. [*Audience laughter and AJG chuckles*.] If you use the scientific method and not some other mechanism of getting there. [*AJG drinks*.] We don't know the answer to most things yet. As a matter of fact, we never shall; that's part of the theory. Because every time you become smart enough to know the answer to something you didn't know the answer to before you become suddenly bright enough to ask questions that never occurred to you or anyone else and now there's an entire new vista, an entire new science, an entire new domain of the universe that has never even been considered before. So therefore we have no problem about ever getting stale.

First Step of the Ideological Program Never Recognized in the Economic Sense

But it is only in the domain of the species time scale that you can have a long duration, significance to a life's living, the living of a life. The ironic thing is, those who have significance to their life usually were seldom properly compensated while they lived it. Therefore incentive under market conditions—market conditions in the economic sense as described by economists of quality such as von Mises, Böhm-Bawerk, Carl Menger—these people who understand market concept. Nevertheless, their theory, though valid in the second, third and fourth steps of the ideological program, has never affected the first step. And yet that's the source of progress. It's the source of knowledge and it's the source of integrity, and that's what the disclosure barrier means.

Extending the Significance of the Disclosure Barrier Concept

Each time I give a new course I extend the significance of that disclosure of barrier's depth and importance. I've already identified it as the source of all coercion in V-201. I'm also pointing out it's also the source of the prevention of anyone, except those in the first step of the ideological program, to have the capability of integrity, as of now. That disclosure barrier not only does not permit those who have developed the ideas to live in a market concept, which means they get paid for what they do and not punished for what they do. In other words, they get positive and not negative compensation. Not only is it a barrier to them, it's also a barrier to civilization to receive the concepts of integrity, flowing smoothly into the second, third and fourth steps. Because integrity is a long duration concept and the durability of it does not have the slightest importance to anyone else.

The Commodity Concept of Money

Those who are interested in money cannot get it because money is an unacceptable medium of transmission of integrity because it has a low time scale in its present usage. Now the present concept of money, of course, is geared to something trivial. It's geared to the state and/or to a commodity. What does that mean? It means simply that what is called money today is something that is desired by many or all, and gold has been a traditional such commodity because of its various character-istics described already in both V-120[1] and V-201. It has many desirable and stable characteristics which make it desirable. They're stable and that makes them desirable. Stable chemically, stable physically; it's rare, it's dense, it's easily divisible, it's attractive, and so it's a very desirable commodity and people have focused their attention on it. That's a com-modity concept of money.

The State Concept of Money

There's also the state concept of money, which is: we issue you chits which are exchange mechanisms to lubricate the economy. They issue chits, whether they're made of copper coins or wooden slugs or shells in more primitive cultures or beads or polished stones, or more re-cently, paper with something printed on it, like a number. One unit of whatever it is. Two, six, twenty, a thousand, a million, ten million, a thousand billion. That's called inflation. [*AJG chuckles with the audi-ence.*] That's state money. Of course, economists such as von Mises or one of his disciples, would recoil and say, "That's not money! That's fiat money!" That means money by decree. Of course! But money is what you call it. The state calls that money and von Mises calls, let's say, a commodity money, such as gold. Who is right? What's the definition? The question is, which one is useful? Answer is: gold is useful and fiat money isn't.

Therefore, from a point of view of economics, there is no question that von Mises is right, and the state's got garbage passing under the name of money. I call that the hot-air standard. [*Light audience laugh-ter.*] The other is the commodity standard. The hot-air standard and the coercion standard are the same. You accept hot air as a commodity only because you have no other choice. It's called coercion. So the hot-air or coercion standard of the state is the nondurable, trivial time scale money that disintegrates in purchasing power and produces the infla-tion effect.

---

[1] V-120: "Seminar on Money." A one-day lecture comprised of three sessions.

Gold Can Stabilize an Economy—It Cannot Have Any Production Effect in the Long Term

Gold is necessary to keep politicians honest, which is why quality people throughout the ages have tried to stabilize unstable countries by requiring gold as a currency because you can't counterfeit gold as you can counterfeit numbers on a piece of paper. Well, it's no question that from the point of view they're looking at it from, which is short term personal time scale, they are quite right. Gold can stabilize an economy.

It cannot have any production effect, however, in the long term, because it's not based upon something more important than the relatively trivial desirability of gold. When you put gold in the scale of things, which would you rather have, hot air or gold? Well the answer is gold unless you mean literally hot air and you've got an engine. [*AJG chuckles.*] I mean that's very hot air and much hotter than even the Capitol Hill generates under higher pressure [*some audience laughter and AJG chuckles*] and then it can drive an engine.

Gold Will Not Attract Major Innovation

I'm not trying to be funny about this, I'm simply pointing out that hot air simply is under the colloquial meaning of the present usage of slang, hot air means baloney—of course that's a meat too, and it's also a city in Italy, Bologna—it means that it's nonsense being distributed under the guise of authority of the state. [*Galambos speaking in dramatic, whispering tone.*] "Bow down. Face Mecca. Face Washington," which is the local Mecca. [*AJG pauses for a drink.*] Anyhow, that's a trivial time scale concept of money. It does not last for a durable civilization of even a couple of centuries, when it becomes a hot-air standard. The gold standard can last longer. How long? As long as the particular individuals who sought it are around and possibly an immediate successor or so, if they can find one. Will that attract major innovation, gold? No, it will not. It will not.

Disclosure Barrier Is for Both the Innovator and the Rest of Mankind

That's why there's a disclosure barrier; there's a disclosure barrier in both directions. Which I have not discussed before in V-201.

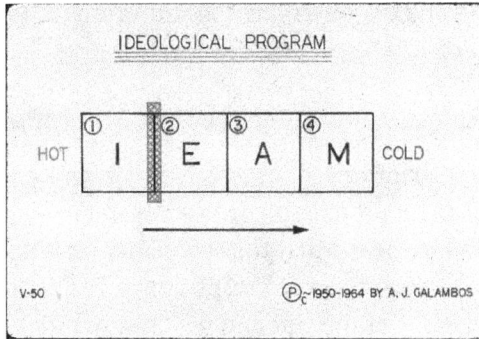

The disclosure barrier does not only prevent the innovator from properly being compensated, recognized and treated decently and in a proper humanitarian way by the rest of mankind—it's not only a barrier for the innovator who is set apart from the rest of people and treated harshly—it's also a barrier for civilization, which is the rest of mankind, the rest of the people, to get a proper form of flowstream into their culture to make their work and their civilization durable. That's why there are no continuing enterprises lasting for thousands of years.

Cultivating a Mutual Psychological Relationship at the Hot End

I know you all have wondered about it, is that right? "Why do companies last such short times?" You all thought of this ten years before you took my course, I'm sure. [*AJG chuckles with the audience.*] Liars will say yes. Those people who have, in the past, failed at having a durable business measured in the species time scale, have failed simply because they have not had the proper mechanism for handling the propagation of new concepts. V-201 shows how to do that. Wherein the first and second steps of the ideological program are recognized both to be production. They're both the hot end, except one is hotter than the other. One is primary production; the other is secondary production. The two together produce the full production. Secondary production cannot exist without the knowledge. The knowledge has nowhere to go unless it's applied, and the application of knowledge is called production. The application is some practical use to which it can be put and which will then produce a marketable product.

So, they both need each other and they have been in opposite ends of a personal or psychological relationship. To one, one thing was a goal and the other one was not, and the other reversed. The one, the long term counts more and the other one the short term counts more. There was no proper reconciliation of this.

Producing a Primary Mechanism for Production

V-201 is the mechanism proposed to counter this and to produce a primary mechanism for production, which doesn't mean there is no secondary production, it means that the secondary production adopts the time scale and the flowstream concept of the first step. In short, the ideological trend is set at the hot end which is natural. That's why the choice of thermodynamics as an analogy is so appropriate. It's more appropriate than I knew it was when I made it. It gets more appropriate with each passing year and course. The fact is that the larger the scope of knowledge this covers, the more certain this is right. Not only was this originally right for volition, it's now shown to be right for psychology. It has been shown to be right for biology. It has been shown to be right for all aspects of volition from the trivial time scale to the longest of term time scale which is cosmic, which we haven't even entered yet but have conceived.

Only the Species Time Scale Considered Long Term

Well on this basis, those who have a short-term view—and short term means either trivial or personal time scale, now. In the theory of volition only the species time scale can be considered long term. In V-30 I discussed long term in terms of investments made in a period of time that last decades, because long term there is defined to be a large percentage of your life expectancy. That's why V-30 describes the present world of defective insurances and investments. And V-201 is the same subject extended into a longer time scale and therefore has enormous superior improvements to offer.

Psychology Dealing with the Shortest Time Scale of All: The Subtrivial

The question is, in psychology, we're really dealing with the shortest time scale of all, and even make a new name for it: the subtrivial. Because in psychology we're talking about what motivates human behavior. In psychology, we're dealing with the determination and influencing of the motivation of volitional action. Well, what you do at any given moment is motivated by some kind of thoughts, which are either rational or irrational, articulable or non-articulable, but something motivates you to do whatever you do. When something motivates you to put the foot on the brake, when to put the fork in your mouth and what's on it, what you do at every moment, small things and big things. Habitual, instinctive things like tying your shoelaces, getting dressed, putting your car in the garage, taking care of your personal functions of life. These are all, basically to a large extent, habitual and instinctive. Only a few things that people do are basically, deliberately conceived on the moment. What do I have to do now? Normally, when you're evaluating something in your business or profession, or occupation, you have to make a conscious decision as to what to do at any given moment. Those are, however, determined, to a large extent by immediate reactions, which you've conceived of at the moment in some form of a thought.

Higher Quality Thoughts Produce the Greatest Yield

The quality of such a decision, however, depends upon how much background there is to that thought. Did that just at that moment spring forth from something you felt? Which means you have no ability to explain and have a direct conception of what it means, and you have no ability directly to determine what caused that action. What makes it right in the sense of V-50—true, valid and moral. Now most people wouldn't even bother, wouldn't even have the tools to think about this with, but those who can and do, they have a higher quality type of behavior. In turn, quality refers to yield. How much do you get for what you put in? What is the ratio of what you get out with respect to what you put in? The higher quality thoughts are the ones that produce the greatest yield.

That means the longer this yield lasts, the higher quality it is because the yield gets bigger because some thoughts produce actions or articulated thoughts which are ideas which are either said or written down, which will influence not only your behavior but someone else's behavior and become a pattern and a style and a functioning mechanism which will be used again and again, by yourself or others. These are things that are worth doing, because the return, the yield, is good, and it gets better and therefore the quality of it gets to be recognized. The quality thoughts are those that do not come from immediate emotional attitudes, but rather they come from a long experience factor which includes drawing on your memory cells as to what related things happened in your experience, or that you knew about that has a bearing on what you are about to do. And you integrate in your mind not only what you're just thinking about, but a whole background of what there is to think about it with. What other things have happened either to you or to someone else that you know about that has a bearing on it where there is a logical connection?

A Facsimile Form of Logic

Now most people don't know much about that concept called logic, and many people will do the same thing but do it in a facsimile form. A facsimile form is they will indeed connect it up with previous inputs, but the connections will be invalid and they will form what are pseudo-hypotheses. They don't know it's a pseudo-hypothesis because they don't even know what a hypothesis is, let alone a pseudo-hypothesis. They will form connections between other thoughts, other memories from their past retained experiences or past retained knowledge of other people's experiences, correlate them invalidly, and come up with a behavior which therefore will produce results that are other than what they expected. In business that would be known as failure or bankruptcy or insolvency, which leads to bankruptcy if you're dishonest or if your creditors harass you to the point where they won't wait. In love, it leads to a thing called divorce or non-marriage. If marriage was never entered into, which is now a growing fad and where people try each other on like a suit of clothes and see if it fits before they undertake any longer-term commitments which in general, most people—by the way, that shows the quality of a civilization. The fear of long-term commitments and the failure at making good ones. That's characteristic of a decadent culture too.

Most Thoughts Carried Out in the Subtrivial Time Scale

In short, the emotional background of people and their rational background both influence your immediate actions. These are usually done in the subtrivial time scale, but there is a distinction between the quality of thoughts. All thoughts are basically done at the time of their being thought up and it's not always an articulable thought, it just might be some hazy concept. I'm sure most of you have not got words associated with why you turn the steering wheel left when you come to a left turn. You don't consciously say, "Now I must make a left turn," and see the sentence written out in your mind. You just come to the corner and you know you're going this way and so you turn the car. But there are things that your past experiences are connecting this up with. Now this is done in the subtrivial time scale where the most important thing is what you're doing at that moment. And indeed, it's necessary because if you don't keep your wits about you while you're driving you could not have much time to think about anything else later. Because it's very urgent that you consider it because the rate of change of property could be enormous [*AJG chuckles with the audience*] if you don't keep your wits about what you're doing.

The same is true in a business decision. A businessman comes up with a point where he has to say yes or no to somebody on some issue, it makes quite a bit of difference whether it's yes or no, and he should know that the consequences that will be faced by making the wrong decision could adversely affect his future. [*AJG takes a drink.*] Yet the actual decision is made more or less abruptly. Now some people are slow ponderers and they go on and on, you talk to them, "Well, I haven't made up my mind yet, but..." And you think that it takes a long time to come up with yes or no. Now these people are simply indecisive. But when it finally is one or the other, it happens at the moment he says it. Because up to the moment of saying it, he could say the other one. Except it took him a long introductory period to lead up to making the final decision, and basic thoughts are carried out in a very short time scale, which is basically subtrivial.

The Distinction between Positive and Negative Output

Now the question, what is the quality of the output? For most people the answer is: dismal. [*AJG chuckles with the audience.*] I should have said, what is the output vs. the input? I should have said, positive output. If there's a very big negative output that's low quality. In other words, when I said quality is defined as the output divided by the input, the larger is the output with respect to the input, the higher is the quality of the product, or worth of the thought in this case. But this of course should be modified to say it's positive output. What does that mean? It simply means that it has to add to property, no different than it meant in volition.

151

Psychology:
A Downstream Concept of Volition

You must realize that in all of this theory that I'm discussing here we have nowhere lost sight of the prior existence of volitional science, and which is not only chronologically first in terms of the development of what I'm talking about comes from volition, but also not only chronologically earlier, it is epistemologically earlier which means that it's closer to the source of everything, which is physics. Physics includes all of the universe. Biology includes the living part of the universe. Volition involves the choice-making capability within the living part of the universe and that part of the universe which has this choice-making capability. Psychology deals with the motivation of that volitional action and therefore it is a downstream concept. Nowhere here have I therefore in any way ignored or eliminated this connection, nor would I even try, it would be wrong. Therefore, in defining things in psychology it is not only proper, but even necessary to use volitional concepts to do this with, and where we extend the subject farther it becomes psychology when it deals with that part of volition which motivates the actual things that are done.

Quantity and Durability Both Affect the Quality

Therefore, to say the quality of a thought as well as the quality of acts which are beyond thoughts, in other words, when it actually becomes action, and ultimately products—this applies to products too— the quantity and the durability both affect the quality. The more quantity there is and the longer it lasts, which increases the utility of that entity, the more there will be of it and improves the quality of it. Now this of course means that the output is positive. If it's negative, that's destructive. Negative property is destruction. Well, then of course, we have a reduction of quality because anything negative is less than zero. The lowest quality in the positive sense would be no output, no output with some input. But it could be worse than no output. That would at least have a neutral yield. It has done no harm, it just hasn't done any good. But if there's a negative output, then it's worse yet. Then it has destroyed property, the output is dangerous and the quality is negative. Then it can go up in a negative sense, which means in the opposite direction.

Plus and Minus; Positive and Negative

As you know in mathematics this is symbolized geometrically

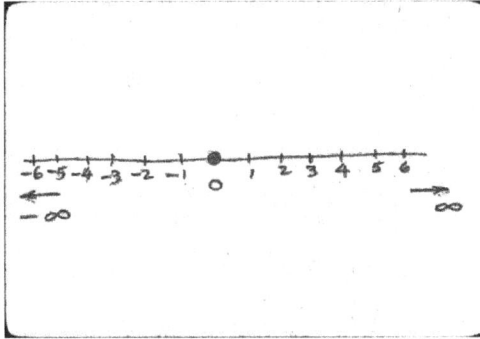

that if we take this point here and call it zero, draw a straight line. You can lay it off in this direction, one, two—that's how you make a scale of numbers. Geometrically represented, you'll see this on a ruler or any measuring stick. In the opposite direction, the same line but going the other way. That's minus one, minus two, minus three, minus four, minus five, minus six and so on. In this direction, ultimately, if you go far enough you approach, but do not reach, something called infinity. And this direction if you go out far enough you approach, but never reach, something called minus infinity.

Well, the output then can be constructive or destructive. They're in opposite directions and they go away from each other. This is of course necessary to recognize the importance of the positive and negative in all of this, which brings me to an important digression, and then I'll go back on the main discussion.

Plus and Minus to Describe Production and Destruction

One of the most important things I have ever done, I think, in the theory of volition which has an effect upon this subject called psychology, very penetratingly, is that I came up with the concept of property being called plus and minus to describe production and destruction. You may say, "Well, everybody knows that." I know. You do now. You all knew this in high school? They call that in high school "positive and negative production"? That's as important as calling charges positive and negative.

Benjamin Franklin's Beautiful Choice

Most people have no sensitivity to the significance of anything. When Benjamin Franklin called charges—first of all, he discovered that there are two kinds of charges, not one, when he could have called them by any other names. Red and blue, or black and white, or up and down, or male and female, or north and south. That's what they did in magnetism by the way: north and south. Any two names that distinguish one from the other, he called these plus and minus. It emerged later—decades later, more so; a century later, still more so—the beautiful choice he made. If he hadn't made that choice, it would have had to have been renamed, it's that important.

The Mathematical Significance of Plus and Minus

Only plus and minus can be handled mathematically and it has a mathematical significance in physics that the charges are called plus and minus. They will obey the rules of ordinary algebra in terms of plus and plus adding up; plus and minus, one subtracts from the other. Plus times plus is plus, plus times minus is minus, minus times plus is minus, minus times minus is plus. And if you use this symbolism it will have mathematical utility. If you call it north and south or male and female it will do no good, because north times north doesn't come up with north, and north times south doesn't come up with south, and south times north doesn't come up with south, and south times south doesn't come up with north. That's meaningless. That's a geographical direction.

The reason it's used in magnetism, for your information, is because in magnetism it was discovered that the earth itself has a magnetic field, and that therefore, the north part of the earth has a certain polarity and the south part of the earth has the opposite polarity, and the magnetic polarities were named for the earth's geographical poles. That's why in magnetism these are called poles and not charges and they're called north and south instead of plus and minus. Yet, I'll tell you this. When you want to deal with magnetic polarity, mathematically, you've got to rename it plus and minus, even in magnetism. Usually the North Pole is called plus and the South Pole is called minus, otherwise you don't get anywhere. [*AJG takes a drink.*]

It's a natural choice once you know what you're talking about, meaning you had some physics and you understood what you had. That's a higher filtration. That property should be called plus, if it's added to.

Clearly, increasing property either in quantity or quality, desirability, utility, diversified usage and so forth, all of these things add desirability to the property and that's called plus. Destroying any of these characteristics is called minus. So if you do that, you have something that's usable and ultimately can be quantifiable.

Applying the Concept of Plus and Minus to Psychology

Applying that now to psychology, which is in the digression, we continue this and therefore the ego is also called plus and minus, which is what I brought up many sessions ago. I believe it was Session 3 it first came in. I called the ego characteristics plus and minus also and I defined as a positive characteristic of the ego which is defined as the self-recognition of oneself or more briefly self-esteem, the opinion you have of yourself, the recognition you have of yourself. If you had done something that has produced positive property, and you recognize it in yourself, then you esteem yourself and you have added a positive addition to your ego.

If you have done something which is wrong, which is equivalent to saying you have injured property, which is equivalent to saying you have reduced property, which is equivalent to saying you have generated negative property, then you have also developed a recognition of yourself as having done something which should reduce your opinion of yourself. That's called a minus ego.

It Takes More than One Action to Build an Ego

Now of course, you don't get a full ego, which is a tensor as I explained in an earlier session, from any one action. It takes more than one action to build an ego. But any one action is typical in general of a person's continuity of behavior. There's a basic continuity about people. When a man is dishonest in Place A with Person One, he is very likely to be also dishonest in Place B with Person Two, because he doesn't change his basic thought patterns as he moves around geographically. Now something can alter that and that's called an understanding of what he's doing and a deliberate conscious attempt to alter his behavior. That takes more than most people have in both ability and willingness to try. Anyhow, the positive ego or ego plus and the negative ego or the ego minus are both inputs which you get from certain actions.

Ego Minus Characteristic Produces Internal Frustration

Now when a person has ego minus as his basic characteristic as I pointed out, he will find that he cannot stand himself. This will produce psychological disorders which are very readily identifiable, and this is largely what produces cases of internal frustration. I will discuss the distinction between external and internal frustration presently. I doubt if I'll get to it this early in the course, but we'll see. This is an internal frustration. External frustration is quite different from internal frustration. Internal frustration comes from within yourself because you are having internal quarrels with yourself. You have contradictory positions, different things that you believe in, and they don't match up in reality, which is the observational world which is corroborable as true or false by observation, doesn't verify your conclusions and you find internal problems which are sources of internal quarreling with your emotions.

Frustration Can Lead to Psychological Disorders

This produces a major catastrophe for some people and some kind of a disorder in a large number of others. These are things that are recognized as psychological disorders in the world today, for the most part. It includes, among other things, schizophrenia which is a split personality, but this is only one manifestation of it.

Majority of People Do Not Recognize Their Own Negative Ego

Now, I'd like to say that the negative ego, which is possible to develop can, in an extreme case, produce suicide. A negative ego, in a less extreme case, could produce a lifetime of morbidity, melancholy, failure in everything one touches. This is the Stone Midas business where everything you touch turns to stone. Or it could be less extreme, which produces frequent failures and erratic behavior. The majority of people, however, do not have a negative ego recognized. They don't recognize themselves as failures even when they are. The majority of people are incapable of being that honest because it takes a certain amount of self-honesty, which is integrity, to be able to recognize that when something goes wrong, you made the mistake. The natural tendency is to find a scapegoat. Which means put the blame somewhere else. "Yes, this certainly is not what I wanted. This isn't the way I planned it. But the reason is..." and then you babble forth some excuse as to why it didn't work, other than, "Well, I didn't know what I was doing," or, "I made a mistake," and then try to find what the mistake was and try to see to it that it does not become repetitive and habitual. And then learn from it. That's called the scientific method. As I say, nobody's infallible, but when you're able to cope with this on a consistently positive basis then you are using the scientific method that produces both the greatest advances in science and successes in business and successes in personal living, including happy marriage and all that.

It Takes Two Rational People to Have a Primary Marriage

It takes two, by the way. Marriage is one game you can't play alone. You have to have someone else to play it with and you both have to be rational to have a primary marriage. Well, that makes the probability very low. It's hard enough to find one rational person, but to have two that find each other, recognize each other and make a contract—well, the probability of that is remote. Which is why primary marriage, though the most desirable form of marriage, is also the rarest. And by the way, calling it that doesn't make it so. A cow's tail is still not a leg.

It's Easier to Make Excuses than to Apply the Scientific Method

Now the point is, in not being able to apply the scientific method and learning where you made a mistake and localizing it, finding out what it was, finding out how to do it right and not doing it again. Instead, it's much easier to say, "Well, the reason this didn't work is something else..." and then you find some excuse. "Somebody cheated me." Or, "Someone else gave me a bad hot tip." Or, "Someone else failed and I was simply caught in his maelstrom." Well, there may be truth to this. You could have been given a bad hot tip. Did you have to take it? This is all covered in the V-30.

V-30: A Much Deeper Course than It Seems at First Blush

V-30 is a much deeper course than it seems at first blush, you know. When you first hear it, it sounds like just deal with some trivia, like stocks and insurance policies. It's a very deep course. It's the only investment course on this planet that's compatible with the theory of primary property. That by itself makes it unique, which is why so many people flunk it. It's a lot harder to pass V-30 than to pass practically a

Ph.D. requirement in economics at Harvard. As a matter of fact, I know at least one person who got thrown out of Harvard and is now making a good noise for the presidency. It's a high qualification to be thrown out of Harvard for cheating. It shows very high political potential. [*Light audience laughter as AJG takes a drink.*]

Majority of People Not Accepting Responsibility for Their Errors

The majority of people, as I say, find it the most difficult thing in the world to absorb the blame for anything that they do. Whether they have made an error by negligence, by lack of inputs, by acting on inadequate inputs, by developing a wrong thought process which led to an invalid hypothesis which did not lead to the conclusion they expected. Or whether they plundered someone else either unintentionally out of error, and then when it was called to their attention, "Oh, no, I didn't plunder. You're crazy. It's not yours." You think that can't happen? Just think back in V-201.

Why the Marketplace Has Always Been Inferior and Short Term

When a person cannot admit an error, there's no retracing of his movements and start up again. You can't go back in time, you can't actually wipe out what you have done. But there's a beautiful mechanism developed in the market theory applied to all four steps of the ideological program which starts with step one, which has previously been excluded from the market. That disclosure barrier, as I say, keeps the innovator's higher standards and higher quality out of the marketplace, which is why the marketplace has always been inferior and short term, and attracted not only producers, but crooks, and more of the latter than the former.

It Is Not Possible to Go Backward in Time

Now, because of this, the fact that you cannot go back in time, that's not possible. That violates the second law of thermodynamics. There is no backward motion in time, you say, "Okay, let's go back in time, do it all over again." That is not available in nature. All fiction stories on time machines to the contrary, notwithstanding. That is a violation of the second law of thermodynamics. It is not possible to go backward in time. That's more than we do not know how, it is not possible.

Internal Frustrations Caused by Inconsistencies in Oneself

All right. On that basis, you cannot thereby eliminate the mistake and say it didn't happen. Now that's the reality that makes a psychotic! When he doesn't recognize that reality. When he does something that produces a result he didn't want, there are only two things he can do: compound the disaster by saying it didn't happen or if it did happen, he recognized it did happen and it wasn't his fault. There are several variations of this. Deny that it happened, deny that it was his fault, and various variations of the latter. And all of these lead to further disasters and produces further failure in volitional behavior and internal frustrations—*internal frustrations*—which are caused by the inconsistencies in himself.

The two kinds of frustration are internal and external. Internal frustrations are caused by internal inconsistencies you cannot live with or you cannot cope with. External frustrations are your inability to cope with what the rest of the world is doing, and you have a mismatch in you and the rest of the world in some way or another. Which I'll have further discussion of some other time. But when you have internal frustrations because you have failed to recognize you cannot go back in time and undo it. Or you think that it didn't happen, or you claim it didn't happen or you claim that it did happen but it wasn't my fault—you are not solving the problem, you're adding to it.

An Error Is an Error, with or without Coercion

Not only have you left yourself open to making the same error or crime—a crime is an error. It's an error with malice. An error with deliberate interference with property, intentional interference with property. But it could also be a mistake, which is not criminal but at least it is still an error. The results are not what you expected. So with or without coercion, an error is an error. An error with coercion is a crime when it's successfully consummated. When a person has done this, and he fails to recognize his blame, his error, then he is open to doing it again and to justify how right he is he will not only do it again but expand his horizons. And then he tries to restructure the world in his own distorted and psychologically erroneous image. He creates a dream of glory about what ought to be instead of what is.

Apology: A Necessary Minimum Prerequisite

Now of course the right way to do it, as I say, is beautifully furnished within the theory of volition, which shows integrity as emanating from the hot end. If you do make an error and there's no way of going back and undoing it and that's an observational physical fact, the obvious thing to do is, on a primary level, apologize if you injured someone. If you injured yourself you say, "I won't do that again." You injure someone else you say, "I'm sorry."

That's not enough! I have a discussion of "I'm sorry." That's not enough. "I'm sorry" is not a justification of continued blunder. That's just the first introduction of the rather sorry expression, "I'm sorry." That is a necessary minimum prerequisite. An apology. But the apology must be followed up with right behavior, not merely a continuation of the same incompetence or crime. But that isn't enough.

Restitution

If you have interfered with property of someone else, it's called restitution. Pay back property that is measurably a market replacement of what you have lost for the other person, if it has a contact with someone else, of course. Sometimes the error is internal, then you don't owe anybody else anything but you owe yourself at least the courtesy of learning so next time—courtesy my foot! *Necessity* for your own survival. You owe yourself the necessity to learn so you won't injure yourself in the future and make the same blunder and get yourself off the emotional hook by saying it's somebody else's fault or it didn't happen, both of which are observationally false.

A Mistake with Property Loss to the Other Party Is a Crime

Now there are cases where it is the other guy's fault. For example, if A stabs B and B dies, then it is A's fault. That is a true statement. But if A and B have a contract and the contract fails, then they're both at fault. Because a contract is bilateral and volitionally open for selection without coercion. You don't have to make the contract. If a bad contract is made, they're both at fault. If, however, a good contract is made and B fails to satisfy his end to A, but A has satisfied his end to B; A delivers to B what he promised and B fails to deliver to A what he promised, then it is B's fault and there is no way of successfully blaming it on A.

I'll repeat that. If A and B make a contract and A has fully developed and produced and delivered what he has promised to do for B, and B has obtained this value, or it has always been available to him, "It's ready here, ready for pickup. Pick it up. It's there." And B doesn't deliver what he has promised to do in exchange to A, then B has made a mistake, and if it involves A's property, it's also a crime. A mistake with property loss to the other party is a crime. In that case, not to admit that is not only criminal from the volitional point of view, but a psychological disease from the point of view of subvolitional subject of psychology. And that is called the ego minor.

The Ego Minor Behavioral Characteristic

That is the ego minor characteristic. When you do not have the ability to recognize that you have done something wrong which would subtract from your ego and is a negative characteristic, an ego minus characteristic, and then you don't recognize it to be minus—you claim it is okay, it is plus—then you have invented a pseudo-positive image, a false positive image by deluding yourself to believe something to be true which is not true.

True is not a relative matter. Anything that is observationally corroborable on a general basis is true and it's an absolute. Reference V-50. And therefore such a person who has developed a minus addition to his ego, which means a subtraction from his ego, but fails to so recognize it and tries to justify it by claiming it to be a positive concept, that's the ego minor behavioral characteristic which is what I call the universal psychological disease which happens to people all over the world.

The Ego Minor Characteristic:
The Basic Source of Instability of Mankind

That's the basic source of the instability of mankind, the ego minor characteristic. That's the source of man's failure as a species, it's the source of man's failure in business, it's the source of man's failure in marriage, it's the source of man's failure just in developing a proper behavior for his own internal satisfaction. Ego minors always are frustrated. Because you see, the fact that they think of themselves positively, at the same time it is not true. Even if they're not bright enough to know it's not true, the reality will always surface from time to time, and something they do produces a different effect than they expected and so everything they want to do doesn't come out right and so they become frustrated.

One of the Biggest Ego Minors of All History

Take one of the biggest ego minors of all history, Adolf Hitler, whom I will refer to many times as a supreme example of everything that is psychologically defective in mankind and which also, of course, ties in with earlier courses which I like to build upon and add to. Take Adolf Hitler. And you can have my share of him any time. [*AJG chuckles, then takes a drink.*] He had one of the most colossal cases of ego minor in history. Yet, he is not known by historians—the early historians in the negative sense, the late historians in the positive sense will remember him differently—he's not thought of as a man who has a negative ego, but a positive.

As matter of fact, he's called a megalomaniac. He's called an egoist. As a matter of fact, he wasn't. He's called an egotist. That he was. Egotism is conceit. Egoism is a positive drive, developed by one's self-esteem. Hitler was not a megalomaniac. Well, let me put it this way. I'll withdraw that statement. If you define megalomaniac as a man who was psychologically unbalanced, hence maniac, and who thinks of himself as enormously large, then, yes, he was a megalomaniac. But most people look upon it as that he had a positive, inflated ego. No. He was totally inadequate in coping with his own elementary, vital survival functions. He was incapable of earning a living in flatland. That's observationally true from his biographical inputs. He was incapable of earning an honest living in flatland.

The only thing he liked or claimed to have liked, was his experience in war, in World War I, when he was only a private, and later a corporal. But he claims that this was to him a great thing, because he was fighting for Germany's glory and he used that corporal business very effectively, the fact that he participated in World War I, which so did many millions of other Germans. He wasn't even German, incidentally, he was an Austrian. That's another frustrating thing for him. He was an Austrian and he hated the Austrians.

He loathed his own country. Did you know that? Those of you who didn't take V-113 may not have known that. Those of you who had V-113 may know this. Do you know he destroyed Austria? Did you know that? He destroyed his native country. He eliminated it. Austria has existed for much longer than Germany. Austria is a much older country. You may say, "Well it exists again. Yeah, well, he lost the war, fortunately and they reconstituted Austria politically. But he destroyed it. It was eliminated from existence. Did you know that? How many didn't know that? You say, "He annexed Austria into Germany." That everyone knows, I presume. 1938. More than annexing it, he eliminated the name. It was not called *Osterreich* anymore, which is what Austria's name is in German, in their own language. For those of you who haven't learned it, Austria is a German country and the native language there is German and it was called *Osterreich*, which means eastern empire. That was eliminated. It was called, when it was annexed to Germany, *Ostmark* and that means eastern province. In other words, it just became the eastern province of Germany. There was no such country as Austria.

And he loathed Austria because he tried to make a living in Austria and failed and he blamed it, not on his own inadequacies, but on his own native land. They didn't give him the proper opportunities. Have you heard that locally? [*Audience laughter*.] People who are underprivileged and they haven't been *given* the proper opportunities. *Given* the opportunity. You get the contradiction? Opportunities are manufactured. Anyhow, the long and the short of it is, this person who failed at everything he tried in terms of honest employment or honest endeavor, he tried to be an artist. He was not accepted as one. That, by the way, doesn't make it bad. A lot of great artists were also not accepted and they surmounted it.

The Market Rejections of Van Gogh, Hitler and Galambos

I know other people who didn't get recognized in their own lifetime. For example, van Gogh, who didn't get recognized as a great artist in his lifetime and didn't sell any paintings. The only painting he ever sold was one to his own brother in his entire lifetime. I would say that's market rejection too. I don't notice van Gogh going around conquering the world. It's possible, and I can't evaluate Hitler's artistic talents, because I'm not in the art connoisseur business and I don't consider myself an art expert. So I can't tell you whether he would have been an acceptable artist or not. But the fact that he was not accepted by the market—which does not necessarily say that his drawings were bad. I've seen a couple of his drawings, I don't think they're too bad, but then I'm not an artist. The fact that he couldn't cope with the market rejection is what I mean to say. He needed market support. Now I've been rejected by the market plenty of times, by the way. F-201 was a total market rejection, but I was able to cope with it. I was rejected in my childhood. The same age that Hitler was growing up, I was just as much alone as he, or more so, and I wasn't trying to rabble rouse amongst my cronies. The point I'm making is he was not able to cope with market rejection because he didn't have a big enough self-image, an ego plus.

The Persuasive Tongue of Adolf Hitler

He did not have enough things going for him that he had a positive enough image of himself. He had to do something else that would magnify his ego. Now he tried his hand at coercion because he happened to have a very persuasive tongue. I won't say he was a great orator, but he was a persuasive one. He seemed to have had a hypnotic effect upon those he talked to because he could spellbind people by the tens of thousands or a hundred thousand at a time, when he finally got a big enough audience. He could get these people to say, "*Ein Volk, Ein Reich, Ein Führer!*" Which means, "One People, One Empire, One Leader." And he would get them in unison to put that chant into motion on a horrifying basis. You can see this in old newsreels. We showed some of these in V-113 in the live version. This man had a spellbinding effect upon people. Now that was something he was good at. But please note it was coercive. He interfered with other people's property.

He found his niche in life, coercion. But did he actually do something positive that could identify with accomplishment? No, it was negative. He attacked other people's property and assumed control over it. In the short run, he assumed control over the destiny of Germany. It made him think he was the top of the world, the greatest and strongest and most powerful man in history. He thought of himself in such a fashion that when they think of Alexander the Great and Attila, Genghis Khan, and Napoléon Bonaparte; these men will all be looked upon as second-rate conquerors compared with himself. He bloated his own self-image. Was it based upon, however, production or his ability to coerce? He used his ability to get other people to follow his political leadership as a way of magnifying his own capabilities, but none of his capabilities were positive, and that's an ego minor in a supreme case. That's a man who made a mistake a way of life.

His own failures at production became his passport to the dictator-ship of Germany and an attempted dictatorship of the world. On the other hand, it blew up. In the long run, it didn't work. The thousand-year Reich was the twelve-year Reich.

Magnification of Man's Self-Image Based on False Inputs

Now, the proper evaluation of what happened there is necessary both historically, which is covered in "Positive History," also in V-113. Also in psychology. All politicians basically have an ego minor drive of which this was a supreme case. Ego minor will have all kinds of effects. This will produce the effect of the Ptolemaic hypothesis, the religious blunders of history. All of these are ego minor concepts. Why people fall into them. Not just because they are earlier primitive concepts, co-ercion being more primitive than production, and believing the earth to be the center of the universe to be more primitive than the earth to be moving around the sun. It's true. Primitive in the sense of earlier to think of it, it's true chronologically. But it also happens to be a magnifi-cation of man's self-image based on false inputs. And that's the ego mi-nor.

Ego Minor:
The Only Disease There Is in Terms of Man's Thinking Processes

The ego minor is the only disease there is in terms of man's thinking processes. Every form of error is an ego minor behavior. Those people who are habitual failures in anything produce the most damaging re-sults. However, some people are able to confine the failureship to themselves, and don't commit any coercion. These people are sad to behold, but they're not dangerous. But of course, when they transmit this disease, by their erroneous and falsely conceived contacts with other people, it becomes dangerous because it affects their property and then they turn to coercion and if they are successful at getting away with it, which is defined as not being punished for it, then they will try for bigger stakes.

In the previous session I discussed the development of how one learns to generalize. I would like to continue that, but I will do that after the intermission. [*Applause from the audience for sixteen seconds.*]

PART B

In Session 7, I discussed the concept of gratitude being dependent upon rationality which in turn is dependent upon a long-term view. Gratitude of course is a concept which is closely identified with integrity. All people who have integrity are grateful. Grateful to all values they have received from any source, whatsoever, and act accordingly.

Gratitude Does Not Fade with Time

Gratitude is not an emotional concept. It has nothing to do with personal liking or disliking, although it can be congruent with it. It's possible—not only possible, it's desirable—if you like the person you're grateful to, but that's not necessary. You can be grateful to someone who has done for you any form of value whatsoever whether or not you like that person. Gratitude does not fade with time. Only thankfulness fades with time because of the law of logarithmic stimulation and the weakening of the remembrance and the effect being diluted.

Gratitude Not Available to People with a Short-Term View

Gratitude, however, is not available to people who have a short-term point of view, for the very obvious reason that they don't have a long-term memory. I don't mean they can't remember, for example, something that they consider important, like who hit a home run when they were eight years old. I don't mean that they have a defective memory. They don't have a proprietarily acceptable memory. They don't remember the things that are of proprietary value and significance to them. Therefore, gratitude is an obviously rare characteristic. The concept of gratitude, for example, never crosses the minds of the vast bulk of the people in its truest form. How many people are grateful to the inventor of the wheel? Whoever in flatland even considers it? "Wheels are here. Well naturally, how else could it be?"

Inventor of the Wheel Cartoons

And somebody, somewhere will see a joke about the inventor of the wheel. For example, cartoonists sometimes cash in on the unknown identity of the inventor of the wheel. They have cartoons of varying kinds, I have a whole collection of them which partly came from my own not throwing away those I've seen and some came from some graduates who thought I'd like to have one. I have a whole collection of inventor of the wheel cartoons, including the guy who is sitting with a round object between his knees in a patent office, a modern patent attorney's office, and he's sitting there next to 20th century types and he is obviously waiting in line to patent it. Another example is a big round stone which has a center missing like a doughnut, and the caveman is standing there next to him holding this up and the other caveman is there listening to him, and he says, "This is the greatest invention of all time, I call it the hole." [*Loud audience laughter and AJG laughs.*] This has a very important ideological significance that even the innovator does not recognize the total significance of what he has done, it gets bigger all the time and outstrips his wildest imagination as to what it can be used for. [*AJG laughs.*]

Anyhow, as I say, most people, the only encounter they have with the inventor of the wheel concept is that people make jokes about his non-recognized identity but nobody's really concerned, "Well, what an injustice it is." Is there anybody here that can tell me truthfully they worried about this as a child, the injustice of this? Well, what did you do about this? One person says yes. [*Student inaudibly answers.*] Nothing. But you worried about it. Well, I commend you for the worry. That's at least a good step. Anybody else can honestly say that they worried about this? It's more than a personal affront to that man, it actually affected the caliber of civilization.

Connection between Nonrecognition of the Inventor of Wheel and Our Present Plight

Is there anybody who now, post-201, fails to see or trace out a connection between the nonrecognition of the identity of this prehistoric personage and our present plight? Is there anybody here who missed that? [*No audience answers.*] Then you missed the whole of 201, really. [*AJG chuckles with the audience.*] I think if there were such a person they would not put their hand up and I rather suspect that might be the case but I'm not going to look into anybody's direction at the moment. [*Some audience laughter and AJG chuckles.*] In any event. [*AJG drinks.*]

Gratitude Is a Component of Integrity

All right, I've discussed therefore the obvious correlation between gratitude and integrity. They are not identical concepts, but they do

167

refer to identical people. They're two different concepts. The larger of the two is integrity. Gratitude is a component of integrity. Integrity includes other characteristics besides gratitude. Gratitude is a portion of the concept of integrity, but it is absolutely correct to say that no one who fails to possess integrity, as defined in V-201, can have gratitude. Furthermore, there is no possibility of someone being truly grateful and not just using the word, misusing it as a substitute for thankfulness, which is a short-term emotional factor which rapidly decays and is forgotten. It is not possible for a durable permanent concept of gratitude without having integrity as a base. There's that connection between these two. They are a mutually compatible characteristic of which gratitude is the smaller, it's a part of integrity, but it's the same group of people we're talking about. They are the successful and major hotenders. Okay.

Possible to Accomplish a Sensitivity to Gratitude, But Not Automatic in Flatland

Now the question is, is it possible for someone who is not in the hot end to accomplish this? The answer is, to accomplish a sensitivity to this. It is possible, it is not automatic. Why is it not automatic? Because the civilization you were born into—which I have called flatland. So we don't have to differentiate between the French and the German and the British and the Russian and the Chinese portions of it. These are political differentiations, also geographic by continent, Europe, Asia, Africa, and America and so on. So we don't differentiate which component of flatland we're talking about, we just use that one word. That means the universal politically oriented, politically dominated structure, which includes theocratic politics, incidentally.

In Flatland People Have No Inputs on Gratitude

Many civilizations have been theocratically ruled. Instead of being ruled by tribal chiefs they were ruled by witch doctors claiming to owe their power to the deity or the deities, singular or plural as the case may be, and claiming authority to coerce on the basis of their brokerage capacity to transmit the deity's will to mankind. Now on that basis, we have a situation wherein we have a mechanism of political management which, as I say, includes theocratic, which is universal for flatland, and in flatland people have no inputs on gratitude. That's only available in areas of extreme long-range concepts. How many people deal with such people? How many people deal with people who deal with long-term concepts that affect the species time scale?

What Makes a Person a Hot-Ender?

Therefore, those who are not in the hot end by innate characteristic—in other words, what makes a person a hot-ender? I haven't determined that as yet. Maybe I never will. What makes Newton excel over other people born in the Woolsthorpe area of Colsterworth? What makes Archimedes better than other Greeks? I do not as yet know. Obviously, it can be sought after, this question. But when such a person emerges he has a natural aptitude, a natural conversance with long-term concepts. He deals with this because that's the way it comes to him automatically. That happens to be the case with me. I can identify where I got it from, partly from my Father and partly from my own. Well, it may be all hereditary, I don't know. In any event, where he got it from, I don't know. I'd like to point out that most people do not have this.

The Integration of Various Human Characteristics

One of the things a person has to do in psychological analysis, is not only to see himself the way he'd like to see himself, but also see himself the way others see him, realistically, and learn to cope with the mismatch. For example, I know what I am and I don't suffer from any false modesty. I also know how various other people view me, with no illusions or delusions. I used to have, but boy, the school of hard knocks coupled with rational evaluation of truth, knocks that out. If you sometimes think I'm cynical towards my own audiences, it's only because of the school of hard knocks. This does not necessarily mean 100% of the people here. As a matter of fact, I know damn well it doesn't. But it's not a homogeneous group of people I speak to. Ever. There are no two identical people, and the spectral range varies enormously, and the quality varies enormously, the comprehension varies enormously, the stability varies enormously. All these characteristics are capable of being integrated, by the way. That's a part of what I'm doing in this course, integrating the various human characteristics.

169

Flatland Teaches the Glorification of Coercion and Destruction

Now, one of the necessities of coping with reality—*reality* is defined as the way the world is, observationally corroborable, and not the way you imagined it might be, ought to be, should be, or will be. Now, it so happens that most people who do not have an innate and thereby, you might say, natural long-term view, they do not have any inputs on this from flatland. Flatland doesn't teach that. Flatland teaches the glorification of coercion and destruction, not production. That's why history is replete with long accounts of wars, conquests, rises and falls of civilization. The heroes, the most mentioned people, are the rulers and the military figures who sometimes they're one in the same and sometimes two different groups of people. Also religious people who have been religious leaders. These occupy the bulk of what is called history to this date in human literature and classified under the category of history. This is all negative history.

Scientists Become Lackeys of the State— Always the Case in Every Flatland Culture

There is a very small, but not zero, literature on positive history. There are histories of science. These are the true histories of the world. Yet, the histories of science that have been written are poorly done, not necessarily from the scientific point of view, but from the point of view of the impact they have had on the world as a whole, including the non-scientists, and they fail to point out this innate superiority of the positive production of science and technology, and the inferiority of destruction and coercion to the very point where scientists traditionally have been the lackeys of the various states. This is always the case in every flatland culture. Always the scientists are working for or through the good offices of the state or some mechanism that is over them, and they provide the value and others get the benefit and also the significant accreditation in history.

An Example of a Significant Positive Achievement Where Ultimate Credit Went to the State

Who, for example—just to name a recent example that all of you are old enough to remember, or you wouldn't be in this class because I wouldn't take you if you were younger than this. Since this happened in

1969, and there's nobody here who is younger than six years old, right? So you're all able to remember this. [*AJG drinks.*] Who got the credit for the landing on the moon? Above and beyond all other people, the then-reigning president: Richard M. Nixon. Also, honorable mention was given to John F. Kennedy because he started the program which said that the United States will put a man on the moon in this decade, and he spent enough money, by God. Newton was able to be harnessed quickly enough to be doing it. Newton, Ziolkovsky, Goddard, and Oberth's work was harnessed fast enough, with enough billions of dollars expended, they made it. And so, Nixon and Kennedy were the heroes of this venture. Also, honorable mention went to the three men who went, which essentially was a piece of piloting, not to be put on a zero-accomplishment basis or level, but certainly not comparable to the achievements of the people who made it possible in terms of the knowledge. [*AJG takes a drink.*]

All right. That's a significant positive achievement where of course the ultimate credit went to the state. Only once did I hear Newton's name mentioned, and not on that voyage. Please note I didn't call it a *flight*. That in itself is an impertinence. It's not a flight to the moon. You can't—never mind. You've heard that before. [*AJG chuckles.*] It was on a different voyage to the moon, which did not produce a landing but a circumnavigation thereof, and it was Apollo 8. Somebody on that voyage gave Newton a mention. Probably the only reason he'll ever be remembered. [*Some chuckles from the audience and AJG.*]

Short-Term vs. Long-Term Effects upon Civilization

In any event, in the long term, you see, what is important is quite different from what is considered important why you live the life you live. Is there any exception to this? Can you think of something that was considered important in the days of Caesar or Genghis Khan? Or Alexander the Conqueror? The Great, he's called. Actually he was not great. He was just very obnoxious. [*Light, sporadic laughter.*] Please note who's called the Great. Have you ever heard the expression Newton the Great? Or Lagrange the Great? Or Faraday the Great? Or Kelvin the Great or Maxwell the Great or Einstein the Great or Galileo the Great? Or Archimedes the Greatest? No, you don't hear such expressions. You only hear Alexander the Great and This the Great and That the Great. They're all kings. All right.

How many of you can think of a single person or event that was considered important in the days of these long-ago times, that still has an effect upon your present way of technological 20th century automated industrial civilization? What effect does Alexander the Conqueror have on this? Or Genghis Khan? What effect do you think Hitler will have on a world of two thousand years from now, if man still exists in two thousand years? Which depends not on Hitler, but on me and those who are on the entire ideological flowstream behind me and those on the same flowstream in the future. The same flowstream. Not one that is diverted into coercive channels. All right.

The Bastiat Significance

Two thousand years from now, be there civilization for man, what effect will Hitler have on that? Only a Bastiat influence. Do you understand what that means? I'm giving Bastiat the primary credit for a major concept: that which is seen and that which is not seen. I call that Bastiat effect, or the Bastiat significance. Only that will be affecting the future which cannot be known, and therefore is not really knowledge, but as a might have been.

It may be that somebody who could have been of Archimedian or Newtonian stature was wiped out by Hitler before he did what he would have had in him to do, and this coercion of Hitler may have destroyed his potential before it became an actuality. And therefore, it might have been that mankind would be farther along and even better off for such a person that might have done it had Hitler not murdered him. In that sense, there is a permanent damage done forever, or more accurately by my own terminology, for all eternity. There's a permanent damage. But that's something you can't measure or evaluate. But what might have been can't be observed, but it's a recognized possibility. So in the Bastiat sense there's a permanent damage.

Perpetuation of the Disaster

But in the actual known sense of not what *might* have happened, but what actually *did* happen, Hitler's effect will be as obsolete as the caveman's activities that are not remembered. The great mighty conqueror who now is called Hitler the Great, not by us, but you may be sure there are plenty of current-day screwballs who still hold that view, and this is a dangerous situation, to martyr such a criminal. These peoples' errors are perpetuating the past. When you say Alexander the Great, you're essentially glorifying murder, and I only say it with contempt or quotation marks. If I were to put this in writing, I would call him Alexander "the Great" and put it in quotation marks. In speaking it's more difficult to say it so I have to circumvent it by caustic remarks of some nature. [*AJG chuckles.*] But whenever a person thinks that that is an accomplishment, that is bad. That is part of the perpetuation of the disaster.

What Historical Inputs Are You Considering of Significance?

Now only a person in flatland who has an innate long-term attitude would think of these things because the inputs he gets in school—especially what passes for history classes—about the past, do not glorify achievement or long term. Do you know any exceptions? There are histories of science and they are, in general, well done. Well, I shouldn't say in general. The ones I have personal knowledge of and therefore I have bothered to read from cover to cover or at least a part thereof, I have found enjoyable and interesting and also very informative and much of the historical inputs of all of my courses come from these things.

The history of what happened in the past is the first step of the scientific method applied to volitional science. What do we know what our inputs are? What are our observational inputs into volition? What has already happened? That's called history. Now the question is, what inputs are you considering as of significance to develop your hypotheses? If your inputs are to believe in the greatness of conquerors and political chicanery and Machiavellian treachery and trickery, then fine. Then your heroes will be of the kind that will make you into a continuation of that flowstream. And if that's considered exalting, exhilarating and heroic, then man's future heroes will tend to do the same thing. That's considered the great thing to follow in the footsteps thereof.

The True History of Man

On the other hand, even where there is a proper input, which is the history of science—which is not of zero literature. It's a small literature, but it's not zero. That's where I picked up my inputs from. It's a more selective literature but it's a much superior one. I consider that the

173

significant one, and this is the true history of man and the other part is the destruction of what is the true history. I came up with the statement you have many times heard, that you cannot destroy what has not first been produced. And therefore, it's more important to discuss how to produce it than how to destroy it and to glorify the production rather than the destruction. All right.

The Psychological Motivating Factors of Pessimism, Optimism, Reality, and Idealism

Based on this, it is clear that most people don't have any such inputs and they don't have the innate drive to pick it up because they're not thinking in terms of the long run. They're not necessarily productively oriented, and the inputs certainly don't favor it which is why it looks very pessimistic. That's another subject that I haven't come to which I mentioned in the last session. That's on the agenda soon. Pessimism and optimism in terms of both reality and idealism. That's several different concepts, integrable. This has partially been done in my history class.

I also want to discuss it from the point of view of the psychological motivating factors. All right. Most people in general have a somewhat optimistic view of the world in the short run and pessimistic one from the long run. That's the exact opposite of my own view and I'll discuss that distinction later.

All Compensation Depends upon a Proper Concept of Gratitude

Right now I'd like to go and discuss the fact that it is in the domain of gratitude that one finds the germ of satisfactory compensation. This subject is well-discussed in V-201 that all compensation depends upon a proper concept of gratitude, which determines *why* to pay. A proper concept of the flowstream, the ideological flowstream, which determines *who* to pay. And the law of logarithmic stimulation, which is a volitional extension of Weber's law on physiology and psychophysiology, which determines *how much* to pay. If you know why to pay, who to pay, and how much to pay, you got compensation solved, and that's the theory of primary property.

Gratitude: The Rarest Human Characteristic of All

If all of this is anchored on the mechanism of motivation of gratitude, you will see why there is a weak motivation for high culture because very few people have the capability of having gratitude. I'd like to state full credit to my Father of having sensitized me to this as a young child when he told me that the rarest human characteristic of all is gratitude. Being another person of major quality who had many contacts with flatland [AJG laughs] which he didn't call flatland. That's a more recent addition to the vocabulary. I would like to point out that it was he who sensitized me to the weakness of mankind on the subject of gratitude.

The Distinction between Gratitude and Thankfulness

What I added to it, among other things, is that I made a distinction between gratitude and thankfulness, which my Father did not, and he didn't make any distinction on these two at all. I said that emotions are responsible for thankfulness, and rationality for gratitude and it fits my general concept of not using synonyms. To most people thankfulness and gratitude are identical in meaning and they're just synonyms. Another word would be appreciation. Appreciation I would lump with thankfulness as a weaker and shorter-term form of thankfulness. Thankfulness might be an inner attitude of appreciation and what is called appreciation, the external manifestation, are the same thing but they are both in the short term.

Gratitude Is Both an Inner Emotion and External Reaction and Action towards Other People

Gratitude is both inner and outer because if you have an inner you will also do something outer to yourself. You will not just sit on it. That's part of the rationality, that you don't introvertedly do nothing with it. And so, gratitude is both inner emotion and external reaction and action towards other people and, although it has an emotional context it also has basically a rational structure, which means it is corroborable, which means it can be handled by the scientific method directly. Gratitude has a reason for it which can be identified, stated and reproduced, articulated. It's not a haphazard feeling of warmth towards someone because he might have done something that you remember him kindly for. It is an identifiable, articulable and durable concept. Now this is very rare.

175

Gratitude and Debt

I was pointing out in the previous session, and here's where I wish to continue from there, where I left off. Not in the previous half-session but in the previous full session, Session 7. I was discussing the emotional characteristics of human beings and also their rational characteristics. I discussed the source of gratitude comes from having the ability to have a rational base to your thinking so regardless of your emotional reactions towards somebody which may be plus or minus or mixed, gratitude doesn't stem from whether you like a person, but from whether you got value from him and recognizing that value or not, and doing something positive in the same scope and caliber.

You're always in debt to some person you owe gratitude to until and unless it's reciprocated in full measure, then the debt is discharged. It's in that sense that I said I'm probably the first man in history to be out of debt. I don't believe anyone else has even recognized that there was such a problem. That everyone is in debt to such people as the inventor of the wheel. "Oh, yes, it's nice that he once lived, but now we got the wheel, who cares?" We're all in debt to him. We are all in debt to Archimedes. We are all in debt to Newton, and so on. But using the product is not a discharge of the debt.

Deeper: Meaning Inward, Inside Your Own Mind

The theory of primary property of course covers this and I'm not going to reiterate what it's in 201, but I want to discuss this from a deeper point of view. *Deeper* meaning inward, inside your own mind. Inside your own self, your own ego. What you think of yourself as your ego. How you come to think of it is the psychology that produces the ego. All right. So when I say deeper, I mean inward. It reduces the time scale and explains the long time scale in terms of your internal structure of your thinking.

Major Generalizations of All History Are the Big Discoveries of Science

Well, let me discuss what I was saying before, in Session 7, and then continue from there. I will just quickly capsulize where I left it off. I said that [*AJG drinks*] in order to have gratitude, you must have first a rational base. To have a rational base implies that you know how to generalize well enough to be able to formulate general structures of thought patterns which are corroborable and to generalize from one thing to another. Now, of course, the major generalizations of all history are the big discoveries of science. Now, most people don't generalize quite that well as to get from a falling apple and a revolving moon, the theory of gravitation, and that's an observational fact. That happened once that that has ever occurred. After that, it didn't need a repetition because you only have to discover it once and before that, man had just an intuitive grasp that things fall down because the earth pulls it.

The Awe-Inspiring Nature of Newton's Discovery of the Law of Gravitation

It is correct to say Newton did not invent gravity, it always existed. It is also correct to say Newton is not the first one to recognize that the earth pulls things down. That's probably an ancient idea. I don't think it takes that much to figure out that an apple falls down because the earth pulls it. That's an intuitive concept and doesn't require even a major base of physics. Intuitively, you can, for example, if you see a horse pulling a cart, you don't have to be a physicist to know that the one is pulling the other, not the other way around. Actually, the cart is pulling back on the horse. That does require physics to know that. Did you all know the cart's pulling on the horse? That's physics. Intuitively you know the horse is pulling on the cart, and that's why it's moving. All right.

But to know how that mechanism works and to connect it with something that's not that obvious, that the moon which is not falling *down* to the earth, but is moving *around* the earth is also falling. Did they teach you that in kiddie school? That the moon is falling around the earth but it's falling the same way as the apple and for the same reason? Well, Newton figured that out. Now it's one thing for me to say this three hundred years later, because I learned it from my physics courses and books, which in turn are Newtonian. Quite another thing for Newton to come up with that magnificent generalization the first time. And one who cannot perceive the grandeur of that brain!

Now that's something you bow down low to, not because there's a whip over your head, but because it's just awe-inspiring. That's awe-inspiring! If you can't see that you haven't heard a thing I said in all of my courses, and that passes over you and you can sit there and say, "Oh well, ho-hum, when is he going to say something interesting?" I said something interesting. That is awe-inspiring. That's more awe-inspiring to me than a sunset, which is plenty awe-inspiring. Newton could see that connection when no one else had, and then do it mathematically yet, and predict it universally.

A True Recognition of What's Important, Beautiful and Awe-Inspiring

Not only did he develop a law of universal gravitation which can be said two different ways. It's a law of nature called universal gravitation. It's also a universal law of gravitation. You can put the *universal* in two different places. It means something different in both places. It's a law which is a generalization of universal gravitation. That means all over the entire cosmos. It's also a universal law that applies to all things, and to have a zero-failure record on that centuries later. You try that with some short-term quick-buck artist. You try that with some politician. You try that with some theologian. If he can duplicate that effect. Then you can have a true recognition of what's important, beautiful and awe-inspiring.

Newton's Continuity of Creativity

I'd really like to know how many of you think it's awe-inspiring, although I wouldn't care to ask you to put your hand up because I don't want any mock heroics here. Nor do I want to embarrass anyone, so I won't ask you. Well, it would be really interesting if I could read your minds. What you think about what I just said, whether it was boring or significant. Because if you don't consider that one of the most important things you've ever heard, Newton's accomplishment—that's only a small part of Newton's accomplishment, by the way. That man's achievements went on and on. This is a lifetime characteristic of this man. It's a continuity of creativity. This is one example of it. This is, you might say, the step one.

The Process of Generalization: Deduction and Induction

All right, next thing I'd like to point out. This process is called generalization. Now damn few people can do that well. *Generalization* is getting a universal conclusion from particular inputs or, on a more limited scale, it would also include getting a conclusion, which is a proposition that you derive from earlier inputs. A *proposition* is a statement with a truth content. It's either true or it's false which is observationally corroborable. You obtain a conclusion, which is a proposition at the end of a logical sequence of inputs wherein you get a large-scale conclusion from small-scale particular observations. That's the reverse of what most people do to arrive at a conclusion, and that is deduction. Deduction is to arrive at a particular conclusion from known general truths. But to get the known general truths in the first place is not deduction, it's induction. It's you build up instead of get particular results. The inductive generalizations of history are the major laws of nature. All right, most people obviously are not doing this for a living or for anything else, and most of the people who do do it, don't make a living at it, I might add, as a corollary. Those who have done this in the past usually did this as a personal goal for their own sake. They didn't make any money at it and they didn't support themselves through this. They got their support from mooching or some external business or some inheritance. All right.

Generalizations Have to Be Made by Infants

The generalizations, however, that most people can do—everyone has to know something about generalizations or he won't survive, and this is where I left off in the last session. That some generalization is necessary even on the lowest level of volitional life. For example, I discussed generalizations that have to be made by infants, if you recall. Generalizations that have to be made by infants. How does an infant learn anything? How does an infant ever get to know it's a human being? It has to identify in some way with other creatures that are called human beings, which are more frequently referred to as men and women, and they come in two brands. And this infant has to determine that there's a correlation between itself and somebody else that's called a person or a man or a woman. The first person it's likely to encounter in life are its parents.

"Well, who are you?" I mean, that is a more articulated thought than an infant is likely to make when it first opens its eyes, and I really am not quite sure what it thinks because I myself did not remember my thoughts when I first opened my eyes and neither do you, and that's one of the shortcomings. It shows how poor our memory system is. We don't remember our earliest inputs. No one remembers his own birth. I don't think that's even conceivable. So you somewhere along the line begin to remember a few things that have happened and I know the specific things I remember the first and they were rather dramatic things like a bowl of farina fell on my lap on a ship when I was coming to America when I was two and a half and the scalding of me I remember, and I remember the incident. I think that's the first thing I remember although I have vague recollection of things that happened to me earlier in Europe. Those are vague. This one's a very vivid recollection. [AJG chuckles.]

When you finally have to determine, "Well, what is this thing called me, or I?" And you have to identify first of all another person, your parents will do since they're closest and usually the most interested. [AJG chuckles.] If your parents are dead or have deserted you then whoever has become your foster parent by picking you up and taking care of you will become the substitute one. They'll be your foster parents or guardians or what have you. And you say, "Well, if nobody did that then you wouldn't be alive." If when you were one year old and you got deserted, and nobody who was older picked you up and did something for you, you wouldn't be here to talk about it. But no one-year-old human—cat, yes—but a one-year-old human cannot earn his living or even support himself.

So, on this basis, the first identification that you're a living creature, and so is that other one that you're looking at, and then to make that supreme generalization that you are the same kind of person, that there's a connection between you and that very foreign-looking creature that is one or both of your parents that you might at first fear. So big, and so frightening looking. [AJG takes a drink.] Can you imagine the fact that you could ever know that you're a human being, could not be done without generalization? Just think of that. That it's not possible to know that you are a human being unless you generalize between yourself and other things that walk around and are called human.

Then of course, identifying with the specific model you're in, or brand. I mean, you're either like one or the other of your parents, but not like both. There are some freaks, I suppose. [AJG chuckles.] I mean you're either of the one brand or the other brand, but you're not both, and if you are then they have circuses which can hire your talents. I'm not sure that those are real, though. [AJG drinks.] I guess the other brand of psychology is better able to cope with that, the sexual brand. Freud, I think, could explain that one better.

Anyhow, in terms of the generalization of how you can even identify that you can be a person. How can someone that is not tall enough to reach to the top of a table identify with someone who is three or four or five times as tall and who is wearing entirely different form of clothing which is superficially deceiving to a child. For example, a grown man doesn't wear the same clothing as an infant boy. A grown woman doesn't wear the same kind of clothing as an infant girl except certain weird types today, which we will try not discussing. [*Audience laughter.*]

Today we have all kinds of degenerates of all different domains, but in general, the clothing deceives a kid from recognizing that there's a correlation too. A little girl doesn't wear silk stockings, for example, and a little boy doesn't wear a mustache or a beard. He can't. [*AJG laughs.*] And even if the father is more civilized, even then there are other characteristics you can't identify with, like a necktie. Of course, sometimes children have the ability to identify with their fathers very easily today. [*One man chuckles loudly in the audience.*] That's not intended as a serious remark, just as a caustic one. Just the last sentence.

Generalization by Infants of Growing Up

Anyhow, the situation is that there's an enormous gap. What you have are particular inputs which are observational. A kid can imagine what he or she is. Let's say he, but that includes she, as you understand the vocabulary problems. So I don't go through this he or she routine a thousand times, I say *he* but that includes the *she*. Okay. So a little kid looks at himself in a mirror or looks at himself without a mirror. You don't have a mirror, you can't see your face but you can see your hand, you can see the rest of your body that's at least visible to the part of you that your eyes can have a line of sight to. A cat has a larger scope of this view, because a cat can twist like a pretzel, but a human being can't very well see his own spine. But you can see the front and generally recognize what character—and you look at your hands and look at the parents' hands. It may look similar but huge difference in size. A little kid has a hand maybe that big. Maybe two inches, or so. Along comes that mammoth, gargantuan monster which has a hand of six, seven, eight inches long. It's similar, it has five protrusions from a piece of flesh which is called a palm and it looks the same but look at this enormous difference in size! Also the agility. Now to generalize that "Someday I'll have hands that big," that's the point I was saying.

The Concept of Time:
The Supreme Generalization of History

That's where I ended I believe, close to this point in the last session. That is the supreme generalization of history. Someday, or later, or just to know what it means, wait. The concept of time is the most intuitively obvious and the most scientifically difficult to identify. The concept that with the passage of time the child grows up and becomes big and therefore a full-sized human.

Not Everyone Reaches Psychological Maturity

You may say, why am I dwelling on this? I already brought this up in the last session. I'm dwelling on this, ladies and gentlemen, because that's the steps you have to go through to grow up. You say, "Okay, that's fine. What are we talking about? Everybody here is an adult." I just made the comment that nobody here is six years old, I'm not sure of that psychologically. I am positive of that chronologically because I don't admit people who are six years of age into this class for many reasons of which I will not bother to enlighten you with a long list of reasons, but there are many reasons for it. It's not an arbitrary decision. But on a psychological basis, not everyone reaches into to maturity.

Responsibility Is Not Possible without Risk

I have defined *maturity* as the ability to take full responsibility for one's own actions. And responsibility involves risk above all, and above every other consideration. Without risk there is no responsibility. That means if you accept responsibility, that you take the risk of being wrong and then paying the volitionally proper concept of restitution for error. Restitution is due, by the way, even in the absence of coercion or error, which causes another person a loss of property. Responsibility is not possible without risk.

The Importance of Being Able to Generalize
the Concept of Property

Now, how can you get to that point unless you know how to generalize the concept of property, mine and yours, and property in general, and have an aptitude and a sensitivity to property? You will notice that no characteristic of man's activity is one that a person is less sensitive to than to property, and that is the most universal and most important characteristic of all which is why civilization has always been poor at protection of property. There's a very low sensitivity to the recognition of what property is and what your relationship to property is; and whose is it? That's the most important question to ask in volition and the one to which people have zero sensitivity in the most part and the reason is they can identify with what is theirs because that's easy.

Concept of Property Ownership a Natural Instinct

The natural instinct of, "This is my territory. This is mine." That's the most natural thing in the subvolitional domain to recognize. That's why all communism and all socialism is based on complete defect in observational reality. To say that you can share property and no one owns property, everyone owns property, the state controls the property for the good of all. All of these are completely outside the purely observational fact that every creature which has volition will guard what is closest to it. This is true of the most primitive animals.

I refer you to the books by Robert Ardrey: *African Genesis* and *Territorial Imperative* in which he discusses animals and their territoriality which is an instinctive form of property which is not with the theory of property. I don't endorse, by the way, all of Robert Ardrey's conclusions, but he describes the characteristics of animals and has written them up in those two books I mentioned, in which he recognized that animals have an instinct that they naturally guard what is in their own territory. And that goes with the most elementary observation of the smallest of infants.

The Most Primitive of All Instincts on a Sub-Rational Basis

Is there anybody here who has ever had a child, or has been close to a child of your own or otherwise and has noticed its characteristics that fails to recognize a child naturally is possessive about what's near it? Isn't the cat? Isn't the dog when it buries a bone? And the squirrel which squirrels away for the winter and is smart enough to know that it's gonna get colder later and need something to eat later? Isn't the bird when it builds a nest? It's the most primitive of all instincts on a sub-rational basis, long before there's any rationality. It's the natural characteristic of all creatures to be possessive about what is in their scope, in their own domain, their own area.

183

The Only Way to Have Stability for Property

The place where thinking has to come in is to be able to draw a line of demarcation between what is yours and what is not yours and then say the only way I can have a stability for my property, or my whatever you call it—territory, my things, my whatever—you don't have the word *property*. The only way I can have a stability to have what is mine secure to me so it is truly under my control is if I do not interfere with someone else's equivalent characteristic. If you can't do that and can't see the connection between your protection and the other creature's protection, then you have failed in generalizing. And that's the only security that man can ever have.

To have a system where property is stable, which means he has to be stable psychologically for property which is to be developed, to be stably under his control and that means he has to have a stability. Before he can be that stable, he has to have the capability to generalize. Well, that's the generalization that's harder to do than to identify yourself with your parents and say I'm a little human and later I'll be a big human. It's harder to generalize what's mine and what isn't mine and act differently towards the two different concepts and recognize that the other party, who has property too, whether it is important to you or not becomes important because if that other party does not retain control over its property, yours won't be stably yours either.

Rise of States Comes from Man's Infantile Attitude on Property

This inability to generalize has produced the striving for some form of protection, which other creatures come about to offer you, and this is the rise of the political state which I discussed in V-201 and also in V-50. The political state arises as an answer to security problems, security being your ability to take care of your immediate desires and wants, however short-term they may be, but you know you have to eat, you have to be warm enough or cool enough depending on where you live. If you're living in a cold climate you have to be warm enough; if you're living in a very hot climate you have to be cool enough. Whatever it is, you have to be temperature stabilized. You have to have for that clothing and shelter against the elements, against the temperature changes. You have to have food, you have to have water and all of these require effort to acquire and thereby someone comes along to offer you these goodies, survival minimum requirements called security, and you abdicate your ability to choose to him in order to get this security. And the rise of states comes from man's infantile attitude on property, and coercion is accepted as a protection of property instead of an attack on property. It's an infantilism in human beings as a group, collectively, that has produced the demand for and the opportunity for the creation of states.

Emotion:
An A-Rational Attitude

I also discussed in the previous session in the context of this, emotions. I discussed that it is incorrect to assume that emotions are bad, per se. An emotion of course, is a non-rational attitude one develops which is either—I'm sorry. I shouldn't say non-rational, I should say a-rational. Do you understand what a-rational means? It's not rational and it's not irrational. Irrational is non-rational. A-rational is not directly associated with rationality. It's a feeling, in general, which comes to you inwardly much the same as your external feelings such as touching, tasting, seeing, hearing, and smelling.

Feelings You Get from the Five Senses

These are the five senses and then you will react to these things. When you see something which you like you react favorably, when you see something that is disgusting you react unfavorably. Ditto with sounds. There are certain melodious sounds which you enjoy hearing. There are also screeching on a blackboard, for example. Chalk on a blackboard is for most people a very painful sound, actually. Anyone experienced that? Okay, this is called, certainly not a beautiful but a rather disharmonious, also actually a displeasing sound, displeasing at the least, possibly even physically painful.

You can touch something which is pleasant like cat's fur, unless you have a mental blockage against cats, what to me is one of the most pleasant touches you can have, is stroke a cat. That's worth the having of a cat if he had no other characteristics. On the other hand, touching a slimy fish is to me [audience laughter] the most nauseating thing that you can conceive of. It not only wouldn't produce pleasure, I'd be almost ill with the thought and have to wash my hands immediately thereafter and I don't think there is a soap that can erase the memory.

[*Laughter in the audience and AJG drinks.*] Touching a hot stove is a memorable but unpleasant experience, but you won't forget it. You learn from that, by the way. I've given that lecture before on the subject of you learn from experiences like that. You don't have to be a thermodynamic genius not to touch a hot stove twice.

Now the point is, these are feelings you get from the five senses. These are not creative hypotheses. Therefore there's no rationality involved. It's simply a reaction to what inputs you had from the physical contact with the physical universe.

Psychological Feelings

A psychological feeling has similar characteristics but it comes from internal reactions to things that are going through your brain. You have feelings which are either positive or negative, which means that you like it or you don't like it; you react favorably or unfavorably. And these can include things such as liking or disliking, anger or pleasure, and feeling of harmony and feeling of well-being. These are emotions. This is what an emotion is. It does include, actually, the responses to the five physical senses, such as touching, hearing, seeing, and things like that. That's an emotion too.

If you react favorably to stroking a cat, that's a positive emotion. If you react unfavorably to stroking a fish, that's a negative emotion. That's for me. You may have the reverse and you may have your own problems. [*AJG chuckles.*] In any event, I'm sure the fish like each other. I can't say that I disagree that they have the right to do it, I just don't wish to share it. I'm sure that crocodiles think they're beautiful and we're probably the most hideous things in the world. And I can't say I disagree with that. [*Audience laughter as AJG drinks.*]

Your Own Personal Taste: An Emotion

I think cats are more beautiful than people anytime. One of the characteristics of mankind is not his beauty. Anyhow. There are some exceptions, but it doesn't go as far as a cat. I never met a cat that was ugly. Just funny looking. [*Loud audience laughter.*] I've seen some funny looking cats but they were still elegant, but I've never seen one that's positively revolting. But of course you say, "Well, that's your own personal taste." Exactly. That's the point I'm making. You say, "I don't agree with you. I hate cats." Well, if you're in that group of unhappy people [*chuckles in the audience*] then okay, I recognize that this is possible and I certainly will not interfere with your disliking of cats. That's your business. On the other hand, that is an emotion. That is an emotion. How you react to a cat, how you react to fish, how you react to anything. It's also an emotion as to whether you like Person A or dislike Person A. It's also an emotion whether you enjoy this type of music or that type of music or that type of cacophony that is called music, and so forth. These are emotions.

Rationality and Risk

Now, what's rational? Rational is taking observational inputs and coming up with predictable conclusions based on a form of arriving at conclusions by an intellectual linkage of thought processes which produce a hypothesis, which when extrapolated beyond the domain of the previously observed will produce predictions as to what you can observe in a domain where there has not been previous observations. And if the corroborations take place without exception then you have, with the fourth step of observation for corroboration purposes, substantiated the strength of the hypothesis that was formed, that if it does not ever fail someday you or someone will have the courage to take the risk, which is a matter of responsibility. I call this a theory. Of course, your reputation rides on whether there is anything wrong with that. And later, if there's a failure to corroborate, that injures the reputation. That's one of the things that scientists in general up to now have totally disregarded, the market risks they take in formulating their statements and as to how positively they have made their statements or how weakly they've made them. The stronger the statement is, the bigger is the risk.

Rational Differs from Emotional

And so, I'd like to call your attention that rational differs from emotional in the fact that although they're both stemming from the way your brain produces thoughts. In the one case, they are predictably corroborable if it's rational. What is called the scientific method, which I've discussed in all of my courses, is the technique of arriving at conclusions that you will adhere to and believe in. And if you do that, to the extent that you do do that and you do it correctly, which means that all your predictions are corroborable, your achievement is rational. Your thinking processes are rational. Whereas if you believe in certain things which you cannot thus corroborate and you cannot show how you got those conclusions, these are emotional. It is not correct to say, even for a rational person, that emotion should be scrapped and man should be rational 100% of the time, in the sense that we have no other function but as rational creatures. Yes, that is perfectly true.

All Irrationality Leads to Error— Ultimately Leads to Failure

All irrationality leads to error and irrationality that leads to error ultimately leads to failure in whatever you're doing. In other words, you fail to succeed in the pursuit of happiness in however you have chosen to define happiness and most people don't have a rigid or formulated or articulable definition. They have some wobbly thing, "Well, I feel like doing this or I'd like to do that or I'd like it if thus and such happened." Or, "I'm going to try to do this and see how it comes out." These are not very specific goals. But if whatever they felt like ought to happen didn't, the dumbest of people can realize they failed. Okay. But they might not accept it. That's where psychological diseases enter. They will recognize that they didn't get the goals they sought, well then, they pass the buck. That's when the disease starts. This discussion, by the way, started in V-30, that you must take responsibility for your mistakes and not pass the buck. All right.

There Is No Mass Communication without Emotion

So getting back to the discussion at hand. The emotional characteristics such as fear, anger, love, hunger, how you react to a cat's fur, to fish's scales; these characteristics are emotional. There's nothing wrong with having emotions or expressing them. It is not correct to say a person should be unemotional and always speak in a purely routine, provable mechanism where everything he says is automatically explained as he goes along. That's nice if you do it, but most people will lose interest. If for example, a television salesman tries to sell television sets by explaining Maxwell's theory to his potential customer, the customer will think he's a victim and he will walk out. I already told you that this is no way to sell television sets. He has to use emotional approaches such as, "Oh, this will fit in with your living room décor." Or, "The sound and fidelity is greater" or "The picture is clearer" or "The programs you get will be delightful." These are emotional responses. He is stimulating to you to think so you shall covet it. Damn few people will buy it because they want to have an illustration of Maxwell's theory. This we've discussed before. So the emotional attributes are necessary for, basically, mass communication. There is no mass communication without emotion. There is no possibility of communicating with people who lose you after the second sentence in a rational dissertation.

Doesn't Have to Be True That Rational People Are Dull and Emotional People Are Exciting

Most scholars write so poorly and with such dull style that they bore the hell out of me. For example, I have publicly stated, I have never read John Locke except a few pages. His style bores me. Yet, from what I understand he did from people who did read him, he did quite a lot of good things. Not everything. I got him secondhand. I mean, second, third, fourth generation transmission, which may lose some of the flavor, to be sure, I have to recognize that. But I tried reading him in the original and he's dull. Well, that's the way it is. Rational people are dull and emotional people are exciting. Yeah? Well that's not true. It can be true, but it doesn't have to be true.

Thomas Paine: Both Rational and Exciting

For example, Thomas Paine was both rational and exciting. When he explained to the people of the American colonies why independence is a favored course of action over reconciliation with His Majesty's government in London, he didn't refer to His Majesty in quite that way. He referred to him as the sullen tempered Pharaoh or the Royal Brute of Britain. You say, "Now a true scholar would not resort to such name-calling." He would criticize Paine and any intellectual snob would denounce him, "This man is not an intellectual, he is writing for the masses." It is true he did write for the masses, but it is not true that he was not a scholar. He was more of a scholar than most of those pompous self-opinionated intellectual snobs that think they are scholars. If you define a *scholar* as a man who thinks well and explains rational conclusions properly, I think that would be to me an acceptable way of referring to a scholar. I would say that Thomas Paine had better thoughts than most people who think they're scholars. That can be substantiated with rational arguments because he wasn't dull. Some intellectual

189

snobs look down upon him, "He was just a political pamphleteer." Yeah, he didn't write two thousand pages of dull writing which hardly anybody else will ever read.

Newton's *Principia* Does Not Read with an Emotional Response

Well, as a matter of fact, you know, here's Newton's *Principia*. [*AJG holds up a copy of Newton's* Principia Mathematica.] I have mentioned this to you people not within the scope of this course, but in between intermissions and other courses. This is not an easy book to read. This, unlike Thomas Paine, does not read smoothly. I'll withdraw that—it does not read easily. Smoothly yes, but not easily. This is not the lightest reading. You cannot read, for example, a random page as we can in Thomas Paine and come up with a tremendous immediate response as to whether you like it or don't like it. If you read Thomas Paine for one page you will either violently like it or violently dislike it, depending on your past biases, or prejudices would be a better word.

For example, let me just read you a random sentence out of Isaac Newton's *Principia Mathematica*, and I just opened the book at random and it happens to come to page 154 of Volume One. I will just take a random sentence, I didn't pre-select this. He's referring to a drawing which he has letters on it to identify points:

> For let the thread PT meet the cycloid QRS in T and the circle QOS in V and let CV be drawn and to the rectilinear part of the thread PT from the principal points P and T let there be erected the perpendiculars BPTW meeting the right line CV in P and W. [*Light laughter in the audience.*]

That does not read with an immediate emotional response as a sentence about the Royal Brute of Britain.

Newton's Scientific Work Is the Principal Input for Both Paine and Galambos

Yet, would you believe it that the man who wrote about the Royal Brute of Britain, his principal input—same as mine on science and on thinking—was Isaac Newton? Because he did in fact understand Copernican astronomy and Newtonian gravitation and that was his principal interest. And the same pattern of thinking led him to try to apply this to human conduct. As one of his biographers W.E. Woodward said about him, and also another one earlier, Moncure D. Conway, who was his greatest biographer, the man who saved him from the falsehoods of his enemies. [*AJG drinks.*] He said about Thomas Paine that he thought about constructing a constitution with the same precision as an engineer builds a bridge. He was the original iron bridge builder, you may know. To him it didn't matter whether some people disliked or liked his ideas. The only question is, are they right? When he didn't like slavery, it didn't matter what the economic interests of slaveholders might be. Slavery is wrong, period. So somebody has his feelings hurt about it, it's too damn bad. Because it's wrong. That's provably wrong.

Rationality Plus Emotion in Communication Is Nice but Not Essential

So I'm saying that if a person can express himself in a way such as Paine, and still be right, meaning he was rational, that's a very nice combination. Not essential. Newton is no less the Newton because he was not always exciting in his writings. Then again, very few people ever read Newton. Nonetheless, even though his books have never become a best-seller, and even when they will be in the future, they will be because people will wish to have, you might say, a monument to rationality even though they don't personally understand the book. It's a better thing to put up on your pedestal than an icon.

Others Popularized Newton's Work

The point is, if Newton had also been able directly to be his own popularizer, which he was not able to be, he might have had a somewhat quicker response. As it is, he needed other people to bring him to be understood by lesser intellects. For example, you know, it was Voltaire who brought Isaac Newton to the European continent. Voltaire and his mistress, the Marquise du Châtelet. His mistress translated Newton into French from Latin, not from English. What I was reading you was the English translation and even though he was English, that's not the original. It was written in Latin, as you well should know by now and it was translated to French by the Marquise du Châtelet and that was Voltaire's mistress and Voltaire himself popularized Newton's writings into French. And between the two of them this made Newton known on the continent of Europe.

Until then he had this rather, you might say, mighty obstacle to acceptance of his work because of Leibniz, because Leibniz was a scientific rival of his. They both invented the calculus and until Voltaire popularized Newton into French and his mistress translated his book into French, he was not known in Europe very well except as to who he was but not what he did. After that, everything changed. Right after that came Lagrange, Laplace, and the other mighty intellects of the 18th century that continued Newton's work. But first they had to be able to know what he did. Now Eddington, for example, was his own self-popularizer. Unlike Newton he was able to write both the kind of writing you can't, in general, understand, and also the kind that you can very easily understand and recognize the significance of.

Emotion Can Be Compatible with Rationality

I'm just pointing out that it is not necessary that someone who has the ability to write or speak interestingly, has to be irrational. It does not follow that a person who is emotional is irrational. However, the reverse is not the case. It does follow that if a person is being irrational, he is automatically emotional. Irrationality is always emotional. Emotion does not always stem from irrationality. Emotion is capable of being either. There is emotion which is compatible with rationality, and just adds flavor and spice to life. It doesn't make it as dull as dishwater. It doesn't necessarily follow that everything that is right and scholarly has to be dull. If it's flavored with occasional, and sometimes more than occasional, strong but provable statements, it has an emotional impact on the masses as well as a rational impact on the intellectuals.

Thomas Paine: Making an Emotional Impact on the Masses; Making a Rational Impact on the Intellectuals

Now Paine was the kind of person who could do both. As a matter of fact, he was probably the single world master at it. It was Paine who converted the intellectuals of the American independence era period to independence from mere rebellion. He is the one who changed the minds of such people as Jefferson and Adams and Washington and Franklin. They were not for independence before Paine. It was he alone that created the American Revolution. That's a unique accomplishment of one man, which nobody in the bicentennial celebration recognizes out of context.

Thomas Paine in the Context of the Theory of Primary Property

By the way, you can say it out of context but it will make no sense at all, and if any of you people have the bright idea to say it out of context and blow everything you know on the bicentennial year, "Well, I know something you don't know." All you will accomplish is to make a jackass out of yourself and injure Thomas Paine's reputation. Or a jillass, as the case may be. [*Laughter from the audience*.] It depends on who. [*More laughter*.] You cannot put this out of context. You cannot have Thomas Paine out of context, both upstream and downstream. Thomas Paine makes no sense without the context of the flowstream of ideology. It makes no sense in the 20th century and the welfare state philosophy without the theory of primary property.

The Wrong People Have Endorsed Thomas Paine for the Wrong Reasons

Today Thomas Paine is dead, not just physically but historically. Right now, he's dead without the theory of primary property because the wrong people have endorsed Thomas Paine for the wrong reasons. For your information, Thomas Paine is a hero with the communists, and he was not a communist. If that image persists that's the worst thing that could happen to Paine. He's a hero with the welfare-staters. Even Joseph Lewis was a welfare-stater. I know this from a personal discussion with him. The man who wrote the book that proves he wrote the Declaration of Independence. It didn't prove it particularly well, but it proves it. To my satisfaction I've eliminated the parts in my own dissertation, which I don't consider strong arguments and strengthened the ones I think are right, added some of my own. But that's a book that shows he wrote the Declaration of Independence. Mr. Lewis was a welfare-stater. He liked Thomas Paine for the wrong reason. He was also an atheist. Thomas Paine was not. He wrote *The Age of Reason* to show that atheism is wrong. Somewhere some people got the wrong message.

Paine's Major Significance in World History

Then there are some people who take some things out of context from what I have developed and think that they understand what Paine has done. They have not. They don't know what it's all about. Paine's significance in world history is that he is an in-between significant point on the ideological flowstream between Newton and myself. He's the only person who did something significant with Newtonian physics and applied it to volition before me, my only acknowledged antecedent in this domain. That's the significance Paine has. And it's a very major significance.

That, by the way, is not even connected directly with the American Revolution. The American Revolution simply happens to be a partially short-term successful achievement which later derailed. You do know it has derailed? Two hundred years later, this country is approaching its termination. I discussed the longevity of civilization and showed you this one had the highest productivity and the shortest durability. The reason it derailed is because it did not include the theory of primary property, and a number of other reasons, but that's the main reason. Okay.

Thomas Paine: The Exception, Not the Rule

The emphasis I'm making on emotions, however, referring to Paine, that he's a classic illustration. Probably history's best. Certainly the best one I know of. Not the only one I know, but certainly one I know of that I think is the most prominent example of a person who was basically right in his way of thinking and yet was able to reach even the masses with it. However, that is not the significant characteristic of major achievement. He is the exception, not the rule.

Most Innovators Never Really Try to Reach the Masses

Most innovators never reach the masses and never really try, and it doesn't matter. Maxwell never tried to reach the masses, neither did Newton. These books have never entered the mass circulation nor will they ever. As I say, the only way they could ever be best-sellers is if somebody recognized them as, you might say, monuments to rationality and they kept unread books on shelves. [*AJG chuckles.*] That's the only way it could be popular. It will never be popular in terms of frequently read literature. There's not enough people to get anything out of it. And yet these people have reached the masses indirectly through the second step of the ideological program.

Paine did both. He actually penetrated the barrier for himself. It was on both sides, both at the hot end and the cold end. But even hotter end than he and higher up on the flowstream was Newton, and the Newtonian physics and the Copernican astronomy. And that more motivated him than the history of the Roman Senate, which influenced the lawyers of the revolution such as Madison and Hamilton and Jefferson. Paine conceived of what is right and wrong, not what is historically a precedent, what is classical or not.

But the significant and major achievers, and I do not consider that Thomas Paine's stature in history even comes close to Isaac Newton's. I have a personal affection for Thomas Paine exceeding Isaac Newton by far. I mean, I look upon Newton less affectionately—that's an emotion—than I do Thomas Paine. If I had known both of them, if their biographies are at all true, I would have liked Thomas Paine very, very warmly. And I would have respected Isaac Newton, but probably from a distance. But if you evaluate their importance and the value they created, and that's where gratitude comes from, the gratitude to Newton is by far the larger and he did it without having the reaching of the masses as a part of his repertoire.

The Theory of Primary Property Increases
the Mechanism of Rationality

It's a very important point to differentiate between the emotional and rational characteristics of man and know which is which, and neither is to be condemned, and rationality is to be put on top. It's the higher characteristic of man. Emotion comes to all. Rationality comes to some. Civilization depends upon those to whom rationality comes easily and is a durable and strong and innate characteristic. My theory of primary property has increased the mechanism of rationality by a very large leverage factor, and it showed the mechanism of creating a second step to the ideological program which does not steal.

Entrepreneurialship Capable of Entering into
the Species Time Scale

Therefore, the business concept of flatland emerges to the primary entrepreneurialship of the primary mechanism, or the integrity machine, of V-201. That form of entrepreneurialship is not innately but by acquisition, by earned acquisition, capable of entering into the species time scale. And for the first time a mechanism which is productive in the secondary domain can be operated, actuated, from the first step and therefore will be harmonious and compatible, rather than disharmonious and incompatible with the first step. The barrier, thereby, is removed.

As I said in V-201, the barrier can only be removed from the side of the second step, which is the downstream area. The second step is the place to remove the barrier from. That means the teaching of the significance for long-term profit, the much higher profit to be attained through long-term objectives than short-term objectives.

Internal Stability Leading to Longer-Term Objectives

I don't expect this to be greeted with mass adulation and it cannot. There are only a few people who can even understand what I'm talking about. I hope some of them are present this evening. Those are the people that can build a civilization that is durable. This course's addition to that theory is the internal stability that comes about from a grasp of the concept of long term. And the long-term view actuated by short-term, subtrivial time scale thoughts, which lead to longer-term objectives is a large part of what's coming in this course. This is all introduction to this.

Personal Stability Developed on a Conscious and Purposeful Mechanism

It is the ego minor which is the universal destroyer of stability. The ego major, which is the builder of the first step of the ideological program in the first place, and the second step through the theory of primary property in the second place. It is the concept of personal stability that can be developed on a conscious and purposeful mechanism by the harnessing of thoughts, including emotions, to longer-term objectives and always putting the order of things, the priority schedule, in favor of longer term than shorter term. A person who can discipline himself to do that can acquire the long-term view, but that increases his capability to generalize from where it was before. And if a person cannot do this, it cannot be done within that person.

As I said, this theory of psychology explains all kinds of people from the shortest- to the longest-term objectives, but only those can acquire the value from it who are in the longer time scale appreciation capability zone.

External Discipline:
The Unacceptable Alternative to Self-Discipline

Those who have not got the ability to comprehend this—I don't mean hear it, I mean understand it and act accordingly—will not get the direct primary advantage. It will explain the others. It can easily explain the Hitlers. The Hitler types can't enter the hot end of the ideological program. Neither can the short-term quick-buck artists. Nor the people who find any member of the opposite sex as a gay adventure for the night to start a new one all over again with a different personage the next day. These people can never get the supreme value of the stability to be afforded from the personal inward ability to harness one's own thoughts so that the higher priority ones are the longer time scale ones. That calls for something called discipline, self-discipline. Where there are no people in a civilization that have this self-discipline, the tyrants will furnish an unacceptable alternative called external discipline. The Prussian army is the substitute for stability. I'll have a lecture on the Prussian army in a different course some time. It won't be this one. You'd be surprised at how much you can learn from all kinds of things. From fish to Prussian army. [*Light audience laughter.*] Cats.

A Man Who Cannot Learn from a Cat
Is an Intellectual Weakling

By the way, I also resent intellectual snobs who tell me that I waste a lot of—and if this fits you, wear the shoe. [*AJG drinks.*] I resent snobs who tell me that I waste a lot of the class's time in discussion of cats, so let's end that subject. I have learned more from my late cat Christopher and my present four cats than I have from the combined totality of almost all of humanity with the exception of maybe a few hundred. And a man who cannot learn from a cat is an intellectual weakling. Because a cat who is not bright has certain characteristics that can well be emulated by a rather weaselly creature called man. The lowest cat has more self-esteem naturally than all but a few human beings. No cat can be enslaved, and I've never met a deceitful cat. I've never met a cat that pretends to be something other than what he is, or pretends friendship prior to killing you. So if anybody is harboring any resentment against my discussion of cats here or any other time or place, then be it known to you one and all that I resent you, because you're not capable of

197

participating in the progress of man. And it's also an observable fact that only tyrants hate cats because they don't like anything they cannot control. All right, I just brought that in because I know that there's always this kind of an undercurrent going so I might as well face it openly. [*AJG chuckles.*] It's much cleaner. All right.

A Preview of Following Sessions in This Course

I will discuss starting in the next session the discussion of pessimism and optimism connected with short- and long-term concepts. Realism and idealism. All of these are connectable and although part of this discussion stems from what's in V-282, the inward aspect of how it determines personal motivation and what kind of a person you are and how this can be connected with ego problems such as plus and minus, major and minor aspects of the ego, I will bring that in as another major topic.

The Inability of People to Generalize

I also would like to point out that many problems in psychological misadventure come from the inability of people to generalize as I've already said, that is to argue beyond their own local domain to some area remote from them, either in person or geography or time. Time is perhaps the greatest dimmer of memory and therefore produces more errors than all the others. It's the thing that makes it look as though most people do not have gratitude which in fact is true, because time fades memory, fades appreciation which is short term, and the gratitude perishes because there never was any. In other words, what appeared to be gratitude turned out to be only short-term, emotional and nondurable response such as appreciation and thankfulness which is an inward aspect of appreciation. The reason people cannot maintain their gratitude is because they never had any. They had a short-term facsimile and that's because they cannot generalize in time, from one time to another. If they can, time does not produce a logarithmic stimulation in gratitude, merely in total impact.

Gratitude and Compensation

The logarithmic stimulation properly comes in compensation in that meanwhile there have been more inputs and therefore you have to prorate it to all of the inputs, the earlier and the later ones, but not quantitatively the gratitude is reduced in percentage, but qualitatively it remains durable or even grows. I should say in general it grows qualitatively and also quantitatively if it's measured, for example, in P_2 in the theory of primary property. For example, if we applied the theory of primary property to Archimedes then when he lived, and now, it is incontrovertible that from a quantitative point of view, percentage-wise, he'd be getting far lesser royalties today than someone who lived right after him. But in dollars, or in monetary units of any kind, he'd be larger because of the enormous expansion of the production of civilization.

The Ego Minor Characteristic: Producing Short-Term and Unstable Activities

Now, the ego major and the ego minor concepts will explain both the hot end mechanism, not how it came about but what actuates it and how it affects the rest of civilization. And the ego minor aspect will produce the understanding of all the obstacles that mankind has had both in civilizational structure also in personal structure, why an individual cannot succeed at anything. The same characteristic that prevents a person from having gratitude to Newton or to any other significant input is the very same characteristic that will produce in that same person the inability for himself to have an adequate self-esteem. Therefore, since he cannot take it, that he hasn't got adequate self-esteem, he manufactures a false self-esteem which is an ego minor which is what produces his short-term and, in general, unstable activities.

Stability in Relation to Survival

Stability which is the next major course to be offered by me through FEI in a year from now will follow smoothly from this discussion of psychology because it follows smoothly from the financial, the emotional and also the motivational aspects of man. And stability is a generalized concept, which is totally unknown to the average person. Most people have never led a stable life, neither from an external point of view, because political civilization produces so many upheavals and mishaps, and before that they were subject to even the elements, even now they are. Many people die even today as a result of hurricanes and tornados and fires that they themselves did not set but because of natural and also artificial disasters which are not properly controlled through lack of adequate technology. Illnesses that have not been solved through knowledge.

Of course, that's an ultimate universal killer. Ultimately, death comes to all in today's structure of civilization. There's no way you can avoid it at this time because we don't have the knowledge to know what to prevent it with, namely, a biological development which has not yet been integrated but is integrable.

The Universal Disease in Relation to the General Instability of Man

The general instability of man comes from external sources beyond his control and internal sources. It comes from his own internal sources largely because of his improper evaluation of the reality that he is immersed in. He is in the real world, but he doesn't know how to recognize it. He cannot differentiate between real and imagined things that he believes in. The ego minor is the mechanism to have a lesser appreciation of reality. His own shortcomings are translated into a hostility against those that he has not been able to cope with and integrate into the reality structure he has never recognized. This is why I call it the nearly universal disease.

Almost all human failures are caused by the same thing. This is as much of a generalization as that coercion is the source of all civilizational collapse. As I said, the ego minor and the coercion are quite related. I've already covered that. I'll continue from here. [*Applause from the audience for sixteen seconds.*]

SESSION 9

PART A

Emotions Have a Positive Role to Play in Human Health

Good evening, ladies and gentlemen.

In discussing the nature of emotions which can be rational, it is clear that the existence of emotions has a positive role to play in human health. As long as the emotions are subordinated to rationality and do not overpower it and produce a superseding of the rational characteristics. As you know, most people's emotions are not subordinated to rationality, if for no other reason than they haven't got any. Rationality is too rare a characteristic and therefore the concept of emotion has a somewhat jaded image in terms of the standard attitude prevailing as though emotional characteristics represented irrationality. This is not true.

Desirable Emotions: Warmth, Compassion and Kindness

As I have pointed out, some of the most potent achievements of human beings come from people who are capable not only of demonstrating but having rather strong emotional action. Among the desirable emotions are warmth and compassion, kindness. These have rational characteristics. Persons who lack these characteristics, as far as I'm concerned, are deficient human beings. I look upon coldness as a human deficiency and failing, people who are not capable of demonstrating warmth.

On the other hand, there is another side to this too. Warmth also requires a return action called compensation in volition. There's no possibility to be warm to someone who is not warm and keep it up for very long. It produces frustration. This is a limited discussion of frustration here. There will be a much more general discussion of frustration later in the course on a more generalized basis. But to be warm or kind or compassionate or just plain decent to any other human being and have it not reciprocated, produces frustration and it could also produce cynicism. It could produce misanthropy on the part of people who have been burned. The best people in the world are the ones who are burned the most because they have both the highest exposure hazard and they also have the greatest sensitivity.

Sensitivity:
The Measure of a Reaction to a Given Amount of Stimulus

I don't know how many of you have heard V-212 where I have a discussion of sensitivity at the very beginning of the course. Sensitivity represents the reaction to a stimulus. Sensitivity measures how easily there is a reaction to a given amount of stimulus. The larger the reaction, the greater the sensitivity.

Facsimiles and Counterfeit Substitutes
to Deep Concepts

There is a counterfeit to sensitivity I have pointed out in V-212, and that is touchiness. Most deep concepts have short-term counterfeit substitutes or, let's say, facsimiles. I pointed out that the facsimile of gratitude is thankfulness. The facsimile of curiosity is nosiness. The short-term facsimile of sensitivity is touchiness. People who are touchy are rarely sensitive. What's the difference? Time scale, and depth of the thing that the reaction is to. Touchiness usually deals with trivia and superficialities. Sensitivity is to primary matters, more than to secondary.

A Number of Reasons for Not Demonstrating Emotions

I could care to point out that the characteristic of not demonstrating emotions does not mean the person doesn't have any such reaction, but the inability to express it is both a form of inarticulation and therefore inability to communicate. Or it could be an unwillingness to communicate because of repressions that have occurred in earlier life that have produced a fear to express oneself. This could come from, let's say, parental authority which is abusive, or being exposed to certain forms of coercion including, let's say, certain kinds of teachers who are in school who are coercive. This is especially true in countries where the educational system is more stern and strict than in this permissive and rather decadent educational structure that this country has. Where, for example, corporal punishment could be inflicted by teachers. This could produce a negative effect upon an impressionable child, such that upon growing older with years he becomes unwilling to communicate in his feelings in any form. It's a form of fear. Fear obviously is an emotion, fear of consequences.

Repressing Emotions Leads to Frustration and Psychological Problems

I might also say that a person who is seemingly unemotional, that's on the surface, internally he probably is worse off than a person that expresses his emotions. By repressing them he's essentially building up the internal frustration which produces an emotional stress or pressure and like a pressure cooker, if you do not release the steam and keep the lid tightly shut it will produce a higher pressure head and if it becomes high enough it will blow the lid off. This produces, I'm sure, much of what in clinical or medical psychology produces outbursts which end people up in various conditions that end up as cases for the psychiatrists and the mental hospitals. This comes from people who are incapable of expressing their emotions.

A Proper Expression of Emotions Is Necessary to Maintain One's Stability and Health

I think that a proper expression of one's emotions is necessary to maintain one's stability and health. It's safer. It's by all odds the long term, better thing to do as long as one thing is kept in mind: the emotions are not a license to be irrational. Now of course, I'm talking about what is a right concept. This doesn't apply to most people because they do not know how to evaluate rightness. I'm talking about the ideal, not the average case. The majority of people's emotions are totally turbulent and uncontrolled, and if they're rational it's incidental and occasional and usually coincidental to intention or desire which still happens, maybe occasionally they're not in the wrong.

Improving the Hot End to Create Stable Surroundings for the Cold End

Since I'm discussing this course not just in a general context but also specifically for the purpose of improving the hot end of the ideological program, and the basic value that is in this course can reach mainly hot-enders. It is not capable of dealing directly with the cold end. To understand the cold end, yes. To improve them, no. I don't think there is a short-term solution for that. The only way to improve the cold end has not changed one iota from the original discussion in V-201. Any improvement in the cold and must flow in the normal four-step sequence in the ideological program. It must come from example and product development and product improvement.

When the social structure has better products and better means of keeping track of who owns what, and better ways of compensation and better ways in which human beings are capable of getting what they want out of life, namely to pursue happiness in whatever way they choose so long as it does not interfere with other people's property, so long as this mechanism in the volitional domain is available and ultimately becomes operational, the cold end will be benefited, not just economically, not just socially, not just in terms of the achievement of freedom, but also psychologically because when they have a stabler surroundings they are part, they respond to their surroundings. Their surroundings include other human beings, not just their working conditions or their compensation but includes other people and the general structural stability will have an image component, an image development.

People Are Basically Habit Behaving Animals

People are basically habit behaving animals. Most things that are done are done by habit or routine. Most people do not even think about most of the things that they do which is of course compatible with what they have to think with. Most things that they do are relatively routine. For instance, the normal functions of living including the preparations to get up in the morning and get dressed. Subsequently when later in the evening to get undressed, prepare to go to bed, brushing one's teeth, taking a bath. These are the things one doesn't think about: what is the next thing to do, what order we do it, what is the significance of everything I'm doing, what is the quality of the soap, what is the nature of the toothbrush [*AJG chuckles*] and what are the mechanical interactions between a toothbrush and the surface of the teeth? These are things dentists worry about and annoy the patients with. I use the word *patient* deliberately, by the way. This is not something the average person thinks about. It's a very routine thing of life.

Customary Daily Rituals and Routines

One lives to pursue happiness, not service his teeth or his fingernails. [*AJG laughs with the audience.*] I've tried to explain this to several people in the professions. I once had a fellow who took my course back in San Diego in the early days. He was in the men's clothing business— suits, shoes, shirts, neckties, belts, you know, this kind of crap. [*Audience and AJG laugh.*] He was always annoying me about how important the quality of one's clothes is. A suit is what you wear to cover yourself for both social reasons and also for comfort, and temperature control and things like that. And then he said, "A man is what he wears," and I said, "That's so much baloney." [*Laughter from the audience and AJG.*] A dentist thinks that a man is what his teeth are. [*Audience laughter.*] The hairdresser thinks that the man or the woman is what his hair is. That's ridiculous. [*AJG drinks.*] Everybody has his own little bailiwick that he wants to put up into the kingpin role of civilization. [*AJG chuckles.*] This doesn't mean you don't wear clothes; it doesn't mean you don't brush your teeth or whatever other ritual is customary. But it is customary, that's the whole point. It is a ritual and it's the customary thing to do.

Almost Everything Is Done by Ritual and Routine

There isn't anybody outside of a dentist who could explain to you what tooth brushing does. I doubt if he could even make a good case for it and that comes from long experience. [*AJG takes a drink.*] Eskimos don't brush their teeth; they don't need dentists either. I know, they eat whale blubber. [*Loud audience laughter.*] That's good for the teeth and sugar is bad for it, and I get the long spiel about how bad sugar is, as though sugar were the enemy of civilization. [*Audience laughter and AJG chuckles.*] If the shoe fits, too bad. [*Audience laughter.*] This is getting around basic issues, it's circumnavigating the issue. People all have small routines and rituals they perform. Almost everything is done by ritual.

205

Most People Live for Superficial Trivia through Habit and Routine

Driving a car is a ritual. The same thing goes for cars, that fancy cars is what makes your reputation. That's great. That's why I use that tin can I drive. [*Audience laughter and AJG laughs and takes a drink.*] Because if my reputation depends on my car that's pretty damn sad, and yet that's what the average cold-ender believes. That's why they buy the fanciest car to impress their customers how successful and prosperous they are. I have a course called V-30 which deals with the subject too, so I won't go into that. But most people live for superficial trivia and in so doing their lives are consisting of trivial rituals which are done by rote.

They are no more part of their thinking than how to swallow food. You don't have to be a genius to know how to swallow food. Cats can do that. That's a habit too. You get into it because you're hungry and then nature helps you out. Tying your shoelaces is not quite natural, you have to learn the mechanics. Once you do it, how many of you think about how to tie a shoelace? Or do you do that by ritual? Can you do it by ritual? [*"Yeses" from the audience.*] That's how most people do their daily work, by the way, which is why it's so high quality. [*Audience laughter and AJG chuckles.*]

Most People Cannot Subordinate Their Emotions to What's Necessary to Be Done

Most people can't handle a job if it has more than three things to do. If they have something in which they have to do discretionary thinking and judgment, where you have a hundred or two hundred or five thousand different things to think about or do and juxtapose them in the proper order, priority scheduling, and determine what comes first, how much time to devote to this, what's more urgent, what's more important, how much effort to put into this and how much of this can be done by ritual? Most people don't think about these things. That's why they are one-shot people, they can do one thing at a time and then if they get into the groove, and they have not been displaced by some emotional upset, then they can function. An emotional upset makes them even wobble at the one thing they're in the groove for, and they get out of the groove and they can't even handle the most simple tasks.

Most people cannot, under any circumstances, subordinate their emotions to what's necessary to be done because it's thought out to be needed and rational.

There Is Nothing More Dangerous than to Repress One's Feelings

Now all of this is neither an argument to attack the concept of emotional feelings nor to say that that's paramount. It's neither. It has a role to play in living. Many things can be done by instinctive feelings and those can be delegated to relatively low thinking levels. There is nothing more dangerous than to repress one's feelings and pretend that one feels differently than he does. Then he becomes deceitful and false, a very common characteristic of human beings.

Easier to Be Truthful than to Be Deceitful

I've always thought about how difficult it must be to be a chronic liar. You have to have a mental catalogue of who you said what to. [*Some chuckles in the audience and AJG laughs.*] It's so much easier to be truthful because then you don't have to keep track of to whom you said what lie, so you maintain your posture and your prestige, such as it may be. The people who are open with what they do have both a harder and an easier life. I presume you will want to know what that seeming lawyer statement is. It's easier and harder, simultaneously.

Occam's Razor Speaks for Truth

For flatland purposes it's very much harder to be truthful; for space-land purposes it's much easier. Since we're living in flatland today, there's a conflict. Do you know why it's better to be truthful? First of all, you don't have to keep a catalogue of lies so you know to whom you said what. Therefore, Occam's Razor speaks for truth because you can remember what's true more easily than when you vary the statement to fit the person you're talking to. When you tell somebody what you think he wants to hear, or what you think is good for you that he should hear it regardless of its truth content, then you've got to remember to whom you said what and under what circumstances and for what reason. But if you say the same thing to everybody on the same subject then it's much easier. That's clear, isn't it?

Truth Is a Difficulty in Flatland

On the other hand, truth is a difficulty in flatland because a person who is truthful and doesn't hide what he believes, and lets it out openly, gets himself into a great deal of trouble with flatlanders who cannot cope with such a person, they think he's crazy. I ought to know, I felt this all my life, that most people look upon me as, "Well, what planet did he drop in out of?" [*Audience laughter.*] "How did he get here amongst us?" [*AJG chuckles.*]

Galambos' First Friend as a Child in School

I had that when I was a child in school. That's how I got to have so many friends—exact count of zero. It's very difficult to remember that number, isn't it? [*Light audience laughter.*] The first time I had a friend in school was when I was thirteen years old. It turned out he really wasn't, by my present definition a friend. He wasn't quite up to that caliber that I would call today a friend. However, what he was in fact, looking backward upon it, he was my first student, and I was either twelve or thirteen and he was a year older, but he was my first student and I could give lectures to him, and that's how I learned how to articulate. [*Loud audience laughter and AJG laughs.*] We were very friendly for many years until we grew up and separated because of the army. We went in different directions. [*AJG drinks.*] He was a very fine person, incidentally, in character, which helped. Anyhow, it was easy to talk to him, for me.

Galambos Had Difficulty Communicating with Classmates as a Child

It was hard to talk to everybody else because they weren't interested. They were interested instead in much more important things like games, feuding, spitball throwing, and various other brilliant activities that one undertakes in school, especially in such classically, well-ordered schools as they have in America, where teaching is incidental to monitoring the windows and the aisles. And of course the language helps too—the insane language that we are stuck with—helps because three-quarters of elementary school is spelling. That may be an exaggeration, it might only be 70%. [*AJG chuckles.*]

Only a Foreigner Can Master the English Language

I remember when I was going to school, I had the report cards brought home to my Father, and he looked at it and he saw things that were graded included spelling. He said, "Well you had that last year." [*Both the audience and AJG laugh loudly.*] I said, "Yeah, but they have it again every year." "Why?" [*AJG laughs.*] "Because it goes on for all of the school years." He said, "Why?" [*Audience laughter.*] Well. Why? Because nobody in this country understands his own language. It's largely because it's so difficult of comprehension, only a foreigner can master it. You see, they have to try, they have no choice. A person who comes here from another country, finds this anti-rational language, if he's going to talk it at all he has to learn its rules. And the first rule of the English language, which has no exception, is that there are no rules. [*Audience laughter.*] There are no rules that have no exception. [*AJG laughs.*] Now if you understand that rule, that one has no exception. [*Audience laughs.*] You understand what I'm saying. [*AJG chuckles.*]

So, there are two ways to attack this. Memorize each word as though it were a separate adventure. [*Audience laughter.*] And that's what the American dummies do. The foreigners do it differently. They develop a feeling for the language. That's an emotion. That's not rationality. That's the only way to master the English language, because it's irrational. You cannot apply rationality to where there is none. [*AJG drinks.*] There is a rhythm or a style of irrationality here, and when you master the feeling for it then it more or less comes out.

An Example of an Intelligent and Literate Foreigner

I recently a year ago came across, thanks to two of my graduates, a newsletter on energy which turned out to be written intelligently by a person who understands that there is no energy shortage. He himself is an electrical engineer and professor thereof and he puts out a newsletter on energy. It is also very well-written with both wit and style, as well as intelligence. I wrote the man a letter that it was wonderful to find that there was an oasis in this sea of insanity other than myself, so I said I was glad to see that he had something intelligent to write. He wrote me back [*AJG drinks*] and not to my surprise it turned out he was a Czech [*a few chuckles in the audience*] who had come here. Well, the fact that he was Czech is immaterial, but he was not an American by birth, and he had come here in the 1960s. He was essentially an exile from communism, which he abhors and he's all for free enterprise and for technology and science. The right combination of three things, by the way. It's called pro free enterprise, pro science and pro technology. Within the political framework—which he doesn't understand V-50—he is right. He does a very creditable job at it. All right.

He also wrote it so nicely. So I wrote him back another letter and said, "That explains why your newsletters are literate because you were not American born." You may think I'm joking, but I'm not. I don't know any American English professor who has a vocabulary of English compared with my Father who was foreign born. Now he had an accent but as far as the understanding of the English language, he could give lectures to English professors. That's largely because it's 60% Greco-Romanic, or Greco-Latin. Twenty percent Germanic, I understand, is the background of the English language. Twenty percent whatever else. It's a hodgepodge intellectually as well as content-wise.

Your Language Affects Your Thinking Pattern to a Large Extent

You say, "What's this got to do with psychology?" Well, it has a little bit to do with it in that communication is necessary for human beings to be able to know what each other think and do, and as a result of this, much of their thinking is structured to fit the language that is common to their communications. You think in terms of concepts and the more intellectual you are the more you verbalize your concepts unless you pictorialize it. Most people just think in terms of images, literal images, pictures. And then if you become more articulate and more verbal, your concepts become more capable of proper abstraction and therefore the language that you think in largely influences your thought patterns. English is an especially difficult language to be psychologically stable in. I don't think anyone has given this much thought, but the language you have affects your thinking pattern to a large extent.

Better Ability to Express Thoughts if Bilingual or Multilingual

Since I am bilingual, which is a handicap that I do not have multilingual capabilities such as ten languages. My Father was always irritated with me that I only was able to speak two languages. That's only one better than one. He used to say about Americans that they speak no language other than English, and that poorly. He would say, "Only a dog barks in one language." [*Some audience laughter.*] It's not a very articulate language, I might add to my Father's comment. Well, two languages is a hell of a lot better than one. Three would be still better—four. Well, I have some knowledge of about six languages, but some knowledge is not total capability. For example, to think in a language, I would say you have total capability in a language when you have the ability to think in it, not merely to express yourself. When you're translating from your known or established home language or base language and then you're saying it in a different language you're essentially translating it as you go along. That's not the same thing. If you're not comfortable to think in it, then it's not the same thing.

Helpful to Have Ability to Think in Two Languages

Well, I can think in two languages easily and partially in a third. I can partially think in German but very poorly, but I don't have to translate everything. I can understand it in the original context without translation. That helps. I have some structural knowledge of other languages, at least three others, to varying minimal extents, of course. The fact that I know two languages is very helpful, however, over one. Because it's more than double capacity to communicate in. You can compare to structures of languages, and you have an automatic equivalence factor here. How would you express yourself in two different languages? Now if the two languages are very much related, the advantage diminishes. For example, if the two languages that you might speak are, let's say, Norwegian and Swedish I don't think it would be a very great advantage because the languages are so close to each other. Or Portuguese and Spanish are not the same but they're related to each other and they're very close. Now, it would be hard to find two more different languages than, let's say, English and Hungarian, both of which have large vocabularies.

English Is a Giant Language but Hungarian Is More Colorful

Now you might find two languages such as English and Hawaiian. There's a bad comparison there because one of is a major huge language with an enormous vocabulary and the other one is a minor language with a small vocabulary and a very small intellectual capability, and so you're dealing with a pygmy and a giant. English, for whatever its shortcomings may be, is a giant language. I don't just mean how many people speak it. I mean the magnitude of the language, the magnitude of the expressions and the magnitude of the number of things you can say, the breadth and depth of the vocabulary, the richness of it, the colorfulness of it. It is a large language; it is a very large language. The English and Americans boast that it's the largest language in the world. I would like to challenge that. I don't think that's true, but it's very large. I think that Hungarian is larger but fewer people speak it. There is nothing you can say in English that I don't think you can say in Hungarian as well or better, usually more colorfully.

Language In Relation to the Subject of Psychology

And you may say, "What has all this got to do with psychology?" Well, it has very much to do with it, how people behave, what motivates them, what kind of thoughts, what is the color of their thoughts? Do they express themselves colorfully or dull and drably. English is a relatively drab language in case you do not know it. That is even true for cursing and swearing which is more colorful than ordinary communication. Even here it's not, as you might say, colorful as Hungarian, which I don't curse as well in Hungarian as in English. That's because I have less experience. [*Audience laughter.*] I was not in the Hungarian army; I was in the American army. [*Louder audience laughter and AJG laughs.*] Do you understand what this has to do with psychology? Psychology deals with motivation of volitional action. Well, a person has to think before he acts. Thoughts have to have a certain flavor to them and the language it's expressed in affects the capabilities and breadth of one's desires and action and whether one is imaginative or not. The language, to a large extent, affects that.

The British Empire vs. The Empire of Genghis Khan

Now whatever irrationality I attribute to the English language's structure, it does not affect its magnitude. It is a huge language in terms of its vocabulary and in terms of its experience factor. After all, the British and the Americans have had the single largest expansion on this planet in terms of the effect of their culture. The British Empire was the most world-round empire of history. The sun never set on the British Empire when it existed and it was all over the world and the experience factor that the English obtained in dealing with people in other continents was reflected in the usage of the language, both in its usage in foreign lands and the foreign lands' effect upon the language. The only empire in history that was larger than the British Empire was Genghis Khan's empire. That was larger in territory in square miles, but it was not larger—also probably in the percentage of the world's population that was under the domination of the Genghis Khan hordes vs. the British colonialism. But the British Empire is more diversified and more spread out all over all continents and all portions of the globe, all latitudes and longitudes; better dispersion. Whereas the Asiatic empire of Genghis Khan, to be sure was large but it was concentrated on one land-mass.

Also, durability was fortunately short, relatively speaking. Not for its high production, in their case. The lack of durability of that one came from the high coercion factor which did not survive the death of the strong Khans. There were just a few of them and then they disintegrated. As with all major empires, there is falling out among the successors and they internally disintegrated.

The Diversification Factor of the Hungarian Language

In the case of the Hungarian language, the diversification factor is there also but for a different reason than dispersion all around the world. It was a different reason because Hungary is essentially in the center of an area which has many different nationalities. In one single country there were some several dozen different language groups, all of them alien to Hungarians. In other words, there were different language structures. They were mainly Slavic languages but also Germanic and also various other language types and therefore the interaction between Hungarians and their neighbors produced a more colorful type of language. It had a similar effect as the British diversification all over the globe and more compacted.

Bilingual Thinking Affects the Diversification Capability of Developing Thought Patterns

Nonetheless, they are two entirely different languages and when you basically have a bilingual thinking pattern, I think this affects the diversification capability of developing thought patterns. I have to say that my main thinking is done in English, but I can, when I want to, think in Hungarian too. It's simply because I spent most of my life here amongst people who speak English, or at least they tried to, and I find that the very thing that I had to explain to my Father's spelling [*AJG chuckles*] why we had it so long for so many years. He said, "You had it last year, and the year before."

Americans Are Mostly Monolingual, Europeans Tend to Be Multilingual

I find now, in retrospect, now that the formal schooling is long behind me that I don't think that the average American gets it long enough. However, actually, that is meant to say it didn't do any good. Actually, I think if he had another twenty years' dosage of it, he'd be just as incompetent, he'd be just more bored and more confused. It's a lack of desire for perfection. It's a lack of desire to master but have the pride of craftsmanship to even want to master one's only language, which is the case with the vast majority of monolingual people in this country. I don't know the census count. I don't believe the U.S. Census Bureau asks how many languages you speak. But I believe that it is probably correct to say that 90% of the people in this country do not have more than one language. And that is, I say, more quite imperfectly. This is not characteristic of Europeans, for example. Almost all educated Europeans, by that I mean anyone who has gone to school and isn't a peasant out in the fields, most urban or educated Europeans speak several languages and not necessarily well, but they do have multilingual capabilities. It is not rare at all in Europe to find someone who can speak eight or ten languages.

Louis Kossuth: Multilingual Hungarian Patriot-Liberal

The Hungarian patriot-liberal, Louis Kossuth, I believe spoke thirteen languages so fluently that he was a master orator in all of them. As a matter of fact, one of the languages that he was proficient in was English and he learned that while he was politically a captive of the Austrian Habsburgs. For two years he was incarcerated for his political activities by the Hapsburgs and during that period he put those two years to good use. He got himself, by the permission of his jailers, he got himself Shakespeare and an English-Hungarian pronouncing dictionary and he had some French literature and the equivalent French-Hungarian dictionaries, and he learned and mastered both of these languages during those two years, so that later when he was in America he gave a speech before the joint session of the United States Congress. I might add, he was the first foreigner ever to address the Congress by invitation. That does not count Lafayette because he did too, but Lafayette was an American citizen. Lafayette and all of his descendants were granted permanent citizenship for his help to the American side during the American War for Independence, so he counts as an American even though he was of French birth and citizenship.

Kossuth was the first other foreigner to address a joint session of Congress, and I saw that speech when I was in school. I remember that this was one of the things I did for public speaking. I looked up his speech in the college library. It was in a book called *The Greatest Speeches in the English Language*. And that was by a man who was a foreigner and learned English alone, speaking to no one in a prison cell in a period of two years during which time he also learned French equally well.

A Person with Multilingual Capabilities Able to Analyze His Thinking in a Nonsingular Form

Of course, this is a talent that few people have, to be sure, and that isn't what I'm really referring to, the super talent at languages. Kossuth was exceptional. He was superior and exceptional to practically anyone I've ever heard of. On the other hand, he's probably the greatest orator of all time. Less known because of his nationality being to Americans obscure, but Demosthenes, or more correctly pronounced "day-mos-*tay*-nus," he was a great orator in Greek. Winston Churchill in English, but he lisped. Kossuth did thirteen languages. Now as I say, that is an exceptional situation, but it does point out what I wanted to say here and that is that if a person has multilingual capabilities, he in general has a capability to analyze his thinking in a nonsingular form, but in one that you can show thought equivalences in different language structures.

Mathematics: The Single Supreme Language of Mankind

The strongest language of all, the single supreme language of mankind for which there isn't even a close runner-up, is mathematics. And that's where basically mathematics falls into the pattern of languages. Now of course the amount of things you can express mathematically at the moment is limited. For example, the ritual about tying your shoe-laces, although not that you cannot have a mathematical model for this, the normal person doesn't think of that in terms of equations or math-ematical constructs. Driving a car or tying your shoelaces or brushing your teeth. But basically, as far as communicating rational concepts and thoughts, that's the reason mathematics is the language of physics.

Physics: The Birthplace of Rationality

Physics is the single most susceptible and capable subject matter that yields to rationality. It's the birthplace of rationality. The scientific method comes from the developments of physics. What I have done, to a large extent, is the extension of the use of the scientific method to things other than classical physics. It is a matter of self-defense. The normal place for me to be using physics would be in classical physics which includes all aspects and components of the universe, and the study of the universe as a whole or its various structural parts. But in trying to expand physics to other branches of human activity, one of the frustrations and difficulties is that they are not accurate. In other words, mathematics is hard to apply in biology. It's not impossible. There is be-ginning usage of mathematics, but it's very far behind where it is in physics. In classical physics, as you know—physics is a double usage here. It means classical physics which deals with inanimate matter and now I'm using it in the larger sense of all knowledge.

Mathematics: The Language of Physics

Okay, in the restricted sense, physics has always excelled through the use of mathematics. All the main developments of physics came about through mathematical application. That was true of the Greek mathematician and physicist, Archimedes. It was true of Newton. I might also mention that although there are pure mathematicians, and they're exceedingly commendable for what they do, nonetheless, it is interesting to note—by a pure mathematician I mean a person who deals with mathematics for its own sake as a subject all unto itself for the beauty of it and the study of mathematics independently of what it is applied to as a communications tool and an explanation of the con-tent of something else such as physics. There are pure mathematicians and it's a most beautiful profession and most commendable one.

Strongest Advances in the History of Mathematics Done by Physicists Who Needed New Tools

On the other hand, I have this to say there, that the strongest advances in all of mathematics, the history of mathematics, has been done by physicists who needed new tools. The two most prominent examples of this are Archimedes and Newton. Both of these men did things which are at the forefront of the most important advances in the history of mathematics and they both did it because they needed a way of expressing something that they did not know how to express. So they invented the tool, which is a form of mathematical application.

The calculus was invented by three men independently of each other. Well, let me put it this way. Two of them were independent of each other and these two are both dependent on the third in a subliminal manner. That means that there was a background that they had available to them which was not explicitly within their grasp, but they had an intuitive grasp that this had been done already. The two men, as you well know from V-201, are Newton and Leibniz and they both depended, in the background, on Archimedes' earlier work. Most of Archimedes' mathematical work is not identified as to how he did it or where it came from, but it's clear that some of the conclusions he came to could only come from the use of the calculus. The concept of the limit, which is behind the calculus, is basically inherent in Archimedes' work.

The Usefulness of Mathematics as a Communications Tool

In any event, the nature of mathematics is such that the usefulness of it is as a communications tool. It can be used as a subject in its own right, as a subject worthy of study for its own sake without application. That's basically the attitude of the pure mathematicians, who, on the other hand, do not have a strong enough grasp of the principal function of mathematics aside from the study of its own beauty, for its own sake, and that is that it is the tool to make the most explicit explanations available in all of human communication. All other sciences will benefit when mathematics can be applied to them. Biology has begun the task of being quantified. Volition only to a relatively small extent.

Precision Is What Distinguishes Volitional Science from So-Called Social Science

The only reason volition could even become a science, the reason it has failed ever before to be a science, when it went under the crackpot name of social science, which is neither social nor science—it's anti-social and it's anti-science—the reason it failed to be made into a science is the concept of precision was lacking. Now, I claim some personal advantage in having made a distinction between precision and accuracy. It was believed that human behavior could not be a true science in the same sense as physics because it's not exact and that human conduct cannot be predictable such as the trajectory of a projectile, a stone, a planet and therefore it produces nonpredictable and therefore erratic results. It is not correct to say that this cannot be a science because of its non-exact nature. The reason is that there's a difference between accuracy and precision, which is covered in an early session on my physics course.

Accuracy: A Reduction to a Well-Identified Quantity

Accuracy basically is a reduction to a well-identified quantity. Therefore, the more accurate something is the larger the number of decimal figures of significant places you can calculate the size of something to be. For instance, how accurately do you know a length or a mass or a temperature? One place accuracy, two places, three, four, five? If you cannot put a number on it at all, then it's not accurate.

For example, if you say, "How long is a certain stick?" And you say, "It is 32.78 centimeters long," that's fairly accurate. You have four-figure accuracy, four significant figures of accuracy. That's if they all are meaningful. If you can tell the difference between 32.78 and 32.79. If you're not just guessing, if you can actually measure this. You can say it's 32 centimeters long. That just means that it's larger than 31 and less than 33. Or actually, it's larger than 31.5 and smaller than 32.5. When you say it's 32.0 centimeters long, that means it's larger than 31.95 and it's smaller than 32.05 and you got it reduced to a smaller tolerance, the range of error. Then the larger number of places you can specify, the more accurate it becomes.

Precision Is Required to Make a Science, Not Accuracy

This is not the same as precision, however. For example, in volitional science when you say something is good or bad, that's the basic preference concept. Good is defined what you'd rather have, or you prefer a certain condition over another, a transition from a lower state of preference to a higher state of preference is called *good*, and vice versa, a transition from a higher state of preference to a lower state of preference is *bad*. That just tells you which direction you're going. It doesn't say how much of it there is. For example, you say, "Do you like vanilla ice cream?" Yes. "Do you like fish?" No. Okay. Well, how much do you like vanilla ice cream? State a number. I mean, how good is it? Is it one unit of good or two or eight or seventeen and a quarter? How much of it is there, how accurately can you identify it? Well, this varies with your attitude, your mood, how much you had yesterday or today or what you'd rather have.

What number would you put on fish, your dislike of fish? I don't know. It wouldn't be infinite, but I'll tell you, it would be pretty high. [*Audience laughter and AJG chuckles*.] The only reason it's not infinite is that there are damn few things that you can correctly call infinite, but it's pretty damn high, my attitude of dispreference for it. But it's hard to put a number on it, it's not exact. That means it's not accurate, that you can calculate this and work with it numerically. That's the reason it's of course not considered possible to call this a science.

Yet it is a science because science means knowledge and knowledge is anything that you can corroborate by observation. Any truth that you can corroborate by observation is knowledge. What is, therefore, required to make a science? Not accuracy. That, as far as I know, is original to me. I certainly didn't get it from anyone else. [*AJG takes a drink*.] It's precision.

Volitional Science Definitions Are Precise

Precision is easier to arrive at than accuracy. Precision requires definitions which are razor-sharp. If you have identified as to what is included and what is not included in a definition, then you can identify: this is included in the definition and everything else is the rest of the universe. Then, for example, the first such thing in volitional science was property. That identifies as to what is property, what is not. There's not a thing in the world, not a single thing in the world you can identify or you can point to that, through the definition I've given, cannot be stated yes, it is property or no, it is not property, without wobbling it. In other words, it's yes or no. It's not well, maybe. Or under certain circumstances, or perhaps, or are there certain feasibility studies we can say that it is property or in other cases it might not be? It's yes or no and it's very sharp, and very precise.

Now that doesn't measure how much property, or what its valuation is which is subjective anyhow and varies from person to person and time to time and mood to mood and market conditions which includes desire to obtain, desire to divest, and interactions of people and communications capability. Obviously, market preferences are easier to express when people who are in the same community are easy to get together, there's a marketplace where they can find each other, when they have a language to communicate in. All of this affects their behavior.

Accuracy Difficult in Volition Due to People's Attitudes Changing with Time

Well, it would be a great advantage in volition if quantifiers or exactitude studies could be made, increasing the accuracy. This is not impossible, it's just difficult. For example, there are many variables and many parameters and of course the psychological attitudes are unstable for people which is one of the biggest reasons why there is a difficulty for putting a number on what is called good and what is called bad and how much good and how much bad. The reason is that people's attitudes change with time, which depends upon externalities and other conditions that may not have anything to do with the subject that is being evaluated. They have moods. Moods are emotions. Moods are based upon feelings, as well as rational thoughts. [*AJG drinks.*]

No Systematic Way to Improve Behavior of a Person Who Does Not Possess Rationality

Now, as I say, psychology deals with both a discussion of all people and understanding of all people's behavior, and also the improvement thereof. Only for those who are capable of doing the improving, for that you need rational inputs. I don't think there's any systematic way to improve the behavior of a person who does not possess rationality other than to create a social structure where irrationality becomes a non-successful pursuit of happiness, systematically and not occasionally, but predictably. When irrational conduct produces market disasters then, as I say, it does not require a thermodynamic genius to learn not to touch a hot stove. One bad experience will suffice. A very stupid person has to be burned twice.

After a while, he'll be burned so often that he stops doing it. The dumbest person ultimately will learn to stop touching hot stoves. If it happens often enough where he'll be sufficiently scalded there will be no problem to worry about what his conduct will be after that, because he won't have any conduct, because he won't have anything to have conduct with, if he gets burned sufficiently. You say, "Well that's harsh." Well, there's nothing said that nature is intended to protect habitual offenders of rationality.

Welfare State Concept Is Anti-Natural and Anti-Humanitarian

There's no paradise for lunacy in nature, only in the welfare state. [*Audience laughs and AJG chuckles.*] I might say, only the welfare state offers such shelter for people who are designed for failure. It is believed by the anti-capitalists that the law of the jungle prevails in capitalism and dog-eat-dog, and such things. Actually, quite the reverse is true. What measure of security and safety that has emerged in civilization has come about from an interesting juxtaposition of capitalist production and socialist bleeding-heart pseudo-humanitarianism. The pseudo-humanitarians of the welfare state persuasion believe that affluence should be distributed so that the less competent should share equally with the more competent, and that is the bleeding-heart characteristic. But you see, there would be no ability to distribute the affluence if there weren't any in the first place, and affluence doesn't grow on trees. Affluence is not a natural condition, it's a manufactured one. It's part of high production and when there is a general condition of hunger and starvation, and a general condition of the scarcity of the resources of nature, then there is no affluence, and you cannot redistribute the non-wealth.

In other words, the peculiar conditions that prevail in today's welfare state is a mixture of capitalistic production and welfare state parasitism which requires the socialist concept of redistribution of the wealth, but the whole concept is completely anti-normal or anti-

221

natural. In raw nature there is no humanitarianism. Nature is not humanitarian. Nature is cruel and harsh to living beings. I've already explained that. It is difficult for life of any kind to survive, even the dinosaurs found that out. Especially the dinosaurs found that out, how difficult it was to survive. We can learn that too, the hard way. More accurately, we could experience it, maybe not learn it, but we could experience it. The dinosaur was not capable of coping with nature and perished. Actually, lasted longer than we have so far.

Proper Language and Communication to Harness Knowledge of Nature

Now, the nature of all of this is to recognize that high communication and high precision produces scientific reality, and there is no ability to have any form of systematic stable society in the absence of the production that ensues from a technology that comes from a large understanding of nature and the harnessing of this knowledge. That requires, as I say, proper language and communication. English is a poor language. Mathematics is far better. English is a rich language, but a haphazard language. I'm trying to use English selectively and not totally. I'm using every word in a limited context. Have you noticed? [*Some in the audience say, "Yes."*]

Galambos Writes His Own Dictionary

I have a dictionary I have always here to argue with and in front of you, if necessary. So I look in it and if they have a usage that's acceptable I'll use it and if not I'll make up my own definition. That's one of the criticisms leveled against me, that Galambos is writing his own dictionary. That's correct. I would like to. If I had the time, I'd make a complete dictionary of the whole language. I assure you, I would revise a large number of things including the punctuation, spelling. [*AJG chuckles.*] I don't think I'd have an easy acceptance of this.

The English Language Is Irrationally Written and Pronounced

The pronunciation—atrocious, this language has an atrocious pronunciation. It's totally irrational and unstable and totally unrealistic. In other words, there are mute letters, there are letters that have multiple pronunciations. For example, the letter combination O U G H has many conceivable and even inconceivable pronunciations. The only pronunciation it doesn't seem to have is the phonetic one, which is "oh-oog!" [*Audience laughter.*] "Oh, oo, gay, ha. Oog." As a matter of fact, even the names of the letters of this language are preposterous, come to think of it. I said "ha" for "h". When you say for example, take the word *hill*, which is you know is a rise in the topographical surface of the earth, you say "hill," you don't say "aych ill." Why is it called "aych"? It's "huh"! It's an expiration, it's a rather ugly letter with an ugly sound but it's "huh," like you're exhaling. And "hill," not "aych ill."

And for example, if G is G, then how do you get a garage? [*Some audience chuckles.*] With two GG's in it. [*AJG chuckles.*] It's not "je-rajee." [*Audience laughter.*] "Jerajee." [*AJG laughs, then takes a drink.*] Now how do you get O U G H to come out "oh"? Or "oo"? Or "off"? Or "uhf"? Or "ow"? It's all of these: though, thou, through, thorough, cough, rough. Everyone one of these has a different pronunciation of O U G H. Now is that logical? Is this the great language? Is this the way one learns how to put his organized thoughts into proper perspective so he doesn't go completely erratic in his thinking?

Is this a proper base of language to start with, to communicate in rationally when your thought processes are carried out in an erratic language that can't even be pronounced intelligently? That's why I say Hungarian would be a better base, it's phonetic. For all of its shortcomings, and it has many, I don't mean to say it's a super language, it is not, but it has phonetic pronunciation. Now that's an elementary simplicity that you could expect from any language and you get from hardly any. No language is as ridiculous as English.

The German Language Is Also Irrational

German basically has certain rules, but the same letter can be pronounced different ways. An S is sometimes "S," it's sometimes "Z" and sometimes "esh." S could be pronounced in several different letters, different ways, but there are certain rules. For example, an S is before a T or a P and then it's "sh." It's "shtrasa" not "strasa." It's "Einshtein" not "Einstein." Have you noticed I always call him "Einshtein" and not Einstein? It's not me that's mistaken, you are. Since his name is German. But the point I'm making is, it's not a clear language. If there are three or more different pronunciations of the same letter, but at least there are certain rules however wobbly they may be.

If it's before a T or a P it's one way, if it's an initial S it's usually pronounced Z, such as "Zemmelweis." It's not Semmelweis, it's "Zemmelweis" but it's written with an S. So the average American writes with the a Z because he heard "Zemmelweis." Naturally to most people that's a Z but it isn't. But, of course, German has its own erratic nature, but at least the erraticism has rules to it. It has rules of irrationality. [*Audience laughter.*]

The French Language Pronunciation Is Irrational

French is second only to English amongst the languages I know of. Now, of course, I don't know many languages. I'm sure there are other ridiculous languages. But in French you don't even pronounce all of the word. You pronounce the first part of the word in a gargled and not particularly clear way and the last part you just swallow. [*Audience laughter with AJG.*] For instance, the word *slow* in French is "lahntemah." It's spelled L E N T E M E N T, so that would be "lentement" in intelligent pronunciation, "lentement." But it isn't "lentement," it's "lahntemah." [*Audience laughs.*] And the first part you sort of nazalize— "lahn, lahn" [*more laughter*] and then the end of it you just drop. But you don't drop it in a clear, sharp, precise cliff-like manner, you sort of fade it out: "lahn,tem,oah." [*Audience laughs and AJG chuckles.*]

The Relation of Language to Psychology

Do you understand what this has to do with psychology? I'm a little worried that you may not see the connection. [*Audience laughter.*] Psychology deals with what people think to motivate them and the thinking has to be done in some form of communicative or abstractive manner which they can imagine and which they can articulate if they can communicate at all. A communicated thought then becomes an idea, as I say in V-201.

The Language and Psychology of Cats

Now cats have a psychology too, but I'm sure they have a lower form of psychological analysis. More difficult to understand but lower in stature on the basis that they do not have a verbalization of their thoughts. I'm sure they have thoughts, but their thoughts cannot be verbalized. Meow [*AJG makes "meow" sound like a cat*] is not an easy-to-understand language. Maybe they have variations that they can pass along to each other and maybe an oscilloscope can pick up these variations better than the ordinary ear can. But to me *"meow"* means *"meow."* [*AJG says "meow" like a cat, and the audience laughs.*] And there are different intonations. One means it's hungry, another one means it wants to go out, another one means it just wants attention, another one means it's in pain, and they all may have variations. Maybe if it says, *"MAAAK!"* That means it's in pain. And if it says *"meoh"* that means it's hungry. [*AJG laughs with the audience.*] But I'm not quite sure exactly how it communicates this except with a tone. Now it's not exactly a nonexistent thought, it's just a difficult one to pass along. Maybe cats understand each other better than humans understand cats, that wouldn't take much. [*AJG chuckles.*]

The Language and Psychology of Flies

Now cats actually are quite articulate compared with, let's say, flies. [*Audience laughs with AJG.*] Now, they too must have thoughts, but their thoughts are even more hazy to decipher. [*AJG drinks.*] Now what makes a fly zoom near my ear and another one lands on a windowpane? I can see why they go to a garbage can, that's their natural habitat. They live there and that's where they feed. But what would make my ear attractive to one? [*Audience laughs and AJG drinks again.*] To zoom and buzz around it. Now they obviously do have thoughts because they have volitional capabilities. They can decide which way they're flying.

The Hungarian Language Is Not Perfect

By the way, so you don't get too upset about the fact that I'm always talking about Hungarian which I happen to know, and that you don't get too upset that I'm usually stressing how much better it is, in one area it's awfully inferior to English; it's the most inferior language in the world. I just thought of it in connection with the word *fly*. In English, if you refer to this repulsive creature, the name is fly, which means get out, fly away, fly out. In Hungarian its name is *légy*. And that means "stay." [*Audience laughter.*] And that's unbelievable. [*AJG chuckles, laughs and takes a drink.*] Anyhow, I just thought that would cheer you up, that Hungarian is not perfect. [*Some audience laughter and AJG chuckles.*] But the pronunciation is close to total phoneticism. There's only one flaw in it and it's very minor.

One of the Characteristics of Healthy Communication

That, by the way, is an exceedingly important part of logic, that everything should have only one representation. You have remembered from my many utterances, I presume, one of the characteristics of healthy communication is that every concept should have only one way of expressing it and not two or three or seven or twelve. You may say, "Well that's not colorful." It is precise though. And you say, "Well isn't the poetic distribution of different words meaning the same thing so you can vary it and so you don't always say it the same drab way, well doesn't that have any significance?" Not here, because that confuses. If you have two things to mean the same thing, one is redundant. And therefore you have not only loused up Occam's Razor mechanism, you've also made the communication complicated. It's more dangerous than simply Occam's Razor is violated.

That has to be explained. Why is it worse? Because it produces a confusion, because if there's two ways to say the same thing, then you always have a confusion as to well, what possible slight nuance or slight variation there may be. You try to grasp, what is the person saying? Why did he use A instead of B when A and B refer to the same thing and they mean the same concept? How come he used A and not B? There's that residual doubt. There must be some thought behind the process. How come he selected this one when there are two words available? [*AJG pauses for a drink.*]

A Single Meaning Assigned to Any Word Improves the Language

That's why I say that even within the framework of our present language, English, it is possible to improve the situation enormously by having a single meaning assigned to any word which is why, for example, when I have used words to create the science of volition I have used the word *property* to mean only one single thing and then I gave a definition for it after I gave it considerable thought as to—I think it took maybe a whole shower's worth of thinking, and anyhow. [*Audience laughter.*] I don't remember exactly how long it took, but anyhow it was more than usual amount of thought. I never had to change the definition except in one way. I generalized it from *man* to *volitional being*. That makes it cosmic instead of terrestrial, and not one species but all species. That's the only change I ever put into that definition.

In any event, the definition of property is one which is exceedingly precise. That means you can identify anything in the world, anything in the universe, as to yes it is property, or no it is not property.

Definition of Property in Galambos' "Hippy Dictionary"

Now, if you look in the dictionary you'll find lots of things. Let's do it, just for the hell of it. This is the recent hippy dictionary. [*Sound of AJG turning pages of dictionary.*] The old one has fallen apart so I'm using the present one. I keep the old one available but I don't carry it around, it might disintegrate. This hippy dictionary has much deterioration from the previous editions. All right.

> Property: a quality or trait belonging and especially peculiar to an individual or thing; an effect that an object has on another object or on the senses; virtue. [*That's the third.*]
>
> [*Fourth is*]: an attribute common to all members of a class.

These are four different things which are 1a, 1b, 1c and 1d. Then comes 2a:

> Something owned or possessed.

That's the closest to what my definition is but not quite precise as to what's ownable, and on what basis the ownership can be established. Anyhow, that's essentially what I have.

> 2a: Something owned or possessed. [*And then they louse it up by saying, "Specifically, a piece of real estate."*]

That's the one thing that isn't property. [*Audience laughter.*] I already explained that, why that is. [*AJG chuckles.*] That's a natural resource, land.

> b: The exclusive right to possess, enjoy and dispose of a thing; ownership. c: Something to which a person has legal title.

That's also off the beaten track, in volition.

> 3a: An article or object used in a play or motion picture except painted scenery and costumes. [*Light audience laughter and the sound of AJG closing the dictionary.*]

You will notice that the word *property* has more meanings than is necessary. I might also mention that that's a relatively small number of definitions. I'm sure an unabridged dictionary could improve on the variations for that word. There are many words which have as many as twenty different meanings. Look up a word such as jack, or web, or well, W E L L. Or for that matter, another word, will, which is one letter apart, W I L L. You'll find not three but—well, I'm not going to take the time here to do it, but I'm sure you'll find more meanings for that. This is very difficult to communicate in a language when the same word has so many meanings. Also, the reverse is equally bad. When you have a single concept, you shouldn't assign more than one word to describe it.

Non-Phonetic Pronunciation Is Illogical

Now, in exactly the same way, the pronunciation is deteriorated when the same letter or letter combination can be pronounced differently. It makes it difficult to communicate in a language like that. Without looking at anybody in particular, I know that practically everybody mispronounces some words, including myself occasionally, but I dare say my exposure hazard is larger than most people's and the frequency is lower. By the way, you probably think I mispronounce many words. That's because you do. For example, if I say Einstein [*AJG says "Einshtein"*] you think I'm mispronouncing it. Actually, you're mispronouncing it and you don't know that it's "Einshtein." If I say *Pythagoras* [*AJG puts accent on first syllable*] you say, "Oh, that's unintelligible." You don't even know what I'm saying. If you see it written out, "Oh, you mean Pythagoras." [*With accent on second syllable.*] No, I don't mean "Pyth-*ag*-oras," I mean "*Pyth*-agoras." Or you might say "off-ten" [*to pronounce the word often*] and I say "offen" and the error is yours because it's "offen," you don't pronounce the T. It's is a brilliant language. [*AJG chuckles.*] It should be pronounced "off-ten" but it isn't. Now if you do say "off-ten," then you're like Eisenhower. [*Audience and AJG laugh.*] Because that's the way he pronounces it. It's also the English pronunciation but since you're not English, you're Americans, you're in the wrong country.

English English vs. *American* English

Speaking about English pronunciation of English and English spelling of English vs. American, then you have actually two different languages. Now that's understandable. The two cultures have grown up across opposite sides of an ocean and one is derived from the other. It's essentially a major supremely divergent dialect, the American branch of English. You will find the English have entirely different lingual characteristics.

For instance, their pronunciation is so difficult to understand for each other, or for ourselves for that matter. When I was in England I noticed that the majority of the people there I thought had a foreign accent. I don't mean in English. I was amazed at how many people had a German accent in England considering that the war was over, I figured maybe they had become brotherly love oriented, and I was wondering how come the English all seemed to have German accents. I was prejudiced perhaps because I'd just come from Germany and I said to myself there are more people with German accents here than in Germany. [*AJG chuckles.*] It turned out that most of them were speaking *English* English. And I took it as a German accent. I was a little sensitive to this at the time, having just been a few days in Germany. It turned out that what I thought was a German accent was either an East Indian accent or an Englishman speaking some dialect other than King's English.

I remember when I was in the hotel one morning at three in the morning, which is the normal hour for me to get some food and I found something that was open, I was having my meal and right across from me on the counter there were three hippies and they were talking rather loudly so I could not help but hear them even if I didn't want to, and since I had nothing to do but think and they were in my direct vision and audible ear flowstream, I heard them talking for one half an hour while I was there. I can tell you that I cannot remember one single word that was intelligible spoken by any of the three. It was absolutely unintelligible. And yet, nominally, they were speaking English, I think. [*AJG chuckles with the audience.*]

And yet, the English version of English, which is of course the original version of English, is beautiful. As matter of fact, I think it's more beautiful than the American English. If you, for example, hear in the movies such people as Basil Rathbone or Ronald Coleman or George Sanders, I think that's just about the nicest spoken English you can hear. It's just delightful to listen to them. Absolutely intelligible, clear, precise and very, very easy and delightful for me to hear. But that is intelligently spoken English. Now you hear the average person out in the sticks or even in the cities, I mean, I think it's called Cockney. It's just hopeless. I remember, the whole time I was in England I would say, "What? Excuse me? Could you repeat that?" [*Some audience laughter.*] And they said the same to me. The telephone operator couldn't understand me and I couldn't understand her.

Galambos' Language Experience in South Carolina

I had the same experience in South Carolina [*loud audience laughter*] for different reasons. [*AJG chuckles.*] Now here is essentially the same language spoken in different parts of the English-speaking world. I remember once in South Carolina I stopped in a gas station and there was a detour ahead and a dirt road so I asked, "How do I go around this?" And the guy in the gas station said, "Blublblubblubloo." I asked him to repeat it, please, and he did the same thing [*AJG chuckles*] all over again. This happened three or four times. I finally didn't have the crust to ask him, "I'm sorry, I still didn't understand, would you do it again?" And I said, "Thank you." [*Audience laughter.*] And I left without the slightest idea what he said and I still don't know what he said. Have you ever had experiences like this? [*"Yeses" in the audience.*] A total inability to communicate, within the same language!

Ambiguities in the German Language

You have it in other countries, too. There are in Switzerland, German-speaking Swiss. Seventy-one percent of the Swiss are German-speaking. Now to me, German is German, and I realize there is some difference in the dialect, but since it's hard one place and hard the other place, it really makes little difference to me whether it's Swiss-German or Bavarian-German or Berlin-German, but even to my non-delicate ear on the subject, I could see differences in pronunciation.

I know I have a German-speaking cousin, he's no longer alive, but I have an uncle who lived in Germany and he had two sons whose native language was German and one of them told me that when he went to Switzerland he did not understand anything that was said in Zurich. And his native language was German. And I know myself, and as I say, my German is not zero, but it's weak. Anyhow. For example, I heard "svoe" for "svie." Two in German is "svie," but not in Zurich, it's "svoe." Now how do you get "svoe" out of "svie"? I don't know, but [*AJG chuckles*] and so it goes. I believe the Berlin pronunciation of "ich" which is "I," is "eeka."

Occam's Razor Utilization of Language

Now how does a person communicate within a language where very few people have the same concept of the same language? It's quite important, therefore, to make a simple Occam's Razor utilization of language. Now most people of course are ignorant. The word isn't ignorant, in my sense. The correct word would be *uncultured* and therefore their communication is sloppy, which is how dialects grow up. Sloppinesses which are inconsistent. But even the King's English, or in German the equivalent—*hochdeutsch*, High German—is not a clear language. If you take into account the ambiguities of meaning, the multivalences, both meaning and pronunciation, this reflects itself in precise communication.

Improved Language Will Bring Emotions Under Better Control

I brought this whole subject up in the context of emotions. There's a good reason for it. Emotions are forms of thoughts which are thoughts, to be sure, but they are not as articulate thoughts as a rational thought would be. An emotion is a less rational thought. I would like to point out that most of human emotion is therefore consistent with the low level of language communication that people have. When the language improves, the emotions will be better under control. That's an area where you can show that although you have no direct way to improve one of the masses, either in volition or in psychology, only the hot-enders, or close to the hot-enders, can have deliberate self-improvement concepts wherein they can improve their time scale, their generalization capability, their rational faculties, and their ability, therefore, to develop gratitude, integrity and such rather abstract but very significant cultural attributes, without which there is no culture. You will not be able to accomplish this with the cold-enders, only with the hot-enders, because only in the hot end can you have the capability for self-improvement by direct means.

Thought Processes Are Inherent in the Language Structure

This does not mean that all is lost, as you may have heard, because of the high leverage. The leverage comes into effect that most people are creatures with primitive and simple habit formations and the habits they form will be based upon the environment in which they are immersed. That includes language. That's not the only environmental factor but it's a very important one. Language is a supremely important environmental factor for the psychological domain. And as the language structure is more rational, even though the person may not be rational who uses the language, he is constrained to use the thought processes that are inherent to that language structure because that's the language he's familiar with. That's the routine or pattern he falls into and if the language becomes structurally sounder, there is an evolutionary, ultimate improvement in language is what I'm saying.

Galambos Is a Physicist First and Foremost

It can be started immediately and I'm doing it even with the present language without being a linguist. I make no pretense to being a linguist, a philologist—in other words, a professional at language structure and language communication—any more than I'm an economist or a psychologist, or for that matter anything but a physicist. That's all I am, you know. A physicist. I'm never going to change that. That's all I ever claimed I was, intellectually, a physicist.

Galambos Trying to Improve Other Subjects through Extension and Generalization of Physics

In every course I teach I'm not a historian, I'm not an economist, I'm not a philologist, I'm not a psychologist. But I give a course in other subjects because I'm using physics and I'm trying to improve the other subjects through extension and generalization of the supremely strong and clean atmosphere of physics. Do you understand what I'm saying? ["*Yeses*" in the audience.] Well in the same way, I don't pretend any linguistic talents, which I am not particularly good at. I'm not even a philologist. I'm more of a philologist than a linguist in the sense that I'm able to understand language structure better than I am in learning the language to speak, in its own right. I'm only, as I say, slightly better than the dog that barks in one language. At least I bark it rather fluently and with a better vocabulary than most people. Not only better in the quantity of words, but in the more precise way I use each one. It's not perfect, but I'm improving as I go along and, unlike most people, I'm capable of learning. [*AJG drinks.*]

The Duodecimal Metric System:
The Ideal Numerical System for Man

Anyhow, the point I'm driving at is that the language structure is very important and within the present language—I'm not in a position to undertake everything simultaneously. You'll notice I have not given you any crusades to convert the United States to the metric system, which is long overdue and vastly superior to what we've got. I just make some comments here and there about the metric system such as in my physics course and make some occasional, not particularly well-disguised comments about the English measuring system which is less than acceptable. But yet, why don't I go on a crusade to convert the world to the metric system? It should be. The sooner, the better.

As a matter of fact, it shouldn't even be the decimal metric system, which is what we've got. It should be the *duo*decimal metric system. While we're making a change, let's make a good one, and go to the numerical system based on twelve and not ten, which is vastly superior to ten. That's a double change. You might as well do it in one big jump and jump in the pool all at once instead of going in a little bit at a time. Everybody's always afraid to make the change all at once and doing two hard steps instead of one easy big one. We should go to a duodecimal metric system, the number system based on twelve. It's vastly mathematically superior to ten. It's the ideal numerical system for man.

Improving Language and Numerical System
Can Be Accomplished Evolutionarily

I believe in this, why don't I want to discuss it? Because my main thrust is, right now, survival for the species. These other things can be taken later and they can be done evolutionarily. In the same way, I think the entire English language could be overhauled. It wouldn't even come out as English. For that matter, every language. As a matter of fact, if I had the time to live long enough, I might even construct a language. You may say, "Well that's already been done. Esperanto." Yes, I've heard of it. That's just another artificial language as bad as the natural language in which it's patterned on. I know very little about Esperanto, but I know enough that it is patterned on existing languages and therefore has all of their faults, or most of their faults. It's largely a hodgepodge of existing European languages. It's not a rational language. I'm talking about a rational language where you can adapt it to precise communication.

233

Every Concept Should Have One Usage

The English language could be greatly improved and that's what I'm doing, essentially. When I give one definition it means always the same thing. For example, when I use the word, I mean only one thing by *coercion*. There is only one meaning to the word *crime*; only one meaning to the word *justice*. There's only one meaning to every word. You can say, "Well, that makes it less colorful." No, it makes it more precise. The color can be added by a variation of a number of things that you have precision about. Not how you vary it and wobble the specific words by having many different colorations. But every concept should have one usage. The whole concept of the synonym is contrary to science and to precision. Precision is the basis of science and accuracy is desirable, but not essential. Accuracy follows. I think ultimately volition will be very accurate as well as precise.

Mathematical Precision Used in Physics: Statistical Mechanics

For example, mathematics has been used in areas of physics where precision is possible and accuracy is normally not possible. For example, in molecular behavior. Let's take simply a container of gases and you have molecules moving around in haphazard fashion and molecules have different speeds and different directions of motion and the composite, the whole structure, is a volume of gas. [*AJG takes a drink.*] What can you say about it? What's the speed of a molecule? Which molecule? At what moment? How do you pin it down? There are so many of them. For example, a cubic centimeter of air.

A centimeter, for those of you only speak the English unit system, is less than half an inch. There are 2.54 centimeters to an inch. So a centimeter is less than half an inch. If you take a cube of one centimeter on each side—one cubic centimeter at normal atmospheric pressure at sea level—the number of molecules in a cubic centimeter of air at normal pressure at sea level, atmospheric pressure, is a number with nineteen zeros after it. So you can see that it's an awfully large number of molecules. Now, find some way of describing that mathematically. And what's the speed of an average molecule? Which is the average molecule? Well, as a matter of fact, how do you determine an averaging process?

Well, as a matter of fact, there is a way and there's a whole branch of physics devoted to this, it's called statistical mechanics. You apply statistical methods to handle the behavior of a large aggregate of molecules. This has been very successfully done and it's a very precise subject. It's a very exact subject. And even where you cannot identify a specific molecule's behavior, you can very correctly predict how the aggregate performs as an observable whole and that subject in the macroscopic, that's the large-scale view, comes out as thermodynamics. And what is statistical mechanics becomes, in the large scale, thermodynamics, and you have a direct translation between things that can be easily measured—large-scale thermodynamic systems and small-scale molecular structure out of which it is composed—and there is a very good bridge established now in physics between the two domains, the large- and the small-scale domains. Well, the macrocosm is microcosm here. It's through statistical methods that this has been done.

Statistical Methods Applied to Human Behavior: Insurance Industry Example

Well, statistical methods have been applied to human behavior, also in human characteristics. For example, the entire insurance industry is based upon statistical application of the profit mechanism, not to any given person but to a whole large group of people, let's say, the policyholders in an insurance company. An insurance company undertakes to insure a person, let's say, sell him death insurance which is marketed as life insurance because it's a more, let's say, Madison Avenue has decided that it's easier to sell something using the word *life* than *death*. It's really death insurance. Do you understand that life insurance is truly death insurance? You're insuring yourself not against life, but against death. It doesn't mean you will not die, it means that the company will indemnify your beneficiary a certain amount in the event of your death. It's not if you live will they pay, that's an annuity. If you don't live is when you get paid. So it's death insurance, marketed under the euphemism of life insurance.

Okay, you say, "Well how can they make a profit on it since you will ultimately die?" That's a certainty with our present biological knowledge. Everybody living now will die. Okay, since everyone will die, how can they make a profit on paying you off if you do die? Well, because they calculate how much they must collect from you so that

235

they will still make a profit if you do die on the policies that they sell. Now, please note how I worded that. I didn't say the policy they sold you. I said the policies, plural, that they sell as an aggregate. Now, it is quite possible somebody buys life insurance and dies after paying one premium, the original premium, before the second one is due, he's dead, and in the absence of some fraud in the application the insurance company is obligated to pay off. Well clearly, if the guy paid one premium and then he dies and then the insurance company pays off a huge walloping sum in relationship to the original premium they have lost money on it.

How does a company stay in business with that type of a sale? Obviously, they don't have this happen to all their policyholders. Some die immediately. Some die a year later. Some die two years later, some die three years later, some die four, five, six, seven, eight, ten, twenty, thirty years later, forty years later, fifty years later. Some don't die while the policy is in effect. Term insurance expires, and they will outlive the policy. It'll never pay off. But had you died, what are you paying for it then? In case you did die then you have someone get the money.

All right, now how the hell do they know what to charge you and still make a profit, not on one policy but on the aggregate? That's a statistical analysis that has to be made prior to determining what the premium is to be and as to determine what is the probability that on any given type of an insured, what is the probability of their losing on it in a given term, in a given period of time? Say one year, two years, three years, four years, and they structure the premium rate to fit the expectancy of losses to which must be added the overhead for running the company, the commissions they must pay, the medical expenses they must incur in examining the people who are to be insured, in the administration of the payouts and the claims and the investigations, and of course they wish to make a profit. And all of this has to be figured in and they charge a premium to fit all this, but the first thing they have to determine is what is their risk. That is a statistical determination.

Mathematics Applied to Volition

I cite this to indicate that here is an area where long before my time but not recognized as to its significance to volition, mathematics has already been applied to volition and to volitional beings and not just to their life expectancy, but also the probability of their losing money on theft, burglary, fire, hurricane, earthquake, tornados, marine disasters and, in general, any contingency to which people are subjected to losses. My definition of *insurance* in V-30 is that it's a contractual means to protect a person against losses for specified hazards or specified contingency possibilities of losses under unspecified circumstances and times.

The determination of this is a statistical calculation. The fact that the insurance companies, in general, if they are any good they will make a profit, if they're not they will go out of business. Those that survive for a sufficiently impressive period of time, such as decades, obviously have to have basically sound statistics or they would be out of business. So the point is, here's an area of application to human affairs, which has already been done. Now of course, it gets more difficult with individuals but it's not totally out of the question. It's just a question of time before we learn more.

Accuracy Is Desirable Where Possible but Precision Is Necessary

In any event, in any form of endeavor the more accurate the information can be specified in terms of identified quantities, the better off you are. Quantification and accuracy are always desirable to make a science stronger but to be a science in the first place, precision is quite adequate. This confusion and inability to distinguish between precision and accuracy has long stood in the way of social science becoming a true science. When it did become a science, the bad image of the word *social science* to me personally, and the contempt that I hold people in that have engaged in this pseudo-science, this quackery.

Social Science, Volitional Science— Astrology, Astronomy

I didn't like the term so I looked around and ultimately came up to call it volitional science. It really covers the same domain and no pretenses made otherwise but this being a true science I didn't want to have the stigma of the old name. In the same way that astronomers can't use their proper name, which is astrology. Astronomy is a smaller concept than astrology in terms of the etymology of the word. The word itself, *astrology*, means discourse on stars and it's a larger content than astronomy. But astrology is off limits because quacks have squatted on that word now to the point where it's absolutely disgraceful to use it.

Galambos Can't Use the Word "Cosmophysics" in Today's Quackified Society

As a matter of fact, just in case you're interested, I have always referred to myself since I went to school as being basically interested in astrophysics, not just physics, but astrophysics because that's a larger subject than ordinary physics. Most people who think they're physicists deal with terrestrial phenomena. I'm interested in cosmic phenomena, not just terrestrial phenomena so I call it astrophysics. As a matter of fact, it's even more appropriate to call it cosmophysics. I damn near came to change the name of my physics course to cosmophysics. From three words—astronomy, astrophysics and astronautics—to cosmophysics, one word. Do you know what deterred me? The fact that in today's quackified society, and by the way, if I can't express myself with present-day Webster terms, I make up the word. I just did. *Quackified* I'm sure is not in the dictionary.

On the other hand, I'm equally sure you understand what I mean. Am I correct? [*"Yeses" in the audience.*] So a little imagination, and if you want a precise definition it's always easy to render one in a case like this, but I think you can figure it out, right? In today's quackified society where quackery is the rule and not the exception, the word *cosmic* has become usurped by these mystical, oriental jackasses and jillasses that have imported Asiatic mysticism into the United States. To talk about transcendentalism and reincarnation and cosmic this and cosmic that.

They're not using the word in the physics sense, they're jackasses! And also the other one [*some chuckles in the audience*] as the case may be, they're absolutely out of this real world. They are living in mysticism pure and simple, and they have latched onto the word *cosmos* and *cosmology*, and it's really getting disgraceful to the point where I feel like I'm being pushed out of my own house. [*AJG chuckles.*] That's why I didn't change it to cosmophysics. It's a much more dignified term.

Rationality Is Dependent on the Highest and Most Precise Form of Communication

All right, I really started out this discussion on emotions and I got on the discussion of language to indicate that man in general cannot be improved, just to put it on the track that that was on to begin with so I connect it up and I'll continue after an intermission. I recognize this seemingly covered a larger territory than you might expect. It is one subject even though it sounds like it was many. The discussion was emotions and emotions are basically thoughts and so are rational concepts. The question is, in what way can you reach people to improve what is less than good and what way can you not? Only hot-enders can be reached with rationality. That's clear, is it not? Now rationality is dependent on the highest and most precise form of communication. The supreme communication thus far known to man is the mathematical communication.

How to Communicate with Creatures from Other Parts of the Universe

For example, I have considered what would be the easiest way to have a language that would be understood by somebody else in a different part of the universe. I'm not the first one who thought of it and what I'm going to tell you now actually is I believe due to Karl Friedrich Gauss. The thought was, what could we show here on Earth—this is long before man left the earth, before rockets and things like that—how could we tell someone on a distant planet, Mars or somewhere, that there is thinking beings living on this planet? How could we communicate this to someone elsewhere? And I think it was Gauss who came up with this, and I think it's just beautiful. If you made some kind of a visible configuration on Earth, either by cutting a swath in the forest or maybe a big ditch or something like that inside a big landmass, like Siberia, and made some kind of a visible mark that could be seen millions of miles away, it would look like this:

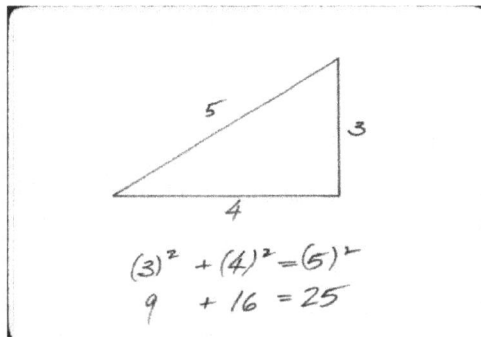

$$(3)^2 + (4)^2 = (5)^2$$
$$9 + 16 = 25$$

239

Making it carefully designed so that this is three units of length, this is four units of length and that's five. You don't have to put the number there, they wouldn't recognize the number anyhow. Just a triangle. Anyone in the universe that has a capability to think would recognize that an intelligent being lives there. That is not a natural formation. That's a right triangle with arms, three, four and five. The Pythagorean theorem will show, that's *Pythagoras [audience chuckles]* will show that the square of three is 9, the square of four is 16, the square of five is 25. Three-squared plus four-squared is 25. Nine and sixteen are 25. The square of five is 25. And a right triangle is such that the square of each of the two arms added up is equal to the square of the hypotenuse.

That's universal, that would be the same anywhere in the universe. And if they see a triangle structured like that, that is not a random accident. That is the work of a reasoning mind and thereby you can communicate. It doesn't matter what language they have, it doesn't matter what form or shape or biological structure they have. If they can think rationally, they will understand that.

Try that with English. Try that with German, or even Hungarian. But you can do it with mathematics. There are any number of things you could do mathematically. It's so simple and it's so beautiful. I might also add, in these unbelievable farces on television and the movies with science fiction dramas about man communicating with other creatures in other parts of the universe, it is always interesting how usually they speak English. *[AJG laughs.]* Have you noticed that? *["Yeses" from the audience.]* They explain, "Well, they're smarter than we are."

No, I think the smarter you are the less you can understand a language as irrational as this. You have to be born into it or environmentally affected such as, as I say, a foreigner who was transplanted here, he has to adapt himself to it to survive. Okay.

The connection of languages to all this is basically we're discussing emotions and emotions have to have a groove to fit into and they have to have some form of communication that is intelligible within, and possibly without, that individual. I'll continue from here.

[Applause from the audience for twelve seconds.]

PART B

This Theory of Psychology Deals with Both Emotions and Rationality

To continue on the discussion of emotional behavior. Psychology actually, if I understand it correctly, in the domain of classic psychology deals mainly with emotions and not with rationality. The nature of psychology as taught here deals with both. I don't think that rational people normally consider that their behavior is part of psychology, and I don't believe psychiatrists have too much occasion to deal with normal people.

Well, let me put it this way. [*AJG chuckles with the audience.*] People who think they're normal usually don't think they need the services of psychiatrists, and probably they don't. I don't mean to say this as a general statement but I understand that many of them are in need of their own services. [*More audience chuckles.*] As I understand it, if it's correct and I believe it is, Freud recommended that anyone who is a psychoanalyst be first required to undergo psychoanalysis himself. Possibly not for therapeutic reasons [*AJG chuckles*] but just to understand the process, although it's quite possible you need the services, I think that's generally true.

Normality: Non-Deviation from Rational Conduct

I don't look upon the subject of psychology as basically one whose function it is to understand only abnormalities, but normality. Now normality of course doesn't necessarily mean average. It does not mean mediocrity. It doesn't mean that which is normal is that which is standard or average or mediocre. I think in a case like this normality refers to non-deviation from rational conduct, which actually from a standpoint of the masses is super-normal. There are damn few people who are not suffering from some form of disorders, mainly emotional.

Emotional Thinking and Rational Thinking Are Not the Same

Now there again, when we're talking about emotions and rational thinking, we're not talking about the same type of thing at all. It does not refer to the same thing. Rational thinking involves logical analysis, by definition. True and valid are involved in the development of

241

rationality. Emotion does not deal with, in general, with conclusions that come to you from logical thinking. They are conclusions that come to you based on your instinctive reaction to things. When you say "ouch" when you get burned, that is not the result of logical analysis. It's your exclamation of pain or despair. When you feel comfortable basking in sunshine, the pleasure you emote is not necessarily a thought-out analysis of the beneficial aspects of solar radiation. It's that you enjoy lying on the beach and resting your can in the sand [*AJG laughs with the audience*] with the warmth of the sunshine upon you. Also, if you get blistered, the pain of the burn will be emotionally reacted to. [*AJG takes a drink.*] Rather than an analysis of how long you should be out in the sun being equated with the burn. Now I understand some people can stay there all the time and never get burned. I'm not one of them. I remember I used to get blisters this high as a kid [*some chuckling in the audience*] then I would spend the time in the bathtub trying to sooth the agony of it with lukewarm water. [*AJG chuckles then pauses to take another drink.*]

Most People Do Not Analyze Themselves in Terms of Rational Concepts

Anyhow, emotional concepts involve not only instinctive reaction to the inputs one gets from the five physical senses but also from the feelings that one gets from reactions to experiences that one has been subjected to. How he reacts to interactions with things and people, the behavior of other people, their own reaction to themselves under certain prevailing circumstances. Also, attitudes of despair, how it could have happened, how it might have happened, how it should have happened, not rationally thought out but the feelings are not the same as the rational developments. Most people do not analyze themselves in terms of rational concepts. They feel.

Emotional Concepts: More Primitive than Rational Concepts but Not Undesirable

I would emphasize something I said before. It requires emphasis so that you do not overlook this. The intellectual attitude might prevail that emotions are inferior to rational analysis. It is a more primitive level of conceptualization. This is certainly true. It does not make it, however, undesirable. I repeat. A lack of reactivity emotionally has one of two basic interpretations, neither of which is good. One is that the person has a very low sensitivity and he acts more like a vegetable than a person with a brain if he does not react to things and emote them. The other is he might react but he represses the external manifestation of it and then he's considered to be stoic.

The Heroes in Ayn Rand's Novels Are Psychological Degenerates

This is what, in much literature, which I consider bad they represent heroes in this role, the strong silent type. These are in fact psychological degenerates. For example, the heroes of Ayn Rand are, in my judgment, psychological degenerates. It is abnormal and unhealthy to behave the way they do. I don't think real people ever get that bad. They are nothing but wooden dummies. To smile is a show of weakness. To laugh, or have humor is an absolutely indescribable sin. They never have the slightest thing witty to say. Have you noticed that? To laugh or to smile or to show pleasure is a demonstration of weakness. These are not human beings. They're robots. Intelligent robots, but they're robots. [*AJG drinks.*] Real people don't come that bad.

Ayn Rand Followers Act Like Robots

By the way, have you ever noticed the clan that follows them? All put on this severe look. They all are robot-like followers of this. I had occasion to visit three of their lectures in New York, in the Ayn Rand-Branden Institute. I'm not as observant on certain things as my wife. On intellectual matters I think I'm more observant than anybody I know. But on noticing people's personal characteristics, I'm almost as insensitive as a stone and I don't notice anything. My wife called to my attention and when she did it was obvious, but not until she called it to my attention.

Look around you. Look at the hairdos of all the women. They all have hairdos either like Ayn Rand or Barbara Branden. [*Some chuckles in the audience.*] They're all smoking cigarettes in a holder, because Ayn Rand does. [*Audience laughter and AJG chuckles.*] They didn't smile at the intermissions, they weren't laughing, they weren't having a good time. That's a cardinal sin. [*AJG pauses for a drink.*] And when I met Ayn Rand personally, she was not the least bit humorous. Intelligent, absolutely. Humorous, witty, no.

243

The Delightful Writing of Arthur Stanley Eddington

How many of you have read Eddington's books? [*Some replies from the audience.*] Eddington! Arthur Stanley! [*AJG chuckles with the audience.*] Have you noticed how well he writes? [*A few "Yeses."*] Would you mind handing me his great book on *Philosophy of Physical Science*? I was just reading this to some V-50 types [*a chuckle or two in the audience*] in a recent—I had the inevitable question challenging Eddington's subjective philosophy. I just remembered it, I just read it the other day. But I didn't call to their attention that the problem there was philosophical. Here it is. There is no problem, I just want to show it to you, how pleasantly Eddington is able to talk about the most deep and profound concepts. He's talking about the ichthyologist which I discuss in V-50. Just listen to this paragraph, how delightful it is. It's not dry, it's not humorless, it's exceedingly witty. At the same time, he's discussing the most profound thing about the universe, the nature of observation, the role of observation in the scientific method, the supreme role of it. This is the one on the "no sea creature is less than two inches long based on the net." The casting of the net corresponding to observation. The net itself is the intellectual and sensory equipment which we use obtaining the knowledge. The catch, the fish, stands for the body of knowledge which constitutes physical science. I'm refreshing your memory on this, just one paragraph I'll read you. But please note how witty it is and what a non-dry presentation it is for a very serious subject:

> An onlooker may object that the first generalization is wrong. [*That's the one that says no sea creature is less than two inches long.*] "There are plenty of sea creatures under two inches long only your net is not adapted to catch them." The icthyologist dismisses this objection contentiously. "Anything uncatchable by my net is *ipso facto* outside the scope of icthyological knowledge and is not part of the kingdom of fishes which has been defined as the theme of icthyological knowledge. In short, what my net can't catch isn't fish."

Now isn't that a nice way to write that? It's clear, it's simple, it's witty, it's not pompous. And yet it's profound and is written by one of the greatest intellects of all time. Then he goes on and makes it even better. He says:

> Or, to translate the analogy, if you are not simply guessing you are claiming a knowledge of the physical universe discovered in some way

other than by the methods of physical science and admittedly unverifiable by such methods. You are a metaphysician. Bah! [*Audience laughter.*]

Eddington Was a Giant and Could Afford to Use Colloquial Terminology

That's in the book! Now can you imagine someone less than Eddington having the courage to have such an undignified expression there as, "Bah"? [*Audience laughter and sound of AJG closing the book.*] But he could afford it. He could afford it. He had the power and the intellectual strength to let that lapse into frivolous and you might say even colloquial terminology. He could afford it because he was a giant and he didn't have to pretend he's a pompous ass because he wasn't a pompous ass. He must have been a very witty man. He must have been a delightful person. Perhaps dry, in the usual British sense, but also witty.

Thomas Paine Wrote with Strong, Witty and Emotional Terminology

Please note how witty Paine was. Please note, in that serious and rather tough writing he wrote, how much wit there is and how much power there is in that wit and how strongly he wrote while he was writing rational concepts in highly charged emotional terminology. But there is nothing in his writing that smells of poor reasoning, or poor ability to apply rationality. Even when Paine made mistakes it was less a fault of logic than a fault of proper observational inputs and inadequate observational inputs. If he ultimately came up with political conclusions, I think it's mainly because he didn't have all of the observational facts organized. He had fewer facts to deal with, for example, than I do. I'm standing on his shoulders.

Paine Did Not Have as Many Inputs as Galambos, 200 Years Later

For example, he didn't have two hundred years of the events that followed the creation of the United States republic, and the political catastrophe that followed. He didn't have this as an input, I do. Also, he was a bona fide humanitarian and he had no solution for old age, other than some kind of a dole that's passed out from the society, which of

245

course is Social Security and is wrong. He didn't like to see poor people growing old and destitute. Now, the objective was humanitarian, the method was coercive. He didn't have an alternative.

Paine Did Not Recognize the Beauty of the Profit Mechanism

Please remember, he was a poor businessman. He failed in the to-bacco shop. He failed Tobacco Shop One. [*AJG chuckles with the audience.*] How many of you took V-76 and heard me say this? That this is the greatest tragedy in the history of freedom. That Thomas Paine was unable to operate a tobacco shop for a profit. I bet you thought I was kidding. I wasn't. Sometimes I put in something to keep you awake and make you laugh, and sometimes I do it deliberately, sometimes it just happens, but in all times it's desirable. But that was very serious, even though it was put in a way that sounded like it was facetious. It was not. That's the greatest tragedy in the history of freedom. If he had known what a profit mechanism was and its beauty, he would have recognized that investments would produce the better result for old people than a dole from the state. Where would the state get it?

Paine Was Sensitive to Primary Property but Had No Mechanism to Apply It

Anyhow, he didn't have as many inputs as I do, largely because I'm later and have the experience of the failure of the American republic in its downstream application. Moreover, he didn't have a theory of primary property. He had a conceptualization of primary property, which was quite admirable by the way. Some of the best things conceived of, the strength and the potency of ideas were penned by Paine. He is very sensitive to the strength of ideas and the durability of ideas. There is explicit references to it in his writings, that the product of the mind of man can endure longer than all the tangible things. He had that completely, correctly two hundred years ago [*AJG drinks*] but he didn't have a mechanism to apply it with, therefore the mechanism he did have, he didn't know how to integrate that.

Galambos Loathed Business but Was Still Successful

I think the reason he didn't is the same reason I couldn't have if I hadn't succeeded at a business. And that wasn't the easiest thing to happen either. I loathed business in my early years, the whole concept of business smelled to me. I had a very close to socialistic attitude on it. I have already said the only deviation I had from socialism was that I did not envy those who had a lot of money. I did not envy them, I pitied them. Do you recall this? [*"Yeses" in the audience.*] I pitied them. That's not the same thing. I did not wish to have their money or want to see them robbed of the fruits of their monetary acquisitions. I pitied them for having a low goal. In other words, I looked upon this that my attitude is superior. [*AJG drinks.*] Which I still do.

Thomas Paine Was Not a Socialist

By the way, on that, Paine is also not a socialist. I want to say that before I drop the subject and before I go elsewhere. I want to call to your attention Paine's magnitude there too. He also was not a socialist. To be sure, the word I don't think was in common circulation. But he does not fit the concept of the socialist either, even though the socialists and communists have tried to make him their pet. The one that they're trying to look to as one of their heroes. They got the wrong boy. He was not a socialist. He was a humanitarian who did not care for seeing destitution in old age or in any other age for people. But he did not share with the socialists their hatred of people who have property, even though he had very little in the sense of P_2. P_1 he was a giant in, but in P_2 he was poor. He could have been rich but didn't give a damn.

Thomas Paine Had the Two Best-Selling Books of the Period, Not Anxious to Make a Profit

Do you realize the royalties he could have had? He had the two greatest best-sellers of the period in English. *Common Sense* is a best-seller of a magnitude that has never been equaled. Do you know that? It sold one half a million copies in half a year in a country that had three million population. A circulation one-sixth of the entire population. That would have to be about thirty-five million people in the United States today to have the same percentage. He didn't make a profit on it. He ended up owing the printer. You say he was a lousy businessman. That's right. He was also swindled. Also, he wasn't anxious to make a profit. He was doing this for a cause.

247

All Causes Are Lost

Now there he and I part company. I don't do it for a cause. Now to be sure, I didn't know this always either. When I was a kid, I had causes too. Then I learned that causes don't have success. All causes are lost. There's no other thing. They talk about the tragedy of lost causes. It's inevitable. Anything that is a cause will automatically be a lost cause because you cannot succeed with the concept. So I dropped the idea in favor of proprietary productive management of developing products and marketing them rather than doing this for a cause. But Paine didn't know that. He was earlier and to him this was a cause.

Making a Monetary Profit of No Consequence to Paine

The cause of the American Revolution was supreme and whether he made a profit on it monetarily was of no consequence to him. The second time he had an enormous best-seller was *Rights of Man*. That sold like hotcakes in Great Britain, also in France in French translation and also in America. There, there was a royalty, which was considerable. He donated it to a nonprofit society for constitutional government in England where its function and purpose was to influence Britain to become a republic instead of a monarchy and he donated his royalties to that. Well, obviously you notice how successful that society was. [*AJG chuckles*.] But the point is, he had opportunity to become wealthy. He muffed it, because it didn't matter to him. The fact that it didn't matter to him is what prevented him from correlating primary and secondary property.

The Function of Secondary Property in the Theory of Primary Property

That's true for all intellectuals, primary property comes first. That does with me too, but I structured it so that secondary property measures the success of the primary property and therefore makes it better recognized in the short term. The function of secondary property in the theory of primary property is to strengthen the short term. Primary property is automatically strong in the long term. You can have a total dismal failure in your lifetime. You can die without recognition, such as Bruno or Aristarchus or Archimedes and thousands of years later or even hundreds of years later the work survives.

Durability of Primary Property Is Obvious, Once It Is Pointed Out

The durability of primary property is probably the single most obvious observational fact of the human species. That is the single most obvious observational fact, after it's been properly pointed out, of course. You understand that. You do, don't you? [*Some replies from the audience.*] I'll give you a refresher quotation:

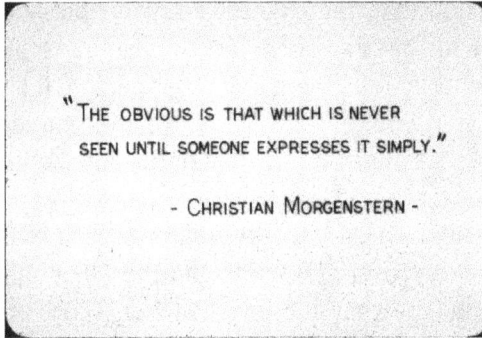

"THE OBVIOUS IS THAT WHICH IS NEVER SEEN UNTIL SOMEONE EXPRESSES IT SIMPLY."

- CHRISTIAN MORGENSTERN -

After it's pointed out by being expressed simply, it's obvious. That's the Morgenstern quote. When it is pointed out it becomes obvious. Isn't it?

You Don't Have to Worry about the Long-Term Significance of Primary Property

Is there anything more obvious, ladies and gentlemen, than that Archimedes' ideas prevail today and are at the base of our present culture? And yet, he was totally ignored by his contemporaries. He didn't even have one person think enough of him to write a biography of him. He's remembered through *Plutarch's Lives*, through the biography of a general who took Syracuse when he was the principal agency of defending Syracuse. All right, the fact is that primary property is the most durable. You don't have to worry about the long-term significance of it. The problem is, the long-termers live in the short term and therefore they don't have the recognition and the incentive to do the maximum they can in the short time they live. So in the short time that they do live, their output is reduced. It's harassed, it's rendered full of obstacles and some probably never even get to be known. It's inconceivable, incalculable, unimaginable how many potential Archimedeses existed of that caliber whom we never even have heard of because they went to their graves without ever being heard of.

Galambos' Father:
A Giant of the 20th Century, a Universal Genius

If it weren't for me, that will be true of my Father, you would not have heard of him. He had the mind of Leonardo da Vinci in diversity and his character was vastly superior, and you never would have heard of him. And you probably, I'm not sure, you probably don't even believe it at this moment. You probably attribute it to filial devotion. If I ask for a show of hands on that, the chances are you would deny it. Subliminally you believe it. That's a very great concern to me, by the way. If he were not my Father, but he were any other person, if I brought him to your attention you would recognize it. It's the relationship that obscures it for you, that this man is a giant of the 20th century, a universal genius.

The only possible way you might realize that it's not untrue, is you might see a small remnant of it in me and thereby it might have some credibility. After all, he's my supreme teacher. These are my theories, but there's a model for everything here. Almost all of it, in some way or another, has an input from him.

Knowledge and Creators of Knowledge Require
Public Exposition to Be Known

Most people have no idea of the durability of ideas that have had some public exposition. Archimedes had some public exposure, not as much as he should have had, but he had some. He had some writings remain in a period before the printing press and only a few manuscript copies existed but a few of them survived, largely thanks to the translations into Arabic, I think they were preserved by the Saracens. It's their principal major value in history. And the rather useless and harmful wars of the Crusades, their only value was that it brought back some of the knowledge of the Greco-Roman period to the Western world, and that started the Renaissance. That's probably the only war that had a good result, but not for the reason that they fought the war. They brought back Archimedes and the other ancient Greek knowledge.

The durability of these ideas is enormous but they had to have some public exposition. Other things that have not been publicly exposed, which never ever got written up at all, are very hard to know who may have done what. We have no idea how many people in China who were of coolie status, but had a potential capability to be an Archimedes, are unknown. The whole culture in China was opposed to this type of a thing becoming preserved. It was a very different kind of a structure than the Western world, which was even more difficult than here. Well, the strength of primary property, if it is recorded and has a preservation, is enormous.

The Secondary Application of Primary Property

The secondary application brings the time scale shorter and makes it possible for it to come to fruition faster and better by eliminating the disclosure barrier by making the application sooner, and basically thereby producing a greater incentive while the person is alive. Quite apart from the justice of it, it also has the practical utilitarian value to people who did not do the innovating. They get the benefits of the innovation faster and better and in larger quantities. And the theory of primary property will also prevent its usage for harmful purposes. Airplanes wouldn't be used for bombing cities. Nuclear energy wouldn't be used to wipe out the population, but to produce cheap and abundant energy. The whole concept of an energy crisis is repugnant to anybody who understands this.

Integration of Primary and Secondary Property Benefits Long- and Short-Termers

All of this means that the theory of primary property, which is a natural integration of two entirely different forms of property that are important in two different time scales, becomes of greater benefit to both participants, the short-termers and the long-termers. Nothing in my theory says everybody has to be a long-termer. I don't even think it's conceivable. I don't think it's even possible. I don't think it's even desirable.

How could a Newton have the stature that he has if everybody were a Newton? You have to have a contrast, but it produces a greater value for those who are not Newtons. They have a better world. For people who are not Newtons and have no potency for becoming Newtons, no hope of it, no possibility, nothing in the world would make it even realistic for them to imagine themself in this condition. However, isn't it better for such people to live in a world where they are not subject to civilizational destruction at any moment of their lives, when they don't have to be drafted into service and to war, when they're not rationed and they're not price controlled and subject to depressions and subject to military despotism?

Doesn't that make everyone's life better? Well therefore the profit is mutual and therefore you have long-term and short-term integration of mutual benefit from each other. The long-termer needs short-termers for a market; the short-termer needs the products of the long-termers for survival and for civilization and for progress and for the comforts of a durable and stable civilization.

Many Long-Termers Also Develop Emotional Disorders

All right. All of these things can be recognized in terms of the fact that human beings are of vastly different ranges of temperamental characteristics. To be a long-termer one has to have, above all other considerations, a rational base. Well that immediately limits the number. That doesn't mean that all people who have a long-term attitude don't have emotional disorders. Many of them, of course, develop this. There are all kinds of historical personages of major quality who had enormous emotional disturbances that I guess, by conventional classical psychological standards, could be called mental diseases in the sense that they were nervous wrecks. They had nervous breakdowns, whatever that means. I'm not quite sure what the technical meaning of that is, and I'm not sure the psychologists and psychiatrists do either. I don't think they have a uniform definition, but it can be intuitively interpreted as some kind of an emotional upheaval during which time they are rendered incapable of production.

Newton's Emotional Stress Caused by Linus

It is stated that Newton is supposed to have had one in the early 1690s. It is not certain that it's true. I've seen it denied and also stated. In any event, he recovered from it. Obviously, he was under emotional stress when he went to the mint and he gave up what he really was best at. That must have produced a psychological trauma for him to leave doing something he was so obviously talented at, so naturally suitable for doing, and then he stopped doing it. Just read his bitterness in his letter in Andrade about this clown Linus. That same letter, by the way, is in another biography of Newton in a fuller text. In other words, in the larger context. It's a very bitter letter. It's in a book by a man called Moore, M O O R E, I think it was Louis. I believe it's Louis T. Moore. I read it there also in a longer context. It's a very bitter letter. I use it profusely in my courses to illustrate the damage that such an attitude toward Newton by a Linus type produces, and that Linus has now become a generic name for a leech who seeks his own glory at the expense of a genius, who then becomes constrained to say the hell with it and not do anything about it. Not do anything more about his work.

Linus Has Attained Immortality as a Generic Type of Criminal

Linus has now earned the maximum historical reputation he can ever hope to have. He has attained immortality as a generic type of criminal. A man who is more dangerous than a Dillinger. Do you understand why? The murder he commits is much deeper than the murder that a Dillinger commits by physical assassination of human beings. This man destroyed the *desire* for further output by the most important man who ever lived. Nice way to go down in history, isn't it? That's the correct direction, incidentally. [*Light chuckles in the audience*.]

Linus Might Not Care about How He Is Remembered

But may I point out, I don't know about Linus personally because I don't know of him. I don't know of any biographies of him. [*AJG chuckles*.] But I can say this, it's conceivable he might not mind. Did that ever cross your mind? That he might not mind being known this way. There are some people who don't care how they're remembered as long as they are. Now consider that posture. That's a disease too. [*AJG drinks*.]

Oblivion More Desirable than to Be Remembered as a Criminal

You might have noticed that it matters to me whether my work survives or not. It matters very much. I don't claim altruism and I don't claim false modesty positions, but I'll put it to you openly this way. [*AJG takes a drink*.] I would rather be forgotten and had the reputation of a blade of grass from two thousand years ago than to be remembered like Linus. Because you can have a plus reputation, a minus reputation or a zero reputation, just go into total oblivion. Oblivion, however undesirable, is more desirable than to be remembered as a criminal.

Linus Will Be Remembered Longer than All Other Criminals in History

Interestingly, Linus will be remembered much longer than Genghis Khan, or Alexander the Great, who will ultimately be known as Alexander the Barbarian or Alexander the Atrocious [*AJG chuckles*] or Alexander by some derogatory term. The "Great" will be dropped, so let's start doing it now. He's only a great scoundrel. Linus will be remembered longer than Dillinger, or Al Capone. He'll probably be remembered longer than Hitler. And that's going some. As I say he's the supreme criminal of the ages in the classic sense, conventional sense of criminal. I think Linus will be remembered longer. Do you know why? Because he's inescapably tied in history to the man whom he attacked and whom he injured, and through whom he injured the world. Newton will be remembered, I don't have to explain this to you, I hope. Newton's reputation will last longer than Hitler. That's certainly clear, isn't it? Without further elaboration, right? Well, since Linus is the one who reduced Newton's output, that automatically makes Linus totally immortal for the duration of the species. But immortal, how?

Criminals Who Produce Permanent Damage Do Not Fit on Importance Diagram

He has the dubious distinction of being a criminal who does not fit on my importance diagram:

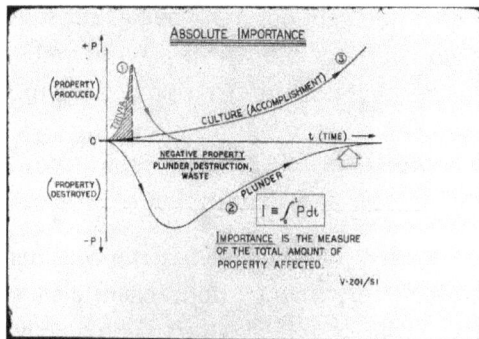

This curve, the negative curve, which is negative property, plunder destruction and waste, where it goes back to zero sometime. This won't. The damage he did is permanent. Forty years of Isaac Newton's life, of productive life, is a permanent loss to civilization. That does not ever get erased.

Psychological Defect of Rather Having a Bad Reputation than None at All

Well, the interesting thing is, there's a form of psychological blunder or defect, let's call it, that makes a person want to have such a reputation, not necessarily for the reason that it's bad, but who would rather have a reputation which is bad than none at all. In other words, it's a person who's basically—let me try to interpret this psychologically. A person who is so anxious to be remembered that he doesn't make a distinction between plus or minus. I'm only anxious to be remembered for plus. If the alternative were minus, I'd say skip it. Zero is better than minus.

The Distinction between Achievement and Dreams of Glory

Let me tell you what this is related to. [*AJG drinks.*] This is related to the distinction between achievement, which I discussed earlier, and dreams of glory, which I have also mentioned but not gone into great detail as yet. I plan to come back to dreams of glory. That's a very major form of ego minor. It's a very large component of the ego minor disease. A person who wishes to have achievement has to deliver. A person who has dreams of glory about achievement but hasn't got the talent, the rationality, the imagination, the stability, the investment of energy and time, and what in pre-volitional science would be called sacrifice, which I call investment.

The Distinction between Sacrifice and Investment

That you sacrifice yourself in the sense that you put everything you got into the drive to accomplish something regardless of what it costs you in health, personal happiness, family life, short-term pleasures, financial considerations, and you put everything into an all-consuming purpose and dedication to an achievement. That's usually called a sacrifice. I'm calling that an investment, if it's done right. If it is done right, I emphasize. If it's done pursuant to what's in V-201. Then what appears on the surface as a sacrifice is in fact a long-term investment. An investment whose duration is, in general, longer than your lifespan.

The Pseudo Long-Termer

Now, a person who hasn't got the *ability* or willingness to put that in, but wants the glory that would come from it if he *did*—then that person is a pseudo long-termer. He wants the advantages of the long term without any of the investment it takes to manufacture products that last into it, and such a person will not care whether he earned the reputation or got it at the expense of someone whose coattails he's hanging onto or, essentially, he's a parasite off of.

Linus Was a Parasite Off of Newton's Achievement

In this sense, Linus is a parasite off of Newton. He has gone into history through Newton's achievement. [*AJG pauses for a drink.*] Someone may wish to come to the defense of Linus here. So I'll forestall it. [*Some chuckling in the audience.*] You say, "Well wait a minute. That was just an honest difference of opinion between Newton and Linus, and Linus was expressing his point of view. He thought he was right, and Newton thought he was right." Yeah. Except whose property was under consideration? Was Newton spending his time discussing Linus's property, or was Linus spending his time discussing and criticizing Newton's property?

Newton-Huygens Dispute Is Not the Same Thing as the Linus-Newton Dispute

This does not have the same effect, incidentally, as the dispute between Newton and Huygens. There is a genuine scientific difference of opinion by two giants, each of whom had his own achievement. They both discovered the characteristics of light and came up with different, and not reconcilable, theories. One was the corpuscular theory. That's Newton's. The other is the wave theory. That's Huygens's. The dispute was on who is right and which one has the right theory and which one ought to be rejected as a false hypothesis. Newton wasn't attacking Huygens's work, he was promoting his own, and vice versa, and they happened to be in conflict. A dispute which lasts to this day. This is 1975 and it is still on, and both hypotheses are corroborated and both are in existence as recognized theories of nature today, and they are both functional.

Newton-Huygens Dispute Is Positive History

It's one of the biggest problems of all time in the history of physics, which is the history of man, if you understand what this means. Because the history of physics is the history of the creation of man's civilization, and the history of politics and militarism is the history of the attack on property. That's the negative history and this is the positive history. This is one of the most important things in all history. That dispute is still on. Who's right? Well, apparently, they're both right, as things stand now. And yet, there's no reconciliation of the two theories. This is covered in substantial length in my physics course, for those of you who had it and for those of you haven't, well, that's where it is. The interesting thing is that they are both corroborated, but in different areas, and this has been no end a problem in physics. [*AJG chuckles*.] This is a much more pleasant problem, by the way, than where the boundary is between two nations, or who should be president, or which way to tax you is the more equitable. [*AJG drinks*.]

Knowledge Production Does Not Have a Destructive Effect

Please note that this, in its worst form of antagonism, does not have the destructive effect. We have learned so much from both theories, that the very fact that they're at odds with one another has produced more knowledge. The corpuscular theory died already and is now resurrected. Non-theological resurrection, you understand, for observational corroboration. It was actually *buried* in the 19th century. The Newtonian corpuscular theory was a failure, it was believed in the 19th century. There was a crucial experiment performed on the speed of light being measured in a material medium.

If the speed of light is greater in the material medium than in a vacuum, the corpuscular theory would win. If it is less, if it is slower, the wave theory wins. The test was made, there is a very clear-cut separation. There is no feasibility studies here, there are no differences of opinion possible, it's yes or no, on or off, which is the faster? Is the speed of light faster or slower in, say water, than in a vacuum? A very clear-cut, sharp as a bell distinction, the experiment was made, Fresnel made it, Kirchhoff made it—Huygens's theory won, and it was considered the corpuscular theory was dead. With due honors to Newton, for once he was wrong. He's not God, but he was the integrator of physical science. By this point it was considered he was wrong.

Blackbody Radiation Dispute Resurrected Newton's
Corpuscular Theory of Light

In the end of the 19th century, some very, very obscure phenomenon which hardly anyone outside of physics had heard of and even most people in physics had not done anything with or known much about, the attempt was made to determine how the blackbody radiation can be explained. Even the discussion of what a blackbody is, is a long subject. It is not connected with a racial studies. [*Audience laughter and AJG chuckles.*] It has nothing to do with human affairs. This refers to blackness as a physical concept. It is so remote from the average person's concept, just to show you how remote it is, I can show you in one of the most dramatic ways, by telling you that one of the closest things to a blackbody in nature—the blackbody is an idealization which doesn't exist, but we have approximations to it, like a Carnot engine which doesn't exist but which is an idealization—one of the closest things to a blackbody is a star, such as the sun. That obviously doesn't fit the popular image of a blackbody, does it? And yet it's a blackbody, or very close to it. It's explained in my physics course, we'll leave it at that here.

What I want to point out is, the blackbody radiation to be explained, failed. All attempts at explaining it in terms of ordinary so-called classical physics failed. It was Max Planck who published in the year 1900 an explanation of blackbody radiation which fit the facts. But in order to determine the characteristic radiation of blackbody radiation and to mathematically derive the shape of the curve that can be drawn to describe it, it was necessary for him to make an assumption, and without that assumption it was not possible for him to do it. The assumption was that light is absorbed and emitted in discrete particles or chunks, discrete amounts of energy. Not half of it, not a quarter of it, not two-thirds of it, not .297 parts of it, but whole units, chunks of energy, and these were called quanta. That's plural for quantum. That's the origin of the name the quantum theory.

Well, you can call it quantum or any other name. A quantum is a chunk and a chunk is a particle and a particle is a corpuscle and here we are, the corpuscular theory is alive and well and is resurrected without any benefit of theological resurrection, and without any intervention on the part of any mythological creature called God. The observational evidence now favors the corpuscular theory and there is no wave theory explanation of this phenomenon. So here you have a second crucial experiment which favors the corpuscular theory.

Photoelectric Effect Also Resurrects Newton's Corpuscular Theory of Light

Then came another one. The photoelectric effect also required an explanation and didn't have one. It was Albert Einstein. The same Albert Einstein who developed the theory of relativity that published that too. That also was explained with a quantum. Therefore, you have immediately within five years of each other *two* corpuscular theory crucial experiments where the corpuscular theory wins. Well, there we are, we're back to that again. [*AJG chuckles.*]

So I just wanted to point out that's a true dispute in the history of science between two giants. I have some discussion of my opinion on this, and how the two can be potentially resolved, which is in the physics course. It's important to recognize that the corpuscular theory always is involved in its success at the time of emission and absorption of light. And the wave theory is always successful in its application in the transmission of light, without interaction with matter, and so it's in a different domain, two different domains where the two theories are respectively successful, and carrying that farther might bring it, ultimately, a proper resolution of this duality.

Newton-Huygens Dispute a Bona Fide Scientific Difference of Option—But Not with Linus

As a matter of fact, it led to something even more fantastic. It turned out that particles also have waves. In other words, a stream of particles such as a stream of electrons, which have always been considered to be particles, they have waves associated with them, and they have wave-like distributions and that really made a hell out of the quantum theory. That made wave mechanics come into existence where the atom is explained as a probability wave. Then everything went to hell after that. [*AJG and audience chuckles.*] So I just wanted to quickly explain to you, here is a bona fide major dispute between two giants where it was not based on hatred of each other or using each other for immortality. It was a bona fide scientific difference of opinion. But not with Linus! Remember, we're back on that. We're still on Linus and psychological disorders and dreams of glory.

Galambos Explains His Teaching Methods

I have to make sure you understand the track is always the same, it's just that we have sidetracks we have to elaborate on and come back to the main issue. Some people tell me, "I can't understand these ramblings." They're not ramblings. I'm bringing in different subjects and pulling it into the same picture. I'm explaining that because I know I'm sensitive to the fact that most people do not know what I'm doing. [*Audience laughter.*] One of the characteristics of rationality is to see yourself as others see you, which is not the same as the way you see yourself, and I'm aware of how negative most people's attitude is on me. So I have to explain this to you. [*AJG drinks.*] Thank you, those of you who are shaking your head negatively [*audience laughs and AJG chuckles*] but you don't know of the rest of the people, and there are more of the other kind.

You know, it's inevitable. When somebody has something new, it's hard to understand it. So I'm trying to explain all aspects of it and I'm trying to explain to you so you don't lose track of the main thread. The whole lecture is on emotion even though it got off onto things like Linus and the wave theory [*AJG chuckles*] and Thomas Paine and Eddington. [*AJG stops to take a drink.*]

How to Determine Who Is Right in a Dispute

It is necessary to recognize that the subject of a dispute, the so-called dispute as the one Linus and Newton had, couldn't have arisen if both of them had property of equal caliber. It is not Huygens that chased Newton into the mint. He didn't enjoy that. Nobody likes any kind of a dispute no matter how basic it may be and how true the observed facts may be. Disputes are not pleasant in any circumstance. But the acrimony, the viciousness of a Linus is quite different. This man had no property of his own that was subject to the dispute. And you can always tell who is right and who is wrong in a dispute when there is such a situation as a Linus, just ask one question. Whose property are we talking about? And who is discussing the other person's property adversely? Was Newton criticizing Linus's discoveries, or vice versa? Or was it Linus attacking Newton's discoveries? That is always a simple criterion.

Si Duo Faciunt Idem, Est Non Idem

By the way, this is related to something that I learned from my Father in a different context, but then again, contexts can be enlarged, and here I'll enlarge it. It is something which is a saying in Latin—*Si duo faciunt idem, est non idem*—which translates into English means: "If two people do the same thing, it is not the same thing." And it is not the same thing. It is not the same status that Newton and Linus have in this dispute. They do not both have property involved. You can see it more simply in secondary property.

Preposterous Example of a Property Dispute Settled by Majority Rule

If for example, you invite a group of people to dinner. Let's say there is a host and a hostess and there are three guests, and after a while, there's a heated discussion on the subject of how the furniture ought to be arranged in the house [*light audience laughter and AJG chuckles*] and the host and the hostess decide that they don't like the people they have invited to dinner and they say, "Your status here is no longer acceptable. So either shut up or leave." They say, "Well, we have freedom of speech. What are you trying to do, interfere with our freedom of speech? We have every right to express our views. What are we, your slaves?" And you say, "Well, the subject matter of how our furniture is arranged is our business and not yours." "Well, let's put it to a vote!" [*Audience laughter.*] Which is why I had three guests. [*AJG chuckles.*] "Let's put it to a vote. Two favor the present arrangement and three prefer a different arrangement. Therefore, in any democratic society, it is clear you will have to conform to the majority desire."

You laugh, do you not? [*"Yeses" from the audience.*] It is funny, isn't it? It's a contrived and rather ridiculous situation, isn't it? Don't you think that the majority of the people in this country would fail to grasp the humor of it? And they would very much say, "Of course it should be rearranged to fit the majority opinion. You mean two people should dominate three peoples' lives?" [*Audience chuckles.*] That's how they would read it. It's their right to express their views. Freedom of opinion, freedom of expression, freedom of speech. The majority should be favored. And if you say, "This is preposterous." It's no more preposterous than the laws you're under right now. Do you get the point?

Latin Phrase Number 32 from Galambos' Childhood Notebook

All right, well what's wrong? What's wrong is a non-observance of Number 32. You say, "What's Number 32?" That happens to be the number of the Latin saying in my childhood notebook that—*Si duo faciunt idem, est non idem*—appears under. My Father gave me all kinds of Latin things to study when I was a kid, and I was a very poor student of Latin. Very good student of geometry and algebra and many other things, but not of Latin, which is to my chagrin now. But at the time I didn't see the value of it, but I sure as hell do now. And that's Number 32 and the reason it's called Number 32 is when I first expressed this to some people I know, including Mr. Lange, they found it difficult to remember the Latin version of the statement, but when they saw the Number 32 attached to it, the thirty-second Latin saying in my list of Latin sayings, in my ten-year-old notebook, he remembered it as Number 32. So I have now, I guess, entered the historical discussion here as Number 32. That's to compensate for the even weaker aptitude toward Latin than I had. [*Audience laughter.*] My Father wouldn't believe that anyone could be a worse Latin student but [*more chuckles in the audience*] everybody I have met since then would make me shine. I at least learned, on the high side, 1% of what he taught me. I have failed to teach even that to others on Latin. [*AJG drinks.*] You see, as everybody believes, everything is relative.

Latin Phrase Number 33 from Galambos' Childhood Notebook

Just in case you're interested, there's a harsher version of it, of Number 32, which is Number 33. That's the next one [*AJG chuckles*] and that says, let's see [*AJG opens his notebook*] it's *Quod licet Jovis, non licet bovis*. That's what it says. That means: "What is permitted to Jupiter is not permitted to the ox." [*AJG laughs with the audience.*] And I remember when I got that as a kid, frequently [*more laughter*] I was very impudent to my Father, and I got an ego minor reaction and I said to him, "Well, I'm not an ox and you're not Jupiter." [*AJG chuckles.*] I would love to have the privilege of apologizing to my Father for that statement. Because of course he literally is not Jupiter and literally I'm not an ox, but the relationship is quite different.

Right Does Not Involve Somebody Else's Property, But Your Own

Let me put it to you this way. Let's get back to the three guests in the house of the host and the hostess. And it's three to two, as to the room arrangement. What's clearly wrong about it in terms of Number 32 and Number 33 is, whose house is it? Is that not simple? Now what does majority rule have got to do with that? Whose house is it? Whose furniture is it? Who has a right to make the determination as to how to place the furniture within the house? And what right does anyone have to express his opinion in someone else's house adverse to that person's proprietary relationships.

Right does not involve someone else's property, but your own. And you have no right to attack someone else's property, reference V-50. That's very clear. As long as you take political coercion out of this and look upon it in terms of rational inputs only. That clarifies it very simply and very easily, does it not? If someone has a proprietary concern, he has a right to handle it any way he wants, as long as it does not infringe on someone else's property. Someone else may not like what he's doing but it is none of his business.

Leonard Read Quote in Relation to Latin Phrase Number 32

To quote Leonard Read, a libertarian of the period before I developed my theory—and he's still alive. I don't mean to say he's in the past tense, but he preceded me and he has many good things to say. One of which is I'd like to quote you, with credit to Leonard Read is: "If there were no meddlers, there would be no socialists." Essentially, a socialist is a person who meddles in someone else's property, and Number 32 clearly says that if two people do the same thing, it is not the same thing. Someone who owns a house may move the furniture around. Someone who does not, may not. They're not in the same status. Property makes the difference. Whose is it? The same thing applies to the Newtonian situation with Linus. Whose property was involved? All of this clearly indicates that those people who attack other people's property are automatically wrong regardless of what their intellectual explanation or justification may be.

Property Flow in Relation to Latin Phrases Number 32 and 33

Another thing connected with Number 32 and 33, which I might mention here out of context and I hope to have the opportunity to come back to it here, I mean somewhere later in this course, is, there is a thing in V-201 which you are well familiar with, I presume, called the ideological flow chart:

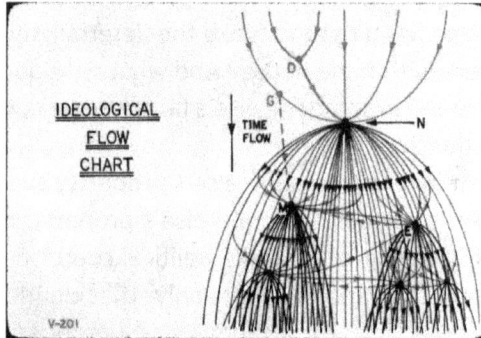

You must always look upon, in property flow—this is not necessarily intellectual property. It can be but it doesn't have to be. If it's Newton we're talking about intellectual achievement, but it can be property in the sense of furniture. You must always look in what direction the property is flowing. Where did it come from and which direction it's flowing. Number 32 and Number 33 are infallible guides as to which direction criticism, and any form of proprietary right to criticize, may flow.

The Distinction between Upstream and Downstream Critiques

Upstream can criticize downstream but not downstream to upstream. Now, you may say that's an interference with freedom of expression. I should have said may, not can. I said upstream can criticize downstream. I should have said upstream *may* criticize. Can is possible both directions. It's physically possible to do it either way. For example, it's physically possible to steal an automobile or shoot somebody and kill him. It *can* be done—but *may* you? No. There's no justification for it. There's no moral justification for it. Okay, in this sense, it is not morally proper to criticize upstream. That doesn't mean you accept errors. It is not proper, however, to attack someone from whom you have received benefit.

Will you please notice that when I explain to you that I do not agree with Thomas Paine on his political conclusions, I have never attacked him. Have you noticed that? I didn't say, "How could he be so damn stupid?" Or, "How could he have gone off the deep end?" I always go to great pains to point out, okay, I don't agree with him, but I'm later and I have gotten inputs he hasn't and I don't *TOLERATE!* [*AJG shouts very loudly*] from someone else a criticism of Thomas Paine for this.

An Example of an Immoral Critic

I remember I had an idiot for a student once, which is actually a very frequent occasion [*chuckles in the audience*] and I remember this was a woman who came to me via the Liberty Amendment fallacy. She was a Liberty Amendment promoter. You know what that is. A repeal of the income tax by political means. Income tax may not be desirable, but the repeal of it by political means is even less desirable, and it could be harmful, for reasons I do not wish to digress into here. That is not germane to this. Anyhow, this woman came to me and took my course and was exceedingly friendly to my ideas, in a rather grim way. She was one of these grim types. You know, one of the grim Randian types. No smile, no friendship, no niceness, no nothing. But she was exceedingly taken with the ideas and she took many of my courses and indicated much interest. And when I gave V-76 she purchased a picture of Thomas Paine and followed my recommendation that no person should have exposure to this and not have a picture of Thomas Paine prominently visible, and she bought one, framed it and put it up on her mantle.

Then one day she read something that she didn't like in Paine. I forget which particular writing it was. It may have been *Agrarian Justice*, which I myself say in V-76 has unfortunate things in it, which I don't agree with. And *Rights of Man* has such things. It may have been one of these two. Anyhow, she read something from Paine that she saw to be in distinct contradiction to what's in V-50. I think it was still Course 100 in those days. And it was political coercion in the form of taxes, or whatever. Anyhow, she got really incensed about this.

She wrote me without a doubt one of the most offensive letters I've ever received, not directed against me, but directed against Paine. I can even more easily forgive someone directing their anger at me, because they know me personally and I can't say that everybody likes my personality, and I'm perfectly aware of that. But she doesn't know Thomas

Paine, and she got a great deal of value from Thomas Paine and then here she writes a vicious letter denouncing Thomas Paine. And so she took down that picture and destroyed it, and "It's an atrocity that he wrote this thing and it's terrible," and made a terrible attack on Thomas Paine and also indirectly on me for having mentioned Thomas Paine favorably.

Well, needless to say, she's not in my market since then. I told her off. I told her that she is totally in the wrong and she has no merit whatsoever criticizing or attacking Thomas Paine and that the only basis she ever had to like Paine was what she learned from me and I'm perfectly aware of what Thomas Paine did that I don't agree with, and it in no way reduces his stature for what he is right on. And I can understand why he did what I don't agree with, he was earlier. He was neither immoral nor stupid. He was earlier and didn't have all the inputs, which is to our detriment. You cannot morally criticize a person for not being God and for not knowing everything and for not being infallible.

Newton, Paine and God Are Not God

Now, let's say the quantum theory did not come into existence and the corpuscular theory was buried in the 19th century and let's say it remained buried. Let's say it became a permanent fact that Newton was wrong on the corpuscular theory. Would that justify an attack on Newton? Absolutely not, because we have the benefit of having the values that are real. And to say, "Well, he didn't know everything," well, so big deal. Who does? He wasn't right on everything. All you're saying is he's not God.

I don't think God is god. [*Audience laughter.*] Because what most people think is God is a false and foolish concept of god. I don't think god is a volitional being. Once I said I don't know. I now think more firmly that it is not. That doesn't mean there is no such a concept as god, but it doesn't make it a creature. It doesn't mean that it's just a Superman, who can do more things and live longer and think more clearly. Anyhow, the whole point is it's an anthropomorphic concept of god and all you're saying is a Newton or Paine are not infallible.

Another Offensive Form of Ego Minor Behavior

Then there is another form of ego minor behavior which is also offensive to me. All ego minor is offensive to me, but another one that comes up in this context. To say, as many so-called professionals in physics do today, "Well, what Newton knew is very little compared with what we know today," and this smug superiority, this pomposity from pygmies [*some audience chuckling*] who are swarming on top of a platform which is a mile above Newton's shoulders, standing on his shoulders, and these little pygmies crawling around on top of that platform which they didn't build. They might have been termites nipping away at one of the grains in the platform that's a mile high above Newton's shoulders. "Well, we see much farther. Look how high we are." Yeah, and what did you do to get that high? You climbed up on his shoulders and other people's shoulders and you're standing on a platform you didn't build.

"Well, Newton didn't know anything about atomic energy, and we do." That's the prevalent view amongst today's great technicians in the field which have gotten so complex no one understands it. It needs a new Newton, the whole subject of the microcosm. Anyhow, I've heard this before I retired from dealing with such people. When I was still at the boondoggle I knew these self-appointed, pompous, technical asses. I had much exposure to this type of person and I was just nauseated by them.

These are people who say, "Well, Newton didn't know anything about atomic energy." Well, fine, add to that ten thousand other things. He also didn't know how to fly. He didn't know anything about gasoline engines or diesel engines. He didn't know anything about electromagnetic wave propagation. As a matter of fact, he didn't know what an electromagnetic wave was. He never heard of it. All of these things came about as a result of his integration of physical science.

No Knowledge Is Non-Newtonian

Every one of these achievements is a post-Newtonian development which is in the flowstream of what we've been discussing. As I have said many times before, there is no knowledge today that is non-Newtonian. There is quackery that's non-Newtonian but there's no knowledge today that is non-Newtonian. It is post-Newtonian, and much of it is vastly beyond Newton, but to attack Newton that he didn't know as much as his successors, well, that's because they came after him and they're standing on his shoulders and most of his successors are not of his stature.

Other Scientists of Possible Newtonian Stature

Only a few of Newton's successors are of his stature. My personal opinion, I would consider men of Newtonian, or close to Newtonian stature, to include just for example, probably Lagrange, probably William Rowan Hamilton. Quite probably Joseph Henry, not quite but, anyhow, approaching that. And Michael Faraday. Definitely Maxwell was of Newtonian stature. Probably Josiah Willard Gibbs. I'm sure that Einstein is of comparable stature, and I'm fairly sure that Eddington is. Possibly Karl Friedrich Gauss, to think of another one. These people are of Newtonian or close to Newtonian stature, but they're still in the post-Newtonian world. You don't get around that.

The Newtonian System of the World Is Getting Bigger

There's still only one System of the World to establish. You can make the system bigger by recognizing a larger world. Basically, incidentally, that's what I'm doing. The System of the World is getting bigger. It includes more than celestial objects. It includes even human behavior. But it's still Newtonian. Everything I'm doing is Newtonian. Let that be clear. Everything I'm doing is Newtonian. And Archimedian. And I'm making no claim that, "Oh, I made a new world here. I built one in the next county." [*AJG chuckles.*] It's Newtonian. It's larger than Newtonian but it's still standing on Newton's shoulders.

Rules of Etiquette Are Arbitrary

And then to attack someone upstream, I consider that the—you know, Emily Post wrote about bad etiquette, and good etiquette. These are sort of arbitrary rules, which hand you hold the fork in and how you talk to the hostess at a cocktail party. Or what is the proper decor at a funeral or a marriage. All of these are arbitrary, and sometimes silly. Sometimes they make good sense. I pointed out in Course 100—I'm not sure if it survived into V-50 because of lack of time—in Course 100 I used to talk about the difference between custom and morality.

For instance, which hand do you hold the knife in, fork in, is arbitrary. Good table manners in America says you hold the fork in the left hand, the knife in the right hand when you cut. Then you put the knife down, you switch the fork from the left hand to the right hand, and then you spear the food with the fork in the right hand, put the food in your mouth, put the fork back in the left hand, pick up the knife and start the whole cycle over again.

In Europe you drop this foolishness, this back-and-forth shuffling. Interestingly, the Europeans who need a siesta after lunch, which they call dinner, and need several hours to rest and they close stores for two to four hours and in Rome they close the stores I think for four hours at lunchtime. [*AJG chuckles.*] They need all this leisure and they have a much slower life. In Europe, where efficiency is less stressed than in the United States, they eat more efficiently. In America they have this asinine custom of back-and-forth play between the knife and the fork. It's utterly preposterous. And yet, it's arbitrary. Is it immoral to have the fork in the left hand? Or to have it in the right hand? Of course not. It's a custom.

It's a Question of Rationality, Not Morality

Is there morality associated with long hair or short hair on men or on women? No. You may say, "Well what are you so down on certain customs then for?" It's not a question of morality. Rationality. We haven't discussed that here, and it's not the course for it. It's a question not of morality. Nobody's questioning the morality of what length your hair should be. It's the rationality of it and what its symbolism is and what its function in society is. Wrongness can be wrongness for other reasons than morality. Rationality came before morality. That's completely arbitrary, morality-wise.

The Distinction between Custom and Morality

On the other hand, certain things about etiquette are not custom, but they are quite rational and proprietarily necessary. For example, it is one of the etiquette features that if you go to call on someone else, you don't barge into his house even if the door should not be locked, you don't just barge in unannounced. If there's a bell, you ring it. If there's no bell, you knock on the door. That's called politeness. Emily Post I'm sure, which I did not read, but I'm sure she would agree that it is impolite to barge in. On this, we would agree.

On the other hand, that's not a custom only. It may be a custom. It's also morality. Because whose property is it? You don't have a right to enter someone else's house unless you have permission. But how can the owner of the premises give you permission if he doesn't know you're there? And he has other things to do than watch out of the window for your possible arrival. So he doesn't know when you're there wishing to enter and you may not enter. You may not enter without permission! If you do, you're a trespasser which is a form of crime. You may not enter without the permission of the owner! But he doesn't know that you're there so you can't ask him permission. By rapping on the door, knocking on the door or ringing a bell, or you can attract his attention some other way like sing a tune or blow a bugle. [*Chuckles from the audience.*] Of course he may say, "I don't like bugles blown on my premises." [*AJG laughs.*]

In any event, the purpose of the bell or the knocking is to say, "I have arrived. May I enter?" And that's why it is a proprietary requirement. That is not custom, that is morality. That is a requirement for moral entry, to get the permission of the owner, and the purpose of that is to announce you have arrived and it's a request to be permitted to enter. With that understanding, that is not only a matter of custom or etiquette. Yet, the people to whom there is no distinction in what is custom or morality, obviously are unable to act properly and behave themselves.

To Attack Someone Upstream of You Is Immoral

It is in this domain, in this context, that it is important to remember the significance of Number 32 and Number 33. That's one of the main functions of the flowstream in determining proper conduct between people. It is more than a development mechanism, it's more than identification of from whom you got what and to whom you owe what—which is of course a basic function you got in V-201—it also tells you what is your proper conduct towards them. To attack someone upstream of you in any form whatsoever is immoral. As long as you're enjoying the value, the attack is not warranted. You say, "Well, he injured me." Step aside.

Number 32 Tells You Which Direction Property Flow Is Coming From

First of all, it's not probable that a Newton could or would injure a Linus. Let's say Newton did go berserk, which is historically not correct. If and when he had a so-called nervous breakdown, he did it privately. He didn't go around attacking other people. He simply had an emotional disturbance over the fact that he was not able to do what he wanted to do and he was undoubtedly frustrated. There was no public attack on other people. He wrote some rather bitter letters to his attackers, but that was defense. That's another example of Number 32. To defend property is moral, to attack property is immoral. Number 32 tells you which direction the property flow is coming from. Whose property is it that you are making use of.

The Distinction between a Criticism in Defense and an Attack

Let's say Newton had gone berserk, which he didn't. And let's say he wrote an attack on someone else, not a defense of his own property which is always moral, but an attack on someone else. To criticize another person for stealing from you is not an attack on him. Do you understand that? If A says to B, or says *of* B, "B, you have stolen from me." No matter what language that's couched in, no matter how angry it is, that is not an attack. He could lose his temper, he can blow his cork. That is not an attack. Do you understand why? Whose property are we talking about? Newton has every right in the world to criticize someone else for attacking his property. The only time it's a true attack on someone else is if the other person's property were in fact interfered with.

The Proper Conduct of the Downstream Person

Now for example, let's say Newton went berserk, which he did not. I emphasize and hasten to add, he did not. And let's say he *stole* someone else's property who stole from him, under the mistaken belief that, tit for tat, he stole from me so I'll take something from him. Now he didn't do that, you understand. It's not even proper to use Newton's name here. This is just talk, in general, but let's say he went berserk and did that. What is the proper conduct of the downstream person in this case? Disconnect himself so he's not exposed to the vulnerability. But to attack his source, that'd be totally wrong. Now that's a case where the Christ turned the other cheek would work. Leave him alone. Step aside, don't have any contact with him. That's a very rare case anyhow. You don't find Newtons attacking other people. You don't find that. I don't know of any such case. This is hypothetical. But if it did happen, just disconnect yourself. To attack him is wrong. Does this make sense to you people? Do you know what I am talking about? [*"Yeses" in the audience*.] Does that give you an idea? [*AJG drinks*.]

The Distinction between a Disagreement with vs. Attack on Thomas Paine

I deliberately emphasize this because you will notice that I have many times had occasion to say that somebody upstream of me, I don't agree with on something. I refer you to my many-times statement that Thomas Paine made mistakes. The entire Course V-111 deals with the mistake of Thomas Paine in advocating a representative democracy wherein you vote for representatives to rule the country to replace hereditary monarchy. I consider this wrong. You will please note that I can disagree with his political conclusion without having attacked him and I go to great pains—everywhere—in 111 and everywhere else, to point out that this is not a shortcoming of Paine's character. I don't accept that from anyone else who may attack him. I told you about the case with this woman who tore up his picture and attacked Paine and also criticized me for having made Paine a hero in my courses.

Actually, I disclaim that I made him a hero, I recognized his heroism. There's a difference. You don't make heroes. Heroes make themselves and I recognized his heroism. And I recognize him as a giant of a person who doesn't happen to be God. And so I accept what he has done which is right. I respect him for it. I'm grateful to him for it. What I don't agree with I don't use and I have done other things which I think are an improvement.

And you know what else I think about Thomas Paine? I cannot prove this, so I cannot offer you observational corroboration, but I can state this as my opinion. I believe that Thomas Paine was such a giant of a man that if he were around and alive today and could hear this lecture that he would acknowledge, having heard V-50 and V-201, that I have a superior system, which I acknowledge is partially resting on his work and I think he's a big enough man to accept that and to recognize such a thing. I cannot prove that, for obvious reasons, he's not around to test this with, but that's the mark of a truly great man.

Galambos' First Encounter with Spencer Heath

I would like to mention a case where something not quite as dramatic occurred but fairly conceptually similar, but not of such a magnitude. The occasion arose, prior to his death, that Spencer Heath and I met each other. He was, I believe, eighty-six when we met, going on eighty-seven, I think, and I was thirty-seven. He heard my lecture, which was an introductory presentation for Course 100 in September of 1961. That was the second go around on Course 100, and he immediately wished to meet me after the lecture, and I didn't know him. I didn't know what I was getting into, whether he was a great man or a crackpot. As a matter of fact, I had reason to believe he was a crackpot from what I heard about him from the guy who brought him. The input I had on him was wrong.

Meeting a Genius

It turned out he was a genius. That's rare in this world. Of course, when you meet one you immediately wonder, "Is it possible I might have met a genius? They're so damn rare." The probability on a random selection basis is if you meet somebody who sounds weird the probability is over 99% you got some screwball on your hands and not a genius. I mean, you don't meet geniuses. I have added up all the total number of geniuses I've ever met in my whole life and I can't fill up the fingers of two hands. [AJG chuckles.] With straining myself I couldn't do that.

Of course, I'm not very gregarious. There are some geniuses who have met lots of them. I mean, I understand Paine knew everybody in the liberal world at the time in France, England and America. I'm not able to do that. In any event, all of whom were inferior to him, I think. I mean, such people as Jefferson, Madison, Monroe, these men were great men perhaps to some extent, but inferior to Paine, and didn't know they were inferior to Paine, for the most part.

Spencer Heath Attended All of Galambos' Early Lectures

Anyhow, to get back to this. I met Heath and I didn't know what to make of it the first night. He gave me his book, he donated it to me. I have it inscribed, I still have that copy. That one's not for sale, the one he gave me with a very nice inscription having heard one lecture from me. It's later that I said I'd sell the book and also that's not in competition with your donations, that must cease. [AJG chuckles.] Well, he agreed and he recognized that I had a better way to do it. More than that, he liked what he learned from me. Now please note, he never got all the way in my theory, he died sooner. He died before 201 existed. Although some of the concepts I talked to him about privately. He was very much enthralled with everything he heard from me. He came to all of my lectures. Even when he heard the same lecture he would come again and again and again, and here's a man who is eighty-six years old.

Galambos Started the San Diego Market
with Spencer Heath in Attendance

I once went down to San Diego to give a presentation for my course. Essentially the same presentation he heard in Hollywood, and he heard it in Orange County every time I gave it there, and when he heard I was going to San Diego, he called me up and asked me if he could come with me. I said well, in principle, yes, Mr. Heath. I'd be delighted to have you come but I don't expect to be back before three or four in the morning. "Oh, that's alright." [*Chuckles in the audience.*] "Well it might be a little strenuous." "No, that's alright." [*AJG chuckles.*] So I took him down. He sat through that thing and he was brighter and stronger and more awake at the end than the people a half or a quarter of his age. [*AJG chuckles.*] Some of these twenty-year-olds were drooping there like this. [*Audience laughter.*] And here it is midnight and he's still fresh and wide awake. Then he comes back with me in my car after not only I finished the presentation, took enrollments for Course 100, and it was our first presentation that we took enrollments for Course 100 in San Diego. That's when we started the San Diego market. And then I talked, you know, the ear-benders, the hanger-onners, and we were there for hours more. And then finally when all this was over we got back in the car and went back to Los Angeles. Well, actually to Orange County, he lived in Orange County and I dropped him off there.

Galambos and Heath Had Overlapping Ideas with Similar
Conclusions but No Controversy between Them

Anyhow, on the way back he was absolutely awake, he wasn't dozing and falling asleep, he was awake and bright and cheerful. It's amazing, really amazing. And then when we got back, I let him out where he lived and I walked him to his house and there he stood on the porch and for another hour he stood—not sat—*stood* till four in the morning! We had a continuation of our talk, we weren't finished talking so we continued to talk, and then I finally felt exceedingly guilty, the man was eighty-seven years old by now. [*Audience laughter.*] I said, for god's sake, he needs some sleep! [*Audience laughter.*] So I bid him good night. [*AJG laughs.*]

274

He was an amazing person, and he liked what I taught, and it was not what he taught and yet it was very much overlapping. We both talked about proprietary management is superior to political or coercive management, and he understood that perfectly clearly. His book was on that subject. It was published before I taught anything. I had never seen it and it's perfectly clear that I did not get it from him. The thought never entered his mind. The subject never even arose between us, that maybe I could have read his book and plundered him. It couldn't have happened. The independency tests, which he'd never heard by the way, it was obvious. He had a real estate interpretation of proprietary management and mine was based on primary property and flowstream of knowledge. It wasn't the same subject but arrived at similar conclusions. There was absolutely no controversy then, now or ever between us.

The Tremendous Ego Major of Spencer Heath

What is more, here's a man who I believe liked what he learned from me—you know what, just consider that. Consider the quality of a man who was willing to learn from someone less than half his age when he himself is a genius. Just consider that. I mean that's a privilege I rarely have seen in this world. As a matter of fact, I think it's unique. In my experience, not unique in history. Unique in my personal experience— no, not quite. Hoiles was also willing to learn from me, R.C. Hoiles, and he was almost as old as Heath. As a matter of fact, I met Hoiles through Heath. It was Heath who had introduced me to Hoiles, and both of them were willing to learn from me, except Hoiles was a tremendous person but I don't think he was a creative genius, but Heath was.

It takes a tremendous quality of person, a tremendous ego major, to be able to recognize that someone else younger than you whose life you have interacted with and you've crossed paths with them, and here he's at the end of his life, he is meeting someone less than half of age. The age difference is adequate for him to have been my grandfather, age-wise. That he was willing to recognize that I had something even bigger than he had, and there was no competition as to who got it from who. I didn't get it from him and he didn't get it from me. It wasn't even the same theory but it had similar conclusions. Now that is the basis of a proper flowstream. When he died, I preserved his work. I'm still doing that. I'm still doing that. Now obviously it would be very easy to just

275

ignore it. It isn't as though he's world-famous. He will be, through me, if my work survives, which I hope it does. He will. But on the other hand, there will be no other way for it to survive because he has not had the proper recognition from others. I mentioned this in V-201 but here I brought it up for a stronger reason, and that is to indicate the quality, the psychological strength, of a person who is able to do this. That's an ego major.

Galambos-Heath Did Not Have a Linus-Newton Type of Interrelationship

He has nothing to fear from me, nor I from him. That is not a Linus-Newton type of interrelationship, and it's especially pleasant for me because he was upstream from me, although I didn't know it. He was not technically upstream from me, in that my theory does not come from his and there's no direct relationship. But in the sense of earlier, he was earlier than I. Not just in age, but he had published his book before I did anything on the subject. But he was able to see something bigger, which in fact it is, because the intellectual flowstream is prior to any form of proprietary community, but it fits in very nicely because a proprietary community is developed from intellectual capability, which he was not even considering this aspect of it. All right.

Integrating the Compatibility of Certain Emotions with Rationality

The nature of the role of emotion which I'm not finished with yet, as a matter of fact, even when I am finished with the formal topic I will still come back to and draw upon it in the later parts of the course. Actually, if this were an ordinary discussion of psychology, most of the course would be on emotion and not on rationality. Here it will not be, only part of it. Most of it will be on the rational requirements for psychologically sound behavior. I restate what psychology is: the science that determines and influences the motivation of volitional action. Well, as you know, most volitional action is motivated by emotions and not by rationality. This theory will give a disproportionate size to the aspect of rational function and rational behavior and rational thoughts, and I'm trying to integrate the compatibility of certain emotions with rationality, but not of all emotions. The majority of emotions are negative and they are capable of being harnessed for criminal reasons.

Hitler Ruled with Emotion and Is Looked upon as a Hero in Germany

For example, Hitler ruled Germany with emotion, not with rationality. Hitler influenced the masses by getting them to be mesmerized. A hypnotic spell existed between Hitler and his subjects and he got out of them an emotional response which was hysterical. It was positive hysterical in the sense that it was favorable to Hitler but it was an hysterical reaction of blind devotion, fanatic support, fanatic followership where people followed him would have followed him through fire to the ends of the earth. And they did. They went all the way to Stalingrad, where they bled to death following Hitler. They followed him through the catastrophe of a war and only became unpopular when Germany itself was destroyed and his unpopularity is not that total. He is still looked upon as a great hero in Germany, and it's gonna get worse as time goes by.

The Popular Heroes of France, Heroes by Emotion

Please note how an earlier man of less criminal fanaticism but still a major military conqueror, Napoléon Bonaparte, wrecked the French Revolution and wrecked France, permanently wrecked France, yet he is the most popular Frenchman in history. Anytime you have a Gallup-type poll in France as to asking the question, "Who was the greatest Frenchman who ever lived?" Napoléon Bonaparte will win, hands down, number one. He wasn't even French, incidentally, he was Corsican, which means he was an Italian. He didn't even speak French perfectly. He had an Italian accent [*audience laughs*] so I'm told. I never heard him speak [*more laughter*] but that's what I read, that he never spoke perfect French. But he will win, without any question, any popular poll in France as to who was the greatest Frenchman. Not Pasteur, not Lagrange, not Voltaire—Napoléon Bonaparte. After he took France to its doom. Led a grand army into Russia which a few remnants of that army straggled back.

Other popular heroes of France you will find high on the popularity list are Joan of Arc and Charles de Gaulle. I don't believe Charles de Gaulle will make it a thousand years from now. Joan of Arc wouldn't either except there's religious fanaticism connected with her. She is Saint Joan these days, you know. It was not just that it's unusual for a sixteen-year-old girl to lead a French army, or any army for that matter. [*AJG chuckles.*] It's also unusual for her to see angels beforehand, which in fact she didn't, but that's what the basis of her popularity is. Anyhow, you understand of course that these are the popular heroes. And these people are heroes by emotion.

A Little Bit of Rationality in Respecting Charles de Gaulle

There's a little rationality to respecting Charles de Gaulle, I myself had quite a considerable respect for him during the war years because I respected him for not giving up to Hitler when the rest of France had laid down and died and were willing to submit to the conqueror. He was a courageous Frenchman who said, "No, we will not surrender France," and he continued the war against Germany, and he resurrected what was left of the French army. I had some respect for him too, but of course when the war was over that turned into pomposity and a love of dictatorship, so my respect for de Gaulle diminished. It doesn't alter my gratitude to him in World War II, by the way. That does not change. I still respect him for that. A man who did not submit to Hitler when the rest of his country collapsed. I still respect him for that. There's a little bit of rationality in respecting de Gaulle. There's no rationality in respecting Napoléon Bonaparte or Joan of Arc or any other general, male or female, religious or military.

The Basis of Most People's Rulership Is Emotional

I'd like to call to your attention that the basis of most people's rulership is emotional. Kings, emperors, dictators, presidents come to power and stay in power by emotional response. That's clear, is it not? ["*Yeses*" *in the audience.*] Now, it is in this area that emotions are dangerous on a civilizational basis. Also, on a personal basis.

Subordinating Emotion to Rationality

All personal failures, whether it's in business or in marriage or in just plain being able to cope with the reality of the world they live in, all people who have failures to record have emotional problems both as the cause of the failure and as explanations for the failure thereafter. This is what is a dangerous characteristic. The theory that I'm teaching on psychology will be able to serve as a solution for those who have the ability to subordinate emotion to rationality. It will not be able to cause an improvement in the people who cannot do that. That's a small class of people, but it's not a zero class. Those people are either in the hot end or can swim upstream.

You Can't Drift Upstream

You can't drift upstream. By the way, anybody who wants to be a hot-ender, that's another point—you can't drift upstream. You can't get on a raft and float upstream. The river takes you down. You must swim against the current. You must have the strength to swim upstream. Now if you are in the cold end of the ideological program, you can get hotter but not without effort. And desiring it won't make it so. That's where the dreams of glory comes about.

Strengthening Not All—But a Few People— to Become Hot-Enders

Saying, "Well I'm gonna be a long-termer from tomorrow morning on, but I gotta have one last fling tonight." [*Audience laughs and AJG chuckles.*] Or, "Next Thursday I'll become a long-termer." Or, "I just think it's great to be remembered in a thousand years." That's not enough. Now these are the kinds of people that have no capability for success. This is the dream of glory, ego minor disease. But those who can do it, this theory will make them stronger and if it has no other effect, that in itself might hasten the time of the natural republic coming about by strengthening not all, but a few people to become hot-enders.

This Theory Also Has Value in the Cold End

Now as far as the rest of it, this theory also has value in the cold end. Not by improving these people directly, but through the hot end, coupled with what's in V-201, which is still very much not only in the picture, but upstream of this. The people who can be helped to become stronger—I shouldn't say can be helped. I don't mean to be helped in the sense of a middleman, in the sense of current-day psychoanalysts helping their patients. This is not a middleman operation. I don't mean can be helped. I don't mean by me personally or by some other agent teaching this to them. I mean that they can harness what's in this theory for their own improvement. This is a self-achievement based on this theory. Who can become stronger in the hot end will hasten the day of the natural society.

The Natural Society Automatically Makes an Integrity Structure for Cold-Enders

By doing so, the natural society will automatically make an integrity structure for the cold-enders where it is simply not possible to be an immoral person and prosper. That's in V-201, I don't have to explain that anymore. If you don't understand that, that's in V-201 at the end of the course. That the whole structure is so organized that if you commit an immoral act you will starve. Therefore, a person becomes honest without integrity. *Integrity* is self-honesty. It's a structure you are part of and you cannot get out of that.

And that's the way the cold-enders will get to be stabler, by being immersed in an ocean which has no turbulence, and where there is not the storm, there are no fanatics to adhere to. Your own failure cannot be blamed on other people because nobody's going to buy it. Therefore, there's no other way but to stabilize yourself, if you plan to survive in the society at all. As I say, the environment becomes just, and not welfare-oriented. And therefore, there will be less need for the concept of externally imposed welfare.

Some Emotions Have Favorable Effects and Should Not Be Repressed

Therefore, the basis of all of this discussion of emotion is that emotion is not a bad concept, per se. Most emotional attributes of human beings end up disastrously and with poor effect, either by followership to political demagogues or by internal emotional upheavals which produce personal failure and end up as acrimony towards those that they wish to blame instead of themselves. These are emotional diseases. That doesn't alter the fact that some emotions have positive, favorable effects and all people have to have emotions in order not to succumb to repression of their natural emotions and thereby injure themselves. You're essentially sitting on a powder keg if you repress your emotions, which is ready to blow up and you're creating a larger and larger pressure.

Emotion Has a Positive Aspect to Stability

The emotion has a positive aspect to stability. It gives you an outlet and a release from things that may otherwise eat you up. It's an ulcer-preventing thing to some extent, and other emotional disorders. I heard it from my Mother. I don't think it's original to her. It may be, it may not be. I don't think it's original to her but I heard it from her. "You don't get an ulcer from what you eat, but from what's eating you." [*Audience laughter.*] Anyhow, I heard it from her. I can't say if she coined it or not. In any event, I think that's very true. I will discuss emotions more.

Connecting Emotions with Rationality

I also want to go into things that are connecting this to rationality. Much of this we have to convert to a subordination of healthy emotions to rational conduct and the unhealthy ones have to be simply transferred to healthier forms of emotion. I told you in an earlier session that many emotions are compatible with rationality. As a matter of fact, I just, so far, only identified one major emotion I can't render compatible with rationality, and that's hate. I don't know of any such thing as rational hate. Okay, other than that, rationality can be the dominant one, and in the hot-ender it must be. Okay, I'll continue from here.

[*Applause from the audience for seventeen seconds.*]

SESSION 10

PART A

Good evening, ladies and gentlemen.

In this part of the course I have been discussing the respective natures of emotion and rationality, although I will undoubtedly have much further remarks to make in this regard later in the course as well, the formal part of it I will try to finish today. Later I will have occasion to return to both emotional and rational concepts and their proper juxtaposition in human motivation which is what psychology is about, the motivation of volitional action.

Emotional Output Has No Long-Term Durability; Rational Output Can Endure

It is quite clear from the outset of this entire discussion that motivation for most people is almost all on an emotional level. It is quite clear from everything I have said, that an emotion is basically a short-term concept. That ought to be very clear, shouldn't it? Is there anybody who thinks that there might be such a thing as a long-term emotion? Is it clear that it has no long-term durability? Let me put it this way. All right, when a person has died, his emotions die with him. When he has died, the rational output of his life in general, if it is properly disclosed and identifiable to posterity, will not die with him.

Archimedes Is the Supreme Example of the Durability of Right Ideas

As you know from V-201, the whole of culture depends upon the cumulative impact of right ideas and the fact that we do not have to reinvent the wheel every generation. We do not have to rediscover laws of nature. Once they are known—every generation—we do not have to reinvent or rediscover anything provided they have been done and properly recorded, or at least recorded adequately that somebody will know how to recall it. This can happen over enormous chasms of time where entire cultures have died and new ones have arisen and they had died and new ones have arisen and they had died and new ones have arisen, and so on. The supreme example of that for a known person is Archimedes.

Akhenaten: The First Identifiable Monotheist in History

Another supreme example of that for another known person is Akhenaten, the first known monotheist in history. Well let's put it this way. The first one whose name is identifiable and who made a permanent impact on civilization, even though he personally was unpopular, even though pharaoh. Did you ever think of that? How few people have ever been kings who did anything other than be coercers. Damn rare to find a man who is in the coercion business who has done something useful as well. And interestingly enough, of all the coercers in history he's one of the most unpopular. His life's work was, you might say, vandalized as soon as he ceased being pharaoh. He was pharaoh only a relatively short time. He was a young man when he died. I think it was thirteen or fifteen years he was pharaoh. Some of these guys went on for thirty, forty, fifty years. [*AJG drinks.*] And then, Egypt went back to its polytheism prior to Akhenaten.

Akhenaten an Example of the Nondurability of Coercion

Here's a man who had all the quote, advantages, unquote, of coercion on his side and he couldn't make it stick through coercion. That shows you the nondurability of coercion. But on top of coercion, which he got by inheritance, I mean, it just was chance that he happened to be born the son of another ruler, and then he got it by inheritance. And even by greater chance he happened to be a genius. Then with all that going for him, in the short-term sense, coercion, on a maximal scale, it didn't last. Where it did last, it came through noncoercion.

Moses Was Akhenaten's Intellectual Disciple

It came about through another people, not Egyptians, but the Hebrews. And Moses was his intellectual disciple. Akhenaten has thereby persisted through the most durable religion in history, namely Judaism, where the founder of Judaism as a religion is a disciple, intellectual student of Akhenaten. And according to Sigmund Freud he was an Egyptian: Moses. I cannot say that is right or wrong. His argument is interesting. Freud, who was a Jew, makes a case for the fact that Moses was in fact an Egyptian. [*AJG drinks.*] As a matter of fact, that'd be interesting if it's true. At the moment I can't say I'm convinced one way or the other. I'm considering it as a possibility.

The Influence of Monotheism on the Newtonian Integration

The fact of the matter is, let's take the case that Moses might have been an Egyptian, and not a Jew. In that case, he was not able to make it go with his own people, so he had to go and make a name for himself by taking a relatively primitive and previously obscure tribe and getting them to accept the monotheistic faith, and in those days, you might call it that. Today, monotheism is not a faith. It's part of the scientific concept. But of course, you have to redefine god from a bearded, superhuman old man.

Nor can you accept the Christian farther account of a bearded young man [*light audience laughter*] who was subjected to what all other people who are not one of the team got subjected to, namely, ridicule, and in his case, worse: execution. God can't be identified on a cosmic scale with such anthropomorphic concepts, but the monotheistic idea—the idea that the source of natural phenomena, which is my personal definition of god—is one. That there's only one source of the universe's explanation. This rather fundamental and basic idea which I consider a major and necessary and, as a matter of fact, it's so necessary that without it, it couldn't have happened, for the Newtonian integration. To identify the idea that everything is attributable to one single explanation.

Moses: A Master Mass-Psychologist

Anyhow, this whole thing came about through a single person's efforts to get it to be durably propagated. You might notice how Moses did it, and this is independent of whether he was Egyptian or a Hebrew himself. He was, in one person, one of man's greatest geniuses. Secondly, he was a witch doctor, *par excellence*. [*Some chuckles in the audience laughter.*] He was a master mass-psychologist. That's what made him capable of [*AJG drinks*] getting something which was a new idea across to a primitive people that neither understood it nor wanted it. He was in a sense a tribal chief, too. He was their physical leader, you know, he was specifically their leader in getting them out of Egypt, their promised land, so-called, which he promised it. He probably didn't know where he was going. [*AJG chuckles.*] He was just looking for a place out of Egypt where he could settle down. He finally found a place and that's called the Promised Land. That's because Moses promised it. [*AJG chuckles.*]

Did you ever think of the supreme talent this man must have had? Do you know the people that he had were one of the most primitive tribes of the ancient world? They had no previous major culture. You can skip the crap in the Bible on the subject, that precedes Moses. Prior to that the Jews had no accomplishment. After that, they are the supreme accomplishers of all history as a people. He didn't even take it easily. If the biblical account has any truth to it at all—and you know, these legends probably have some basis in fact, in the sense that something happened and then they got sort of fairy-tailed up somewhat.

Definition of a Miracle

For example, there's no question that they had to go from Egypt to where they ultimately went which is Palestine, which is now called Israel. Well, this so-called parting of the Red Sea. That has one of the easiest explanations of all time, you know. It didn't happen the way it did in the movies, you understand. [*Audience laughter.*] How many of you saw *The Ten Commandments* in the movies, the Cecil B. DeMille production? I mean, that was a Cecil B. DeMille production in every sense of the word. [*Laughter from the audience.*] Did you see the water stand up, bolt upright, rectangular? [*More audience laughter.*] Did you see that? [*"Yeses" in the audience.*] How many of you believe that happened? [*Audience laughter.*] That's preposterous. That's an insult to the concept of god, to imagine that could happen. Do you know why? How many of you ever gave that any thought? Of any miracle, that would be a miracle. A miracle is something that can't happen. [*AJG chuckles with the audience.*] A *miracle* is something that would defy natural law. That's my definition of miracle. A miracle is something that's extra natural, outside of nature.

Laws of Nature Are Perfect

The argument is, "Well, God is omnipotent." Well, then he must also be incompetent. [*Audience laughter.*] Because the greatest merit to the concept of god is that the universe is harmonious and that it is what man isn't—perfect. And that everything operates the way it's intended to without exception. Well, the laws of nature are perfect in the sense that they never fail to operate, and you don't have exceptions to it. That's the beauty of it. That's the clear-cut beauty of it! That there are no exceptions. When something is a law of nature, it is not subject to: "Well, except on such-and-such a case."

Concepts of God That Allow for Exceptions to the Rules

Well, what does that mean? If a miracle were to take place, then essentially what you're saying is God is so incompetent that the first time he made up the rules of the universe he made enough mistakes that it can't handle certain cases. Something will happen, something will come up, an SCU,[2] and that requires a new way of handling it, different from the normal way, which is called the natural way. The beauty of science is that it is total! And to me the concept of god is the source of natural phenomena. Therefore, it's not hostile to, but inseparable from, science.

Well, if it doesn't need an exception, that's fine, that's perfect. But if it needs an exception, because [*AJG snaps his fingers*] "Oh! We hit a snag." Something doesn't go right, we need to find an exception, and this exception we will now have on this unique occasion only, for the particular purpose at hand, and he who sees it is subject to a revelation. And the laws of nature are suspended for the moment in favor of something super nature, supernatural, extra natural, and this will now take over and will suspend the universe's operation for the moment and then we'll continue.

---

[2] A term from V-50 which is an abbreviation for "something came up"; used as an excuse for breaking contracts.

Belief in Miracles an Insult to the Concept of God

Well, the implication of a miracle is that God is too incompetent to do it right the first time and he has to make up new rules to fix it up. It's like he has a blowout and he has to have a patch on it to fix it. [*Light audience laughter.*] I consider the concept of a miracle, blasphemy. And you can throw that at your local friendly theologian the next time he tells you that you must have faith! [*Audience laughter.*] If this were true, then God would be a blithering idiot. He should be locked up as a dangerous lunatic [*AJG laughs with the audience*] if that were the concept of God that could prevail. Those who believe in miracles, and think it's to the greater glory of God, well I'll just tell you, they have insulted the very concept of god with that. All right.

A Natural Explanation for the Parting of the Red Sea Story

So obviously, either the parting of the Red Sea didn't happen, or there's a natural explanation. Since they make so much of it, what probably happened is he didn't go through the land part, he hit the north part of the Red Sea where there was water and he waited for low tide. [*AJG drinks.*] When the Egyptians came by, it was high tide. They got drowned. That's if it happened at all.

A Natural Explanation for the Water Coming Out of the Rocks Story

Do you want the explanation of the rocks? The water coming out of the rocks? [*AJG pauses and the audience chuckling turns into laughter.*] That's simple. Moses was a big man. I'm not talking about how high he was, how tall he was. He was the tribal chief; he was the ruler of this operation [*AJG drinks*] and he didn't do manual labor. But he was a smart fellow, and he probably was able to do something what today geologists more or less look for, to be able to tell from geological formations where it's likely to find water or oil or whatnot. So he had some primitive ideas on it, and maybe one day he said to some of his tribesmen, "Hey, you three come over here." [*Audience laughs.*] He pointed with a stick and he tapped the rock. "It looks like we have some water here. Dig here." So then they dug, the three of them or the ten of them or whatever it was, and they dug for a while. Maybe a few days, and finally they hit water. In the intervening several thousand years, the manual labor got forgotten, which wasn't very important. You know, the digging, the physical digging, and all they got left was, Moses said, "Here." He tapped the rock and out came the water, and you just forgot the three days of manual labor. [*Audience laughter.*] After all, that isn't the essential part of the story, you know, the digging. That's the so-called tapping the rock and out comes the water. That's not very hard to figure out.

The Ten Commandments and Property Protection

Now the part about the Ten Commandments is very important. [*AJG drinks.*] Because you see, that's what converted the Jews from a primitive tribe to high civilization. The Ten Commandments either were essentially written by Moses, or maybe he was somebody's student and got this good morality from somebody else or else he was the original source. Obviously this is well-known in other cultures too. You know, don't steal. Don't kill. Don't commit adultery. That's a form of a contractual violation. Seven of the Ten Commandments are clearly property protection. [*AJG drinks.*]

The Psychological Value of:
"Don't Take the Name of the Lord in Vain"

I'm not quite certain exactly what's meant by "Don't take the name of the lord in vain." I think that probably means that if you talk about it too much you might question it. [*Audience laughter and AJG laughs.*] I'm not quite sure what the significance of that is. I'll have to think about it some more. [*AJG takes a drink.*] I'm sure it has something to do with property protection too, in an indirect way. It has a psychological value, come to think of it. This is a good course to mention it. If you talk about God too much, then you would say God is responsible for this and God is responsible for that. That's an abdication of personal responsibility. So don't talk about God and just talk about other things and you'll handle your own problems better.

The Bible, the Movies and Brainwashing

Anyhow, the long and the short of it is I hope there isn't anybody in this class now quite so primitive as to believe that he went up the mountain, Mount Sinai, and had a chat with God who gave him the stone tablets. I do hope we are not that behind here. That is the way it happened in the movie *The Ten Commandments*. It's very interesting watching how people's minds are brainwashed. The Bible already did it for thousands of years, but now you have it in technicolor on the screen. [*AJG drinks.*] He goes up on the mountain and he sees some kind of a moving brush fire or something like that. [*AJG chuckles.*] In a booming voice, yapa yada! All of a sudden he comes out with stone tablets for the masses to gawk at. Well that's what happened to his original tribe, they gawked at it.

As a matter of fact, when he came down the mountain, do you know what they were doing, according to the biblical account? His brother, Aaron, had taken over as the acting tribal chief and he was getting them to have some pagan ritual. He was quite annoyed, Moses. And he erupted in a burst of temper. I was discussing temper last time. He erupted in a burst of temper and he threw the stone tablets down and broke them! He was so damn sore. Then he went up the mountain and got a new set. [*Audience laughs.*] Is it necessary for me to point out that either he or one of his flunkies carved it out for him?

If this happened, which is fairly likely that he had some kind of a stone, because everything had to be carved in stone those days to make it look authentic. [*Some audience chuckling.*] That was the early form of printing. [*AJG takes a drink.*] Today everybody believes the printed word. Well, let me put it this way. The masses believe the printed word. That's because the technology of printing is beyond their dimwits, and so anything that gets in print must be true because they don't know how it gets into print. It's sort of a witchcraft beyond their skill. Well, so [*AJG chuckles*] something that gets carved in stone must be authentic. So he goes up and carves this out, either directly or through one of his assistants, comes back, and then he could make it stick. Those Ten Commandments represent the durability of this culture, the Hebrew culture.

Fabulous Theology

To the extent that Christianity has survived two thousand years, despite its fabulous input—and I'm using the word *fabulous* here in Thomas Paine's sense. That's the original sense of the word fabulous. It does not mean wonderful or marvelous. It used to mean pertaining to fables. If you look at *Age of Reason*, on the cover page it deals with fabulous and true theology. Fabulous is theology that's pertaining to fables, and true theology is that which is compatible with observation. In any event, despite the fabulous inputs of Christianity including Immaculate Conception, God is man, man is God. And there's a third party involved called the Holy Ghost. Then they have what is called a three-in-one God, the Trinity, which they also have a three-in-one machine oil. [*Audience laughter.*] So I sometimes would call this the machine oil God. [*Loud audience laughter and AJG chuckles.*] The Immaculate Conception is one of the total mysteries of life. It's so much easier to believe in birth by a normal process, that it takes a tremendous amount of fabulosity and belief without reason to take the least probable explanation in favor of the one that's undoubtedly true.

Anyhow, despite its fabulous, you might say, farther accounts on Judaism, let's remember the basic parts of the Christian faith come from the Judaic one, and the Christians have not been so foolish as to drop the Old Testament, they just added the New Testament. The Old Testament is part of their religion. The part that is durable is not all of the fables about various things like the various angels and the various wrestling with angels. You ought to—no, I don't want to go into that. I was going to discuss some of the foolishnesses of the Old Testament too. [*AJG chuckles.*]

The Torah, the Five Books of Moses and the Law

There is something serious about the Old Testament. Not much, but a little bit. And that's the Five Books of Moses which, by the way, is all there is to the original Jewish religion, the Hebrew religion. The Five Books of Moses, which were not written by Moses, of course. You all understand that? Human beings don't normally describe their own death. [*Light audience chuckling.*] The books of Moses were not written by Moses, but in any event, the Five Books of Moses are called the Torah. That means the law. That is also not a religious, but a political basis to the Hebrew culture. That was the law. That included not stealing and not murdering.

The Durable Portion of the Ten Commandments

You know, that's about all there is to civilization, if it's going to be durable. To the extent that Christianity has had any durability despite its numerous wars and internal feuds amongst Christians, as you know, Christian has killed Christian as much as Christian has fought non-

Christian. The Reformation has produced more religious wars amongst Christians. But despite that, what is durable in Christianity is that they too accept the Ten Commandments. What is durable about Mohammedanism is that they too accept the Ten Commandments. And with all the flaws of the religions, all of the political disasters, all of the religious nonsenses, the really solid part of all three of the Western religions, the smallest of which is the Jewish one in terms of number of people, because Muhammadans and Christians outnumber the Jews many times, nonetheless, both the other ones are based on the Old Testament. The Koran of the Mohammedans is added to both the Old and New Testaments of Judaism and Christianity. The durable portion is, that it's basically against the religion in all three of these, not to steal and kill.

As a matter of fact, with less knowledge, I do know that what is durable in any Oriental religion is they also have at least a moral prescription against stealing and killing. Can you imagine how long a culture would last if stealing was the base of it? [*Chuckles in the audience.*] Universal misdemeanors, universal felonies, universal crime. Well, how long could such a culture last?

Why Monotheism and the Ten Commandments Are Inseparable

Now the political structures that have followed have twisted this to mean that the law has a monopoly on the very things that are prohibited, making it illegal for others to participate in murder and theft. Only the states may do it in the interest of keeping and preserving the order for the others. All of this, however, with all of its flaws, all of its coercion, all of its mystical inputs, has at its base a totally rational concept. A *totally* rational concept: the property protection that's inherent in the Ten Commandments. Which is why monotheism and the Ten Commandments are inseparable.

Monotheism Leads to a Quest for Understanding

Now you may say, "Can't a pagan believe in not stealing who does not believe in a single God?" Yes, he can. But it so happens that's not a religious concept. That has nothing to do with religion directly, but it so happens that the monotheism is stronger because it also leads to a quest for understanding of the sources of what's happening. Now there are two ways to pursue this: by trying to explain things by faith and trying to explain things by the scientific method. The latter is the newer and the more sensitive one and the one that ultimately has produced all of the value.

Commendation and Respect for Moses

In his own primitive age, it is quite certain that Moses was a scientist. He was a very good geologist. He obviously was pretty good at direction finding. It only took him forty years to get out of Egypt. [*Audience laughter.*] Before you laugh at this, please remember, that was not when the world was well-charted. He didn't have modern roads and compasses. He actually did pretty well. He was able to take a rather primitive people who had no past culture to speak of and transformed them to the most durable culture of all time, of all human time. It's the only culture that's almost four thousand years old and still is thriving and has outlasted all of the persecutions. That speaks very well for the structure upon which it was based.

When you can do something that lasts four thousand years you will deserve commendation four thousand years from now too. I hereby wish to state my commendation and respect for Moses. Let me also state my commendation and respect for Akhenaten, the Egyptian pharaoh upon whose shoulders he was standing. That's what I started out with.

The Fatal Weakness of Coercion

Now let's discuss that. Akhenaten, with coercion, was not able to make in the short term, his better-than-normal-for-his-period ideas to be accepted, let alone adhered to, with coercion on his side. Most innovators don't have coercion assisting them, but coercion opposing them. But that shows the fatal weakness of coercion. Coercion just cannot survive the grave. The coercer goes to his doom with his death. You may say, "Well, his work will be carried on by somebody else coercive." No, he wants to make his own rules. Please note that Khrushchev following Stalin did not preserve Stalin's reputation. In general, coercion has its own reward. Oblivion, posthumously. If the only reason we would have to remember Akhenaten is that he was pharaoh of Egypt, he would be remembered as one of Egypt's less-distinguished pharaohs. He didn't last long, his work didn't persist. He built enormous temples and they were subjected to barbarism even within the next generation after he died. He has remained through his primary property through another people, people that have never been, you might say, friendly. The Egyptians and the Jews.

Ethnic and Cultural Continuity

By the way, the present Egyptians are not the same as the ancient Egyptians, in case you're considering that the present Egyptians are not ethnically nor culturally the same as the ancient Egyptians. First of all, the present Egyptians are Semites and the ancient Egyptians were Hamites, and the culture is not a continuous one. They are living in the same geographical territory and since the country was called Egypt then and now, they are both called Egyptians, but they're two entirely different peoples and two entirely different cultures. There is no continuity there. Whereas there is both an ethnic and a cultural continuity between the Jews of three and a half thousand years ago, and today. The same is true for example, in Italy. The Italians are not the descendants of the Romans, neither culturally nor ethnically. There's much change in the cultural attributes of the people. The languages are relatively similar, but then again, so is Romanian, and Portuguese, and Spanish, and French similar to Latin. These are all derived from the Latin language. As a matter of fact, the English language is approximately 60% Latin, and in Greek too.

And certainly the English people are not ethnically the same as the Roman people, but there was a connection. The Romans once conquered the island which is now England [*AJG drinks*] and they brought their culture there, and the language partly came from that period, mainly from the Norman Conquest of Great Britain in 1066. The Normans are French and they spoke a romance tongue. Romance is a Latin-based tongue.

The Durability of Monotheism

In any event, we have a very beautiful example here of the durability of monotheism, which is a most durable idea. It's based upon something that ultimately led to, and is certainly integrable with, the Newtonian scientific integration. It's certainly one of the main inputs into it. When Newton stated that he was standing on the shoulders of giants, certainly Moses and Akhenaten are two more of the giants he was standing on the shoulders thereof.

Coercion Is Always Emotional

Without coercion, the idea is durable. But before an idea can survive, it must be rational. Emotional things do not last. Coercion, incidentally, is always emotional. Not all emotion is coercive, but all coercion is emotional. Emotion is acting on the spur of the moment, based upon feelings, a feeling being an intellectual response to an external stimulus, to the senses. That means the five senses: sight, sound, touch, taste, and smell. An emotional feeling is a reaction to such an input. It can also be a response to your previous intellectual inputs and your attitude towards them.

A Rational Thought Process Takes Time

What is your mind doing? What thought processes is your mind working on to generate to respond to your previous inputs about things that have happened to you in your recent memory bank? You can also draw on your long-term memory to correlate it with what has just happened and it's the immediate response that your mind has in the form of a thought which is brought about by a response without any durable and, you might say, rational thought process. A rational thought process clearly is the one that adheres to and conforms to the method in the scientific method, namely, observation, hypothesis formulation, extrapolation and finally to corroborate what you have come up with, further observation to check out that you have not drifted off into fantasyland, but you are still in the observational, and therefore by definition, real world. Clearly that takes time. To go through the scientific method procedure is not as quick as to what your immediate thought process is in reaction to something that has happened to you on a relatively recent basis.

An Emotional Response Is a Short-Term Reaction

As I say, touching a hot stove, you don't have to go through the scientific method to know what had happened. You got burned and it hurts. As a matter of fact, it just happened to me this morning. I burned this finger. I wasn't thinking about something being hot and I touched it with that finger. It's got a blister on it now. And that does not require any long process of thinking to remove your hand. You think about what you did wrong later. [*AJG chuckles.*] Then that might prevent such an occurrence. I assure you I will not get burned in the same place again. [*AJG drinks.*] This was a new place. [*Audience laughter.*] It was the inside of a new stove and I didn't know what was the hot element. I just found out. [*AJG laughs with the audience.*] But I will remember it. Not a cheap way of learning it, by the way. Now the point I'm making is, an emotional response is a short-term reaction to something that has happened to you recently, and the recently can be either in terms of the five senses, or inputs that you have acquired from observing other people, their interaction with you and your reaction to their behavior, which can be friendly, unfriendly, anger, joy.

When your stomach is growling because it hasn't been replenished recently, you call it you're hungry. That's an emotional reaction to a physical reality. There's not adequate fuel in you to keep you comfortable and so you are responding to certain sensations which are inside of you which you are feeling, and your brain is matching it with a concept called hunger. Now, to solve the world's food problem is a rational thing. To solve the fact that you are hungry is an emotional reaction, which may or may not be satisfied based upon whether you find food. Those who don't find food easily will make up a new way, or they think it's a new way, and that's to steal, and they make that a program. That's a short-term concept. That's because they're hungry so it doesn't matter where they get it so long as they eat. When a person commits a crime, he doesn't care about what the long-term consequences of that will be. All he wants to do is gratify his immediate desires.

Long-Term Negative Consequences of Emotional Behavior

When a desire is a short-term one to fulfill an emotional inadequacy, it can lead to erratic conduct and behavior. As I pointed out in this past series of lectures which I was discussing emotion, emotion is by itself not a bad thing, per se. But if it becomes a supreme dominant mechanism of influencing human conduct, it will be bad in its consequences. Because all of the actions will be short term and not thought out as to their long-term consequences. Therefore, there will be no correlation with observational reality because the fourth step of the scientific method certainly is not being followed, and most likely these are the first three. People are reacting to impulses to short-term attitudes which they cultivate based upon their thought patterns that come about from short-term inputs. Emotions are always short term.

The Positive Factors of Emotions

However, what I have said is that there are two things that can be said about emotions which are positive. One is, it is not an automatic factor, that something because it is emotional is automatically wrong. It is possible to be emotional and still retain your rationality. It's not easy. Under emotional upheaval or disturbance it is not easy, but as long as you subordinate the emotional response to something that is not incongruous with the long-term reality of what you want and seek and can obtain by rational methods, then the emotion is not necessarily automatically wrong. If, through emotion, you can satisfy the release of frustration without injuring someone else then it is not in any way to be condemned. Quite the contrary, it is therapeutic.

A Proper Expression of Emotion Necessary for Human Stability

As a matter of fact, a proper expression of emotion is necessary for human stability. That's why I made rather critical comments about people who are not able to express their emotions properly and who bottle them up and who have a totally seemingly unemotional, or even phlegmatic personality. Such a person is probably committing psychological suicide while he's doing that. It is dangerous not to express your emotions. The question is, can you do so in such a way that you don't commit coercion at the same time? Or you don't produce a long-term effect that is either directly coercive or could lead to coercive misapplications.

Temper, Coercion and Crime

I have discussed the subject of temper. There are two kinds of temper. Superficially they look the same. They are not the same. Not all things that look the same are the same. I have expressed, for example in V-50, something which is a discussion of coercion. I pointed out physical coercion is affected if property is interfered with of another. When a person interferes with someone else's property intentionally, that is coercion if he attempts this. If he succeeds, it's called a crime. For those who would have problems with the word *intention*, and I don't blame anyone having trouble with that one, because how do you define intent? The problem is that most people cannot and do not.

Intent

The most sensible way of defining intent in volition, if you remember, was from its after-effect. If after property has been interfered with the one who did the interfering *voluntarily* is ready and willing to make restitution, and does so, then it's not intentional and therefore it's not coercion and the effect is not a crime because it has been reversed. Voluntarily. Now if you have to track him down and threaten him, "You repay me or else." Then it's a crime. If the person does not have to be tracked down and says, "I injured your property. I'm sorry. It was an accident. I didn't mean to do it. I'd like to pay you for it together with all costs that you bear, including your time lost, the interest value thereon, the aggravation I caused you." That is certainly not coercive. And yet, from an internal point of view, it might have been. This is the time now to bring up this separation.

If you recall, I defined intent back in V-201 originally. I think it slopped over into V-50 by now. I don't know if Dr. Snelson put it in. That's one of the partially good and partially not desirable things. Sometimes from more advanced courses the elementary courses have a slopover effect. For example, I don't know how much of a slopover effect there will be from this psychology course in this year's 201. [*AJG chuckles.*] I'll try to reduce it to a minimum. In any event, you have to define two different kinds of intent: operational and psychological. Remember I made that distinction? ["*Yeses*" *in the audience.*]

Psychological Intent:
A Privacy of What's Going on in One's Own Mind

Psychological intent is what is going on in a given person's thoughts prior to articulation. When you're thinking something but have not articulated it then only you know you have thought of it, and until we have thought-reading mechanisms—which I don't think would be necessarily for the betterment of the human species. That's a privacy I don't like to be intruded upon. I don't welcome that. I think it's probably possible through encephalograph development. I don't think it's desirable. It's something that could be easily abused and thereby dangerous in the hands of coercers. [*AJG drinks.*] But to the extent that thoughts cannot be at this time directly read, that's a privacy of what's going on in one's own mind. If you haven't articulated it, maybe even you don't know really what you believe because you think you know.

A Necessary Prerequisite for Testing Comprehension

But as I have pointed out, you never understand whether you understand something until you articulate it and see how well it goes. Especially if you have an intelligent person to talk with. Now, if you're trying to explain something to a person who is a moron, and he doesn't get it, that is not a reflection on your thinking. But if you're trying to explain to someone who is himself not only not a moron, but actually quite intelligent and he can transmit it to someone else who is intelligent, then you get an idea of how well you understand what you've been thinking. That's one of the things I have explained is a necessary prerequisite for the testing comprehension. That's, if nothing else, one of the excellent ways where the scientific method can be applied in determining your own self-comprehension of anything. So whenever something is a thought bottled up in one's own head, it may be a rational thought, it may not be a rational thought, but if it's unarticulated it is not certain yet.

Getting Even: A Short-Term Emotion; A Counterfeit of Justice

Now let's say a person, out of anger, wants to injure someone else. I gave the classic example in V-201 driving over someone's rose bushes. Of course, this is not a very significant piece of property, in general. It could be far worse. I just used that as an illustration. Well, sticking to that illustration, let's say someone does intend to—he's sore at someone, he drives over his rose bushes. And I say, that will in itself release him from his frustration of whatever made him angry. "Okay, now I've done it. [*AJG pounds the podium.*] I've gotten even." The whole concept of getting even, by the way, is a short-term emotion. By the way, that's a counterfeit for justice. Getting even is not an equivalent to justice.

Pointing Out Injustice in Order to Achieve Justice

I do believe in justice. If someone has committed a crime and nothing ever happens to reverse that crime, that is injustice. For example, I don't think people should get away with crimes. Not Newton, but I— I'm making sure that Linus doesn't get away with his. Linus will be more famous as a result of Galambos than he was as a result of Newton. He injured Newton. Newton was not on a vendetta the rest of his life to punish Linus. Neither am I, but I am trying to create a mechanism wherein the Newtons do not have Linuses in their future. The purpose of this theory is not to punish Linus, you understand. [*Some chuckles in the audience.*] The purpose of this theory is to remove the disclosure barrier and make it possible for Newton types to pursue happiness. It will give a flatlander a distorted conception of the theory of primary property, but it's quite correct to say the purpose of the theory of primary property is to make sure that the Newtons of the world can pursue happiness properly, unmolested. You say, "That's a hell of a minor goal."

Well, when Newton succeeds in pursuing happiness, the rest of the human species will find it easier also. Do you see the connection? If not, I will refer you to V-201. [*Some chuckles in the audience.*] The connection is very clear and simple. This is not altruism for Newton's sake. I never met the man. I'm not even sure he'd like me. The point is, it's justice. The purpose of this theory is to make a positive improvement to a positive person's achievement. Not a person, but persons in general. If in the process, those who did something wrong get remembered for their wrongness, that's justice.

An Act of Vengeance Cannot Have a Long-Term Advantage

Now, the significant fact is that it is not possible to commit an act of vengeance, such as that rose bush episode I was discussing, in such a way that this has long-term merit. It may have a long-term damage, but it cannot have a long-term advantage. The only advantage it can have is for the person who is angry to release his anger in a way that is, from his point of view, an external manifestation of his anger which produces a coercive effect upon someone else. In other words, it injured someone else's property. All right. But I pointed out that this has a therapeutic effect too. The person who was angry, this will now satisfy his anger and in general it will be anti-climactic after that. "Okay, there I've done it. [*AJG slams his fist on the podium.*] Now I feel better."

Crime and Restitution in Relation to Psychological and Operational Intent

Now if that person, after having done so, decides, either because of state justice which exists now, or because of market justice which exists both now and in the future and ultimately will survive and supersede state justice, if for either of these reasons he wishes to make amends before he's caught, and he goes up and rings the doorbell of the person whose house he has destroyed the rose bushes thereof, or if they're not at home, he writes them a letter or leaves a note and says, "I drove over your rose bushes and I plan to make restitution and I'll pay you for it." According to operational intent, this is not coercion because the after-effect undoes the damage.

There still could have been psychological intent, which means that's his own internal motivation which he may not wish to disclose, he might be ashamed of it by this time. He might not say, "Well, I meant to hurt you, but I'm sorry. Now I did it." Then he undoes the effect of it. Or he may not be even willing to say that much, and he says it was an accident when it wasn't which is a lie. But as long as he pays for it, from the standpoint of volitional science he has erased the crime, so long as it's reversible.

The Possibility of Reversing a Murder

This does not include murder and things like that. Although the time will come as I pointed out in some of the speculative aspects of V-201, when we look ahead into the future technology, it's quite possible, probably, to reverse a murder, not now but someday, which will hasten the recovery process. In other words, if it's possible to reverse it, then the murderer might also be the rescuer of his victim by seeing to it that he's brought to safety as soon as possible so that it isn't permanent.

A Temper Tantrum Not Dangerous When It's Noncoercive

But the very effect of the psychological doing of something which is a short-term emotional outlet, as I say, is a therapeutic effect. How many times have you been angry and when you've vented your anger you feel better? Has this ever happened to you. ["Yeses" from the audience.] By the way, that's the effect of blowing one's cork. A temper is not dangerous as long as property is not affected of someone else. It becomes dangerous when in a temper tantrum somebody injures someone else's property. Then it is criminal, especially when it's not reversible. Especially when it's not reversible. When Hitler gets angry, he launches an army into Poland and starts off World War II. That's not reversible, the catastrophe that he committed. Of course, he was preparing for it all along. That was premeditated and I was exaggerating here. I don't think he had a temper tantrum to start the war, but he was well-known for his temper tantrums. As a matter of fact, he was called a *teppischfrescher*. That means a rug chewer. I don't know if this actually happened, but he's reputed to have it been so mad, he got down on the floor and chewed the rug. [*AJG chuckles with the audience and takes a drink.*]

Vlad the Impaler:
An Example of a Coercive Manifestation of Temper

A better example is Vlad the Impaler. You know who that is. Dracula. Not the vampire, the real one. Like in the 15th century, there was a Romanian Prince, Prince of Wallachia, whose name was Vlad. V L A D. His nickname was Dracula, which means "little devil." It's his name that Bram Stoker put upon the vampire character. Vlad, if you can call this an accomplishment, is reputed to have murdered one hundred thousand people, not in battle in war, although he had those too, but by individual killing. He would impale them on a pole. He had a pole, up from the ground, sharpened the top of it, and he would impale a person right on the pole. This to him was a form of release for his anger. He had a bad temper. [*Audience laughter.*]

As a matter of fact, there are any number of stories about Vlad the Impaler. That was his title. *Vlad* was his name, *the Impaler* was his title. You know, like they have Alexander the Great and they have all kinds of titles, So-and So the Bold, So-and-So the Fat, So-and-So the Ruthless, and so on. That's usually a good name for a ruler: "Ruthless." [*AJG takes a drink.*] I'm not quite sure what that means. I know what ruthful means. If you don't have any *ruth*, I don't know exactly what that means. [*Audience laughter and AJG chuckles.*] This language has its peculiarities, as I mentioned.

Anyhow getting back to the Impaler. That was his title, he used to impale people. I understand he impaled one hundred thousand people which is 20% of the population of the domain he ruled. Although many of them were foreigners, they would just come by for a visit and they'd never leave, they'd be impaled. You wouldn't have to do very much either. He would ask you a question and he didn't like the answer you gave him and—I didn't bring the book I have on this Dracula with me, but there were quite a few interesting stories there which indicated how little provocation the man needed to impale someone else.

Well anyhow, this was the way he released his anger. Needless to remark, this is a coercive manifestation of temper. There is nothing excusable about this. For those of you who don't know this, I'd like to mention just as a matter of both historical interest and personal pride, it was the king of Hungary, King Mátyás, who had Vlad the Impaler in prison for twelve years. That's the same king I mentioned in the history course. One of the few quality people who ever occupied a throne. King Mátyás had Vlad the Impaler twelve years in prison. I don't know what got him to the release him. It must have been a moment of weakness. [*AJG drinks.*]

Moses:
An Example of a Noncoercive Temper Tantrum

Anyhow, I gave that as an illustration because that's a very large one, killing a hundred thousand people based on your anger. On the other hand, what would you call it when Moses came down from the mountain, assuming this happened, and when he sees his primitive tribe relapsing into pre-monotheistic paganism under the leadership of his own brother, and he got so damn mad he took the stone tablets and he threw them down on the ground and broke them. Would you call that hot temper? I sure would. Who did he hurt? Nobody. He hurt no one, not even himself. It released his anger, and those stones belonged to him.

It doesn't say in the Bible that he cracked the stones over his brother's head. [*Audience and AJG chuckles.*] It doesn't say he shot anyone. Well, in those days they didn't have guns, but it doesn't say he executed anyone. If he did, that would be wrong. As long as he destroyed only what was his own, it's a question of did he injure himself? No. The release of his frustration was more important than the preservation of those stones, and he could go up in the mountain and get another set of stones. And he brought them down and the rest is history.

Moses's Temper Tantrum

The Jews survived all these thousands of years with what he imparted to them, and so has his memory. However, it's mixed up with fables and legends and myths and various forms of ridiculous superstructures to religion. It is not a myth that the Jews have existed for four thousand years. It is not a myth that they produced people of the quality

303

from Moses to Einstein. Sigmund Freud, while we're on the subject. It is not a myth that the basic effect of the Jews has been to create a new culture. The whole Christian religion, for whatever it has done in terms of durability, the fact that much of Western civilization came through Christian countries, they're using the Old Testament and the Ten Commandments. It's sort of a watered-down monotheism with sort of a three-in-one polytheism there. It is not a myth that the durability has existed.

That temper tantrum of Moses was not to be compared with Vlad the Impaler. You recognize the difference? Was it necessary for Moses to do it? You could say, "Well, if he were a better adjusted person he would have just said nothing," and, "Christ under these circumstances would have preached a Sermon on the Mount or something, or turn the other cheek."

I don't know that that would have done a better job. Maybe that's what got the primitive tribes shaped up. [*AJG drinks.*] They saw the boss was angry. [*Light audience laughter.*] The next time he came down they paid attention to him a little better. Four thousand years better. Those four thousand years of culture are real. Moses was successful. And Olympian comments, "Well, he didn't have to lose his temper. He would have been a better man if he hadn't." I doubt it. I think he would have been a weaker man.

Temper Outburst:
Therapeutic if No One Else's Property Is Invaded

It takes a man who is psychologically unstable to be able to control himself under these circumstances. You may find that a weird thing to say. A temper outburst—not tantrum—a temper outburst is more stable than one that isn't, than one is smooth, calm and collected. A man who is smooth and calm and collected in the case that doesn't call for calmness, might have a release of his temper in a different way, in a slower and more harmful manner. He might do something which would either produce for himself his own extinction. It's possibly because Newton didn't necessarily publicly display his temper that he went to the mint. I would say it would have been better for him to blow his cork. I emphasize: if no one else's property is invaded, that's therapeutic. We are not stones. We shouldn't behave like stones.

It's Normal to React to Physical and Emotional Pain

If we have been injured it is not proper, it is not normal, it is not compatible with the nature of creatures that have emotions not to say that you are injured, and saying it has many ways. For example, when you burn yourself, you either scream out loud, "Ouch!" Or you have an internal, silent effect. This morning when I burned my finger, I was alone so I didn't say, "Ouch!" If there had been an audience, I might have. [*AJG laughs with the audience.*] No, this is clearly done for an audience. [*AJG chuckles.*] What's the purpose of screaming out if no one hears you? I heard it though, inside. [*AJG drinks.*] It's normal to react to pain. That's emotional pain as well as physical pain. A burn from a hot stove is a physical pain. Someone having done you an injustice is an emotional pain. It is normal, it is necessary, to express that pain in some manner as long as it is not done incompatibly with reason. Which means you do not attack someone else's property.

Coercion and Crime:
Not Healthy Responses to Emotional Pain

One of the most frequent ways of expressing an emotional reaction to your being injured is to get even and thereby injure someone else, usually not the same person who injured you. For example, what happens frequently in present-day flatland is, let's say somebody steals from you. So you steal from somebody else. "I'm getting even." So A steals B's car. B then is deprived of his car so he steals C's car. "Now we're even." No, we're not even. You just carried the criminality into another step forward. C didn't steal your car. That's not getting even. That's not a normal response. Let me put it this way, that's not a healthy response to an emotional pain. It isn't even proper to steal from A. If A steals B's car it isn't proper for B to steal from A something else. To recover your own car with restitution for all of the incidental damages, that's justice. To kill A is not justice. Now, if A gets killed in the process of stealing B's car, that's not murder either.

The Quest for Justice in Our So-Called Political World:
Compounding the Problem

That's another point that has been not understood in our so-called political world and quest for justice in so-called political means. For example, this was called to your attention in V-50. The same physical act can represent coercion in one case and not coercion in another case. If one person walks up to another, A walks up to B and clenches his fist and punches B in the nose and B's nose is broken, or just is hurt, that is clearly an act of coercion, is it not? That's interfering with someone else's property. Okay. Even provocation doesn't justify it. You may say, "Well, B flirted with my wife," and A then punches B's nose and he breaks his nose. Well, that's not the solution for flirting with wives. [*AJG chuckles with the audience.*] It might be the thing that many people will do, but that's not a solution. That's just compounding the problem. [*AJG pauses to take a drink.*]

The Same Physical Act Can Be Coercion in One Case and Not in Another

I also gave the illustration, maybe not an everyday occurrence but it can happen. Let's say A exercises on his front lawn, his own property, and in the process he swings his arms around like this. B walks onto the property uninvited and not looking at where he's going, he walks into A's rotating hands, and B's nose collides with A's fist. B's nose is broken. A's fist is what broke it. This is not coercion on the part of A. Believe it or not, this is coercion on the part of B. B was uninvited on A's property. He was not there by invitation, he was trespassing. Trespassing is a crime. To be on someone else's property without invitation and without permission from the owner is trespassing. Any damage that comes to you while being on someone else's property without his permission is your fault, not the property owner's fault.

This is not quite the same as the way the state deals with people. The state would not look at it this way, but the state isn't always right. [*Laughter from the audience.*] You can chalk that up as the 20th century's, one of its monumental exaggerations by underemphasis. This state is not always right. You will please note, however, the physical act is identical. A's fist broke B's nose. In one case it's coercion and in the other case it clearly is not. The coercion is on the other side. B walked into a place where he had no business being there and any accident that befalls him there is his own fault for trespassing. That is clear is it not, after V-50? And you don't even need V-201 for that. So the point is, the same physical act does not always mean the same moral and rational quality.

A Temper Is Not the Same as a Temper

A temper is not the same as a temper. Moses' temper in breaking the stone tablets when he was angry with the primitive and pagan tribe reverting to its paganism when he came down the mountain. I'm saying this as though it were a fact, I don't know if it's a fact, it's a legend. It's a legend in a book of fables. But let's say it happened. The whole world has essentially thrived on this fable so far, so I'll cash in on it for a psychological illustration whether it happened or not. And if it didn't happen, something like this could happen. When someone is angry about somebody else's behavior and he demonstrates that anger in such a way without injuring the other party's property, that is not coercion, and it does not have the same quality. And yet, that's certainly not the same as what Vlad did by killing people when they just looked at him cross-eyed and he took that as an insult.

Moral and Rational Analysis of Events

Nonetheless, they have the same internal effect upon the person who's releasing his anger this way. It has a personal therapeutic effect. But if your therapy causes someone else damage then your therapy is unwarranted. If the price of your recovery from what ails you is that someone else should be injured, you have erred. That doesn't mean this doesn't happen. It means it should not happen. Now there is where we have to take in the moral and rational analysis of events. Psychology, in general, is not a subject that deals with moral issues. This psychology does. Psychology, in general, on a classical basis, was not a study in rationality but rather mainly of emotions. Usually, emotions have led to certain kinds of damages to oneself and possibly to external people.

No Way Man Can Be Unemotional and Still Be Human

There is no way man can be unemotional and still be human, and that's one thing we must start off right away and understand. To take the unemotional person is to take no person at all. Then it's not a volitional being. You say, "Well, man ought to cure himself of having emotions." That's not part of the real universe. It is not a possible and probably not even a desirable attribute. I do not look upon emotions as undesirable. After all, that's what makes people interesting. If everybody looked like a stone, can you imagine how dull it would be to have

307

conversations with people and the most exciting things that you can imagine, like going to the stars and to the galaxies and talking about in in a dull monotone [*AJG starts talking in a dull monotone voice and audience laughs*] and give a lecture on going to the stars and how it is a feasible scientific endeavor to go to the stars [*audience laughter*] and that there is a possibility by applying man's knowledge of physics for man to leave the earth and first to go to the moon. And then after the moon, there are more distant goals we can go to [*loud audience laughter*] and discuss the thermodynamics of rockets and discuss the interesting scientific problems that can be solved thereby. [*AJG switches from his monotone voice back to his normal voice.*] And never even to be glad that you're doing it. Not to show excitement. The beauty, the grandeur of the universe that we have the ability to enter upon and grow with—grow into and with.

Emotion Is Not the Enemy of Man

Gladness is an emotion too, you know. Excitement is an emotion. How can a person live if he doesn't ever get excited about something that is interesting and beautiful? How could you not be excited when you see a beautiful sunset? How could you not be excited when you see the Grand Canyon or Niagara Falls for the first time? If you're not excited at something like that then you're not even alive and you should have been born a stone. [*Audience laughter and AJG takes a drink.*] So I emphasize, emotion is not the enemy of man. What is the enemy of man is when emotional, which are short-term reactions, become the dominant effect upon human conduct and human civilization. Also, in that case, the worst emotions, namely the coercive emotions will then prevail. Not all emotions are coercive.

Rational Emotions Can Produce Progress

As a matter of fact, I discussed a list of sample emotions which are basically compatible with rationality. That doesn't mean they're always rational, but when they are rational they produce progress. For example, the emotional reaction to not having enough food in your belly which is called hunger can produce rioting and food riots. It can also produce among more competent and moral and durable and rational people, scientific agriculture. It can produce the technology of fertilizers. It can produce the harvesting machine. It can produce the various innovations of the recent modern world that have produced more food per acre than in previous centuries.

Why is it that this relatively small country of the United States which looks vast to us because it's a long distance to walk from coast to coast. But in terms of the entire earth, which is a relatively small planet by itself, the United States is a small nation in relationship to the whole of the earth both in geographical territory and also in population. The United States' total population has never exceeded 7% of the human population. [*AJG drinks*.] Why is it that the United States has produced the abundance of food whereas the much larger population in China has produced famines? That's their main product. Are Chinese people basically stupid? As a matter of fact, they're not. They have a rather ancient culture too. And with more people and just as long a time they haven't done as well with it as the Western culture has with the Judaic culture.

The Judaic Culture's Influence on the Western World

Please note how few Jews there are. They were never a proselytizing people. They never sought converts. And with a very small population, they influenced the whole Western world. The entire Western world is Judaic in culture because of its monotheism.

You Can Destroy with Marxism, Using Newtonian Technology

Why has China with a much larger population and equally old culture—and certainly not with stupid people because there are many geniuses who are Chinese, both in the ancient and the present world. What's wrong there? They didn't have a healthy concept towards property, did they? How come that in China there's always a famine? United States, we export food even now. Even now that we're talking about a world hunger problem. Ultimately the United States will be hungry too. Sure, we exported everything! [*Chuckles in the audience*.]

What does the United States have to do about its food, in the absence of external feeding the rest of the world? I mean, it's a little too much. I mean after all, two hundred million people in America have to support eight hundred million Chinese and five or six hundred million Indians, and I don't know how many millions in Africa, and so on. This is getting a little burdensome. To me at least. [*Chuckles from the audience as AJG drinks*.] Why in the hell aren't they doing something? Answer:

they don't know what property is. They worship Marx. That's the wrong White man to worship! [*Audience laughter.*] They picked on the wrong boy. Yet Mao Tse-tung is the leading exponent of Marxism in the world today. How come he couldn't have picked out Newton? I'll tell you why. That wouldn't get him into power. You can't rule with the law of gravitation. I answer my own questions. Sometimes at least. [*AJG chuckles.*] You can't rule with Newtonian physics. [*AJG takes a drink.*] You can destroy with Marxism, using Newtonian technology. That they can do. When you harness the science to be the lackey of the state. All right.

A State Subsidy Is Coercion

As a matter of fact, on the subject of hunger and food and all that, it's very interesting that the United States has had the political policy for decades to subsidize the nonproduction of farmers. They produce so much food that the United States paid farmers not to grow crops so that there wouldn't be such an abundance, super abundance. And how many Americans work on the farm? It's a shrinking number. Whereas it used to be that America was mainly rural two centuries ago, now it's mainly urban and I believe the farm population is well under 10% and it supplies 100% of the people with supermarkets where the variety of foods so far—I have to say so far because they are lousing everything up—but so far, there has been a great variety and abundance of food in this country, to the point where the state felt it was compelled to bribe farmers not to grow food. As a matter of fact, they coerce them not to, and paid them subsidies so they wouldn't go under. By the way, what be the normal thing to do if farmers were not subsidized? What does this mean, by the way, in the absence of state subsidies and state coercion? A subsidy is coercion, you know why? Because what they give away to somebody they take from others in taxes.

Without Subsidies Natural Market Mechanisms Would Provide Stability

What would have happened if they hadn't subsidized the farmers? The price of the farm products would have fallen because of the greater supply over the demand. And the farm products would have fallen in prices. Some of the farmers wouldn't have been able to make a go of it and then they would have to, some of them, the ones who are less profitable on their farm operations, would have to stop farming. Sell their farms for what they can get and go do something else. Go to the city and do something else, which would have produced a natural solution to the problem. The less profitable farmers which in the long run might mean the less efficient ones, would stop being farmers. That would leave the more efficient ones remaining in the field and they'd feed the population. That would stabilize the falling farm prices because after a while, there would not be an overabundance, but there wouldn't be a shortage either. It would come to a natural balance.

And let's say the population of cities expanded and the rural population didn't expand, then there would be a tendency to have a shortage of food. That would mean the farm prices would go up. That would attract people back to the farms so more people would become farmers and they would grow more food. A natural market mechanism would keep this thing in nice balance, but that's too simple. That's too simple. That's harmful to mankind for letting themselves figure out what is the most profitable thing they ought to do with their lives.

Famine in China: The Inability to Adapt to the Natural Mechanism of Solving Problems

Now with such a system where a small minority of the population produced a super abundance of food, whereas in China it's the other way around. Probably over 90% of the people are—I don't know if that's accurate today—but I think it probably still is certainly over 80%, and possibly over 90% of people are in the agricultural domain in China. With most of the people in agriculture they still have famine, and they don't have enough to eat. And that is simply because of the inability of the rest of the world to adapt itself to the natural mechanism of solving problems. It means to leave people alone.

The Self-Esteem of the Chinese Peasant vs. American Farmer

Now in terms of the internal effect upon a given person's self-esteem, who do you think has higher self-esteem? A Chinese peasant or an American farmer? And yet, they're basically in the same business. I didn't even call them the same thing, because they're not both farmers. A farmer is an independent businessman. A peasant is a peasant. A peasant is a slave of the soil and it's a term that stems from the feudal system, which fits well into a country such as China. China is basically a feudal country. Communism has merely aggravated it. What do you suppose is more internally satisfying in terms of self-esteem, called the ego? Who do you think has a bigger ego? A man who can support himself and make a profit? Or a man who works backbreaking labor for unending hours and still has a starving family to show for it?

Emotional Attitudes Prevail in All State Policy

Now the entire mechanism of human conduct is a balance between emotional and rational attitudes. Clearly emotional attitudes predominate in both quantity and also in direct effect upon the structure of mankind in his present social structure. Almost all of social structures are totally dominated by emotional attributes. What do you think this whole Marxian *need* crap is about? "To each according to his need." That influences state policy. All state policy, whether in Russia or the United States—more in the United States, I might add—is based upon the Marxian doctrine "To each according to his need."

The Justice Mechanism for Marxism Is Starvation

The reason I say more in the United States than in Russia, is Russia already had a dosage of this in the 1920s and they got famine as their just reward. That is justice for Marxism. The justice mechanism for Marxism is starvation. That's exactly what they got. When you pay people according to their need you get a degenerate culture. There has never been any social experiment, whether on a grand scale such as Russia, or on a tiny scale such as the Pilgrim Colony in the initial landing of the pilgrims in what is now Massachusetts, or on social experiments subsequently in various 19th century socialist communities where without coercion people voluntarily banded together to form a community where everybody owns everything in a commune. The kibbutz concept in Israel is that. If they weren't bailed out by their Knesset, which is their parliament, and also by the American Jewish contributions to their damned experiments, they'd be starving. All of these kibbutz's survive only by external handouts.

The Direct Correlation between Psychology and Society

Nowhere has ever a nonproprietary experiment, whether with coercion or without coercion, had the slightest success in influencing production positively. And if production is not positively influenced, civilization is doomed. There's a direct correlation in psychology to what kind of a society you have. I think this is somewhat not considered in other versions of psychology, which is not integrated with everything else. It's just dealing with everybody else's own internal mishaps and their inability to adjust and cope with their environment.

That's why this psychology is not meant as a substitute for or even in competition with classical psychology. Plus the fact that it is not interesting to me to argue with other people. I'm putting this forth as a psychology that faces the real issue. Human internal motivation and the society with which it interacts. The only concept of psychological evaluation that is compatible with a durable society.

The Concept of State Provides Unhealthy Psychological Conditioning

There's no point in discussing a person's emotional upheavals and problems if he doesn't interact positively and on a survivorship basis with respect to the rest of culture. If the rest of culture sinks that will certainly interfere with his psychological problems. I imagine that under Hitler's Germany there were an enormous number of people who were psychologically disturbed except that the gas chambers overrode their psychological upheavals and little was heard about the emotional stress they were under, and all you could see was the external murders. But I imagine prior to their death they had plenty of psychological problems that make the average victim on a psychiatrist couch look like he's well-adjusted. Can you imagine what that did to people? What it did to the people who weren't in concentration camps? The knowledge that all you have to do is look at Hitler cross-eyed and you'll be in a concentration camp? Or fail to say *Heil Hitler* when you say good morning? Or you meet somebody in business or on pleasure. If you fail to say *Heil Hitler* for a greeting, nothing more than that could get you into a concentration camp headed for a gas chamber.

What kind of a psychological conditioning do you think the German people were in under this condition? What kind of a psychological condition is it when a parent is afraid to talk freely at the dinner table to his children because the children are taught in school that the greatest glory that can come to a German is to serve the state and there is no way more glorious to serve the state than to see to it that all enemies of the state are punished, and if it happens to be your own father or

313

mother, all the greater glory to you for turning them in. And therefore parents were afraid to tell their children what they thought. A man would be afraid to tell his best friend what he may think. A man might be afraid to tell his wife what he thinks, or vice versa. That's some environment for a man's thoughts to be properly processed in.

The Harmful Psychology of the State Is Resolved with This Theory

Psychology in the classical sense and psychiatry as a profession is a luxury man can afford when he has relative tranquility so that those who have relatively minor problems have an opportunity to run for help. The eighty million people in Germany, including and especially the Jews, didn't have opportunity to get psychological relief. And yet, the psychology of Hitler's Germany is very easily resolved by the theory that I'm teaching here, because the psychology that's taught in this course is fully compatible with the volitional theory from which it came. And I'm stressing that though I am not trying to say that man should not have emotions, quite the contrary, I'm saying it's necessary that he does. It is necessary to regard that emotions are what they are. They are short-term manifestations of human thoughts. And short term is always less important than long term.

Rational Output of Man: Producing Civilization

A rational output of man is what produces civilization, and the emotional attitudes of people are either properly under rational structure or they are not. If they are not in a rational structure, it will be on a non-Hitler mechanism basis. You will produce a psychotic generation, which is what we've got right now, incidentally. Or in the presence of a Hitler, he will harness this reservoir of psychoses. His reservoir of unstable, incompetent, unable-to-cope-with-reality population and they are ready victims for a dictator's unification of will under his command. Unstable people respond to dictators automatically. Stable people do not. Stability comes from a harnessing of emotions to fit rational conduct. You may say, "Well, that's only a few people who can do that." That's absolutely correct, and those are the hot end people. Either civilization stems from the hot end or we're running fresh out of civilization.

I'll continue after an intermission. [*Applause from the audience for eighteen seconds.*]

PART B

The Emotional Component of Man Is Here to Stay

On the subject of emotions and rationality I've already pointed out that the emotional component of man is here to stay because if it isn't then man will have not only lost his ability to be angry, he will lose his ability to be glad, he will lose his ability to seek, to know. The rational content of knowledge is one thing. The excitement one gets moment by moment is a series of emotional pulses to want to continue to seek what you don't know. The two can be very well co-ordinated.

The problem is that most of the damage in the world is done by people who have emotions that are not co-ordinated with rationality. It is not that those who have rationality are unemotional, it is that those who are rational make the emotional component of their life have a proper juxtaposition with rationality so the two are working in harmony rather than at cross purposes. As long as emotion is looked upon as a therapy for negative emotions being expressed, as a form of excitement and therefore incentive when it's positive, not only can you say nothing wrong about emotions, it's a very necessary attribute of a living volitional creature.

It Is Not Possible to Have Dispassionate, Objective Knowledge

I make all this discussion emphatic because there have been many pseudo-scholars that think that to be a scholar you have to be dry as a bone, a fossilized bone. Many people think that scholarship means to be humorless. What they like to call objective, what they call dispassionate, impersonal and therefore you lose the flavor of knowledge. *Knowledge*, by my definition, is the interaction between the observer and the observed in such a way that the observer acquires meaningful inputs about the truth of what he is observing that make sense to him and that he can count on and harness this input for future decisions about what he is to do. Therefore, it is not possible to have dispassionate, objective knowledge. I have gone into that in other courses, so I shan't belabor the point of how foolish the concept of objective knowledge is.

The very concept defies a proper definition and those who mean objective knowledge, or objective view of the universe, what they really mean, if they mean anything—they either mean nothing, they're just babbling—or if they mean something serious, then what they really mean but don't know it and they are not capable of understanding the distinction and the precise articulation, what they really mean by objective is absolute knowledge.

Absolute Knowledge vs. Relative Knowledge

Absolute knowledge differs from relative knowledge in that what you think you know is corroborable but others who use the same tools, the same epistemological tools, which is the intellectual and sensory equipment that you possess, which is in common to our species but we don't all make the same gains with them. We have the facilities. We don't all get the same value from them. To use Eddington's terminology, it's the same intellectual and sensory equipment. We all have pretty well within a narrow range the same range of vision in terms of what wavelengths of light our eyes respond to, what wavelengths of sound our ears respond to. We all have the same temperature requirements. There are slight variations from person to person but the general broad range is about the same for the whole species.

Just as there are differences in the heights of people. Some people are six feet tall, some are five foot eight, some are five foot four. Some are as short as five feet. A few people who are usually called pygmies or dwarves might even be three feet or four feet. They are not the usual. They are, you might say, at the spectral range of improbable, low frequency cases, but not zero. You might occasionally find a seven-foot-tall person. I don't know what the record is, but probably around eight feet, I believe. I do not think you will find any thirteen-foot people, human beings, and you will not find one-foot ones either when they're fully grown unless some accident occurred to them that they're stunted in their growth in which case they are maimed. I don't think you will find that.

Sensory Inputs Are the Source of Everything Man Knows

There's a general uniformity in size, also in the range that we're able to see with, hear with, these are the sensory inputs. This is the source of everything we know. Our direct knowledge depends upon our direct inputs from the external universe. If we didn't have these five senses, how would we have anything? If you didn't have the sense of sight, how would we even know there's a universe beyond the earth? A blind man, a person who does not have the sense of sight can, with major handicap, be able to use the remaining four senses to adapt himself to his immediate environment. He can touch his way around and, in general, people who are physically blind have a much more finely developed sense of touch which is why they are able to, for example, read with Braille.

I daresay that it would be very difficult for a sighted person to do that because he relies on his ability to see. It is the blindness, the absence of this ability, that heightens the psychological need to use a less competent but nonetheless available other alternative way to read, and the invention of Braille is a great boon to those who have this unfortunate affliction. A person who is blind in general has a better sense of hearing if his ears are not also affected by a separate attack on his hearing tools, hearing equipment. In general, he has a more keenly developed sense of sound, but he's more reliant on it and possibly and probably his sense of smell and taste might also be more sensitized. [*AJG takes a drink.*]

The Sense of Sight Is the Only Long-Range Sense of the Five Senses

But let me ask you this question. Not by hearsay or by reading in Braille but by direct observation: How would a person who is blind know there is a universe beyond the universe that he can personally have access to with his other four senses? All of which are short range: touching, tasting, hearing and smelling. The only long range one is sight. How would you know there is such a thing as a sun or a star? Now a very sensitive person, even an ordinary person would be able to tell there's something like the sun just by being out in the sunshine. He'd feel the warmth of it. That's a radiation which affects not only the sense of sight but also your sense of touch. The rays affect your skin and you can feel its warmth. Just as you can feel a stove radiate warmth without actually touching it. You can feel the sun's heat. If you're out there long enough you'll blister unless you have very callous skin. But, take starlight. That light is not that intense. You can't get a starburn as you can get a sunburn. No one's sense of touch is so acutely sensitive that he can, at night, when the sun is not in the sky, it's on the other side of the earth and therefore obscured by the earth, you go out in the starlight you're not going to feel the warmth of the stars.

317

There are sensitive instruments that pick this up, incidentally. They're called bolometers, but I don't think anybody would be able to tell this by sense of touch. We would not know about the universe of stars, the galaxies. We wouldn't even know about the whole rest of the universe. Now you tell me, how are we going to get an objective, dispassionate knowledge of the universe? I'm talking about something about which is pretty hard to get angry about. Let's say the galaxies.

Galambos Is Exceedingly Sensitive to Justice and Injustice

My wife has said a statement to me not too long ago, when I was grumbling about the miserable life I lead that I had to spend my whole life bumping everybody's drawers, taking care of everybody's blunders, and I even bitterly remarked you can call my whole life of error management of the rest of the planet and I feel like shrugging like somebody else wrote a book about. [*AJG drinks*]. I'm getting tired of carrying the world on my back, person by person, idiot by idiot, individual by individual. *Flatlanders of the world, unite! You have nothing to lose but your civilization.* I was making such grumbling noises and said, "Why the hell can't I have just a little bit of my life tranquil so I can get something done that's useful?" She made a rather interesting point which set me thinking. No knowledge is wasted, so it finally gets into a course. [*Some chuckling in the audience.*] This happens to be the right course for it.

"I wonder," she said to me, "if you didn't have all of these disturbances upon your personal tranquility, whether you would have discovered the theory of volition, or come up with the theory of primary property." I thought about it a little while. It didn't take very long. The answer is, she has a very strong point there. I wouldn't have. Absolutely certain I wouldn't have. You have to have an exceedingly great sensitivity to it. It's like the blind man who becomes more acute in his hearing and to his other senses. I am very sensitive to justice and any lack thereof upsets my tranquility to a point that you probably would not have even the slightest empathy for, let alone sympathy.

The Ideological Albino

I have called myself an ideological albino. [*Light audience laughter.*] You know what an albino is, don't you? It's a creature that is totally absent of color pigmentation. Totally white and has no color pigmentation whatsoever. It also suffers great sensitivity to light. The slightest amount of light injures it, as I'm told. Well, in my case any injustice has that affect upon me. So I said to my wife, not too long after she brought this up, "You're right." I would not have discovered anything about human behavior. The subject would normally bore me. I don't even like people, as a mass. I just like individuals when they deserve it, person by person on an individual merit basis.

Dale Carnegie:
One of the Biggest Menaces against Civilization

I'm not, in case you haven't noticed, a suitable candidate for a politician. [*Audience laughter.*] I would not make a good evangelist. [*Audience and AJG laugh.*] I damn near flunked public speaking. [*AJG drinks.*] If I did run for office, I would lose by probably the largest majority against me in history. When I was a kid, not only did I not ever hold any office in any club or society or group as a child, I wasn't even considered as a candidate. Nobody even thought of it. [*AJG chuckles.*] Starting with me. I have never thought of myself in Dale Carnegie terms. As a matter of fact, Dale Carnegie deserves a whole course. That man is one of the biggest menaces against civilization that has ever come about, and yet he has one of the biggest images of nice guyism that you can imagine.

He has influenced the entire cold end structure of our commercial world. He has influenced politicians who are easily influenced by Dale Carnegie techniques. It fits them very, very well. It's an exceedingly successful cold end approach to man's behavior. On the other hand, I would find in my personal case, I would be totally out of touch with reality to have any such type of input affect me except adversely.

If I had a Dale Carnegie attitude, any cold end attitude, I could not be sensitive to injustice. I'd be the perpetrator of it. I would not be sensitive to property, but to need. I would not be sensitive to justice, but to popularity. I would not be sensitive to justice, I'd be sensitive to mass merchandising if I were in the commercial world, or demagoguery if I were in the theological or political field. I only could develop a theory of volition and primary property if I were sensitive to something which is not a mass subject but a relatively small subject in terms of the number of people who participate. Namely, those rare people who have a longer perspective than what is now, but what is significant or important, which means what is durable.

319

Galambos Was Not Psychologically Mature When
His Father Was Alive

I have been reading some of my Father's letters recently. Well, I've been doing it whenever I find some, and I'm doing it deliberately now because as you know, we're planning to put out a commemoration of him on his 100th birthday in 1982.[3] And not only do I find enormous insight into my Father, which of course I'm bringing up to date things that happened thirty years ago. See some of the fantastic inputs I had, but also I see something of myself as a child, or as an adolescent, as a young adult, and I really wasn't an adult when my Father was alive. I was of age, but I was not an adult. Maturity comes when you are self-responsible, totally. And yet I was, in many respects, exceedingly responsible even as a kid. I know how many things were of importance to me that were of absolutely no importance to anybody else. My Father was quite concerned about the fact that I don't have any of what he looked upon as practical goals, like earning a living for instance. I knew I had to. And that's an interesting thing. I knew I had to and yet it was not of interest to me. By the way, that's the standard thing that leads to socialism in other people. When you think that the world owes you a living for thinking.

The Longest Detour to the Stars You Can Imagine

At age nine I wanted to go to the moon. That was no joke. Absolutely no joke. It was not a childhood fancy to be replaced by some other ambition like being a head pitcher for the Yankees or some such thing. That goal never changed. The only thing that changed is that bureaucracy reached the moon sooner. They had easier political and financial access to it, but not greater intellectual access to it. Of course, that's a great, grave disappointment, but nonetheless, it has not in the slightest bit altered the purposes of my life. My wife is completely right. If this were a tranquil world and we had, let's say, what is called a natural society by myself, if this existed prior to my coming into this life and let's say I

---

[3] "JBG Centennial": a two-day commemoration held 1982, March 6 & 7 celebrating the birth of Joseph B. Galambos.

were born in a world where we had a natural and free society such as I am talking about and it had been done by someone before my time, I wouldn't even be in this field. This field would already be part of physics and I'm much more interested in the galaxies anyhow. That's still a fact.

As a matter of fact, this is the longest detour to the stars you can imagine. But it's the one that will get man there. This is to the stars and to the galaxies, it still is. [*AJG drinks.*] The question is, what if I didn't have this problem that I have dealing with the present society? In that case I would deal with something which has no anger input, no injustice input: a galaxy.

How Can You Be Angry at a Galaxy?

How can you be angry at a galaxy? But you can wonder about what it is, how it's constructed, its magnitude, its distance, its attainability. You can see the magnitude of such a thing compared with man's puny extent in both time and space. Well, just to traverse a galaxy if we had the technological capability to go to the stars. The galaxy is a hundred thousand light-years across. First of all, how do we know that? That's a long process in knowing that. That knowledge doesn't grow on trees, that we know how big a galaxy is. Let's say a galaxy is a hundred thousand light-years across. And you say, "Well that's just a number." Yeah. And a trillion dollars is just a number too. Try counting it.

A Galaxy Is a Totally Tranquil Domain

The extent of a galaxy is so large that the time it takes for a galaxy to rotate once about its own axis is several hundred million years. And yet the speed at its outer edge is enormous. If a galaxy is a hundred thousand-light years in diameter, let's say it's basically a pinwheel or a spiral galaxy, then its radius would be fifty thousand light-years. That means that its circumference is about three hundred thousand light-years. It's over that, but we'll call it three hundred thousand light-years. If it turns around, let's say, once every two or three hundred million years, and you calculate that in terms of miles per second or kilometers per second, you will find that the outer edge of that galaxy is going at a speed that you would consider you might say super believable. And yet it takes several hundred million years for a galaxy to turn around on its axis, once. And yet from the standpoint of what we call speed, it's going at a

321

very large speed. And you consider that man is capable of even thinking about this, let alone reaching out for this, first with his mind and later with his hide, and later to expand into the universe. This is a totally tranquil domain in terms of interpersonal disputes, in terms of violence, in terms of war. It's far more exciting to me than learning how to put a muzzle on a Hitler, and yet if we don't put the muzzles on the Hitlers we won't go to the galaxies.

Galambos' Father: A Unique Input to His Theory

I really don't know what makes me happier but it doesn't make any difference since I have no choice in the matter. Whether I were born a couple of hundred years from now and man lived in civilization and someone else did what I did and do the more interesting things, or now and do the unique thing which will only be done once. Frankly, I'm not sure it would be done it all, because I had a certain input no one ever else had starting with, but not limited to, my Father. That's a unique input. In case you think people are interchangeable, you're only thinking of people on an assembly line. Or you might figure Neil Armstrong might have been interchangeable with one of the other robots they could have picked who could just as easily have responded to, "Left foot on the moon first, Jack." [*Audience laughter.*] You could, instead of saying *Neil* you could say *Jack*, and he would have responded equally well. Now if you had Archimedes for a teacher, they just don't have them in twos. They don't come in pairs, or in teams. Well, I had one of that caliber too. My own happened to be even more unique because Archimedes didn't deal with justice and be a scientist too.

Anything That Endures Long Term Must Be Rational

Well anyhow, that's not the main point here, but it was something I wanted to just put in by way of illustration that it is quite a necessary thing to look upon the juxtaposition of the long range which is always rational, or it won't be long. Anything that is volitionally achieved if it is to be still functioning at a time which is large, very large in relationship to the lifespan of a single individual human, or volitional being, if it is still to be operating at a time which is very large in relationship to the lifespan of the creature it has to have rationality at its base. You won't find anything left of thousands of years ago except that which is rational.

Monotheism Is Essential to Science

I started the beginning of this lecture on Akhenaten and Moses. I picked these two men because they come from a period of three and a half to four thousand years ago and they are at the fountainhead of our present civilization, which is monotheistic. And you may say, "What's with all this emphasis on monotheism? I didn't know you believed in theology that much." Well, if you define it correctly, monotheism is essential to science because aside from any fables you might invent about God, the one thing that you cannot invent and needs no inventing, is the fact that we have a universe to live in and that universe requires an explanation. The content of that explanation has been called science. The principal science, the head science, the top science, the total science is physics, of which the others are branches. But that deals with the content of the universe and the phenomena of the universe. The source of it is not covered because the epistemology of physics has not extended to that, and because of that a whole lot of empty yearning, unfulfilled, yearning to understand: What are we here for? Who are we? What it is this all about?

Today's Identity Crisis

In today's vernacular, it's called the identity crisis. Most people have no idea who they are, or why. Well, they have no relationship identified between themselves and the universe they live in. The reason they can't make any progress is they're not looking in the right place. There's only one place to look, and that's the universe. The universe is not knowable except by observation, which brings me back to the point I was making. How can you have objective knowledge? How can you know there's a universe? How can you know there is anything about the world other than yourself except by interpreting it with your brain based upon the input you got with your senses?

Eddington's Concept of Subjective Knowledge Missing in Most Scientists Today

There is no possibility of acquiring objective knowledge. This major input came to me by reading Eddington, which most people in physics don't know anything about. It's not one of the standard subjects in the curriculum of ordinary physics courses, which is to the detriment of the entire profession called physics. They sure missed the boat in the 20th century in physics by ignoring probably the greatest one in the profession in the 20th century, Eddington. Who, I'm sorry to say, himself had much mysticism and faith left in his total intellectual makeup which I fail to understand how he could have, although I know he had a very deep, Quaker, which is a standard theological background.

You may say, "Well what do you mean standard? There are very few Quakers." Standard in the sense that it accepts the usual myths. They're not standard in that they're in the minority but they accept the usual myths. I find that an internal contradiction. I don't want to go into that and discuss it because I have no desire here to criticize my great benefactor, Eddington, to whom I owe the highest intellectual debt, which I can show here mainly by recognizing that it's his excellent concept of subjective knowledge which is missing from most people who think they are scientists, who think they can talk about objective knowledge or objectivity.

Relative and Absolute Concepts Are Both Subjective

This is one of the biggest fictions there is and the longer I live, the more convinced I am how dangerous it is. And yet, the other subject of believing whatever you believe in on the basis of the fact that you believe in it because you know it is true, because you have faith. That's called subjectivity. That is also dangerous. The difference is they do not know the difference between relative and absolute. They are both subjective. The difference is that one is corroborable and the other one isn't. All absolute knowledge is relative knowledge first, but then when it becomes uniformly, universally corroborable, it becomes absolute. It becomes independent of the frame of reference, intellectual frame of reference, mental, intellectual structure that it comes from.

Language and Its Relation to Psychology

I believe in the previous session when I talked about language and its relation to psychology and that the very language you express yourself in increases or decreases your adeptness at having imagination and that the stronger and the larger the language mechanism that you have, the better ability you have to have thoughts in and through. You have much more articulate thoughts if you have a stronger language. If you have more than one, you have a variation which may or may not be to your advantage. If you have no integration ability and you know two languages, it may confuse you and your thoughts might be more jumbled. If you know ten languages but don't think well you might have an exceedingly good linguistic talent through memorization capability but not being able to integrate, knowing ten languages might injure you because you get an intellectual confusion and you can't make translations.

Knowledge Comes through the Connection of Seemingly Unrelated Things

On the other hand, if you can integrate a double language or a triple language or a multiple language capability, this may enable you to have the ability to interpret things in various ways and see the connection between them. That is basically what knowledge is about. To see the connection between seemingly unrelated and seemingly dissimilar things. That's how knowledge grows. To connect up two things that were previously not connectable. Physics teaches us from its past that everything if you look hard enough has a connection with everything else. If we don't see it that means we haven't learned enough. It also dispenses with the need for faith [*AJG drinks*] which you replace with the word *confidence*. You might say, "Well, what's the difference? It's just the same thing, except you have a new word and it's a semantic trick, you just substitute another word." The difference is that faith you take before proof; confidence you acquire after proof.

Faith:
To Believe in Something That Is Not Corroborable

If for example, I am asked to believe that Jesus was born of Immaculate Conception, I have no precedent to compare it with except perhaps another fable in a different culture. I understand several different cultures have had that fable and no one has a corroboration of it in terms of observational, acceptable absolute corroboration. And for me to believe that would be a shear act of faith. I accept something because I'm told this on authority that I take without question, and I take it without question. That's considered a virtue, to have faith. "Why should I believe, father?" "Have faith, my son." "How can I believe that there can be a God when there are wars and there are murderers and there is injustice in the world and good men die without having produced any provocation for their murder? How can I have belief in such a God,

325

father?" "The Lord moves in mysterious ways my son [*audience laughter*] and we are not to question him. Ours is not to reason why, ours is but to do and die." Tennyson.

Anyhow: "You are to have faith that everything is for the better. After all, this veil of tears is only for testing our mettle in how we shall fare in the hereafter which is for all eternity. This is a temporary transition into all eternity." If you think this was a cliché-ridden few minutes, don't blame me. [*Some chuckles in the audience.*] I'm just trying, with my limited acting ability, which has always been weak, to try to portray what the bulk of mankind has fallen for, for thousands of years. That's faith. To believe something that is not corroborable and to give advance credit prior to delivery of proof.

Belief in Law of Gravitation Is Not the Same Thing as Faith

Now, to say I believe in the law of gravitation being everywhere in the universe—and you may say, "Galambos, that's a broad generalization. How do you know that the law of gravitation works in a place that's one hundred million light-years away? Have you ever been there? Have you paid to have an experiment out there? Have you tested every possible matter in the universe as to whether it fits the law of gravitation or not?" The answer is, no, I have not and neither has anyone else and neither will anyone else. It's not possible to test everything. Then you might say, "Oh, well then you have to have faith too." No, I do not.

Galambos Does Not Agree with Eddington on the Concept of Faith

As a matter of fact, I'm sorry to say the great Eddington used the word *faith*. [*Galambos asks his podium assistant.*] May I have Eddington's book please? On *The Philosophy of Physical Science*. Thank you. On the very back cover there's a quotation from Eddington himself: "In the Age of Reason faith yet remains supreme, for reason is one of the articles of faith." I sell this book not because of that sentence, but in spite of it. I say this is the most valuable book I've ever sold in my life, besides from my forthcoming book of my own. And it's not because of that sentence, ladies and gentlemen, it's in spite of it. I don't agree with that sentence: "In the Age of Reason, faith yet remains supreme, for reason is one of the articles of faith." I don't accept that.

The Concept of God

I have defined *faith* as belief in something where there has been no proof but you are asked to believe because it is offered on authority, authority of someone else, even if it's attributed to God which of course is a pretty hard thing to attribute it to because I haven't had any way of chatting with God. Now I know that there are people who claim to have. I have met some. [*Some chuckles in the audience.*] I know that the history of the world is replete with people who are less sane than most of the inmates of insane asylums who claim that they have seen God, talked with God. [*AJG drinks.*] I don't have such talents. I claim no such talents. I claim there are no such talents available, to me or anyone else. I don't even believe God can be talked to. Oh, I'm sorry, you can talk to God [*audience laughter*] but you can also talk to a cliff. [*AJG chuckles.*] But don't expect a cliff to have an intelligent response. I don't believe god is a volitional being, let's start there. I have given a lecture on that in a very advanced course. I do not believe god is a volitional being. I didn't always say that. I've said it recently. I used to say I do not know what god is, but I do not believe god is a volitional being. Yet I don't say there's no god. The concept of god is simply an explanation of natural phenomena. That's the content of it.

The Distinction between Confidence and Faith

Now I'd like to point out to you the difference between confidence and faith. I've already discussed faith. You accept something without any proof. *Proof* is defined as observational corroborability by using the scientific method.

The Magnitude of Newton's Imagination and Grandeur of His Mind

We were discussing the law of universal gravitation. Why should I believe in that when I haven't tested it everywhere? Neither has Newton. Neither could Newton. Neither can anyone else. It's a supreme, competent, successful generalization. Actually, you have all the more to marvel at. The magnitude of Newton's imagination and the grandeur of his mind to recognize how little he had to go on to produce this major generalization which has stood three hundred years without any failure. Don't throw at me the things that relativity has surpassed Newton on.

That's a shoulder account for Newton. He has not superseded Newton. It is a shoulder account that Newton has for Einstein. That's still Newtonian, except that it has been expanded by an epistemological enlargement of physics, which makes Newtonian physics a first approximation to Einsteinian physics.

Galambos Has Expanded Physics in a Different Domain

I might also mention that I have expanded physics in a different domain. I've included first volition, and now biology. I sort of skipped over biology—I came back to it. I jumped the chasm to volition, which actually is after biology and I have now enlarged physics to include volition for sure, and biology quite certainly, that hole to be filled in. It's sort of surrounded now on both sides. [*AJG drinks*.] Also, by derivation from volition, psychology is part of physics now. This is a subvolitional subject and this now fits the universe by the same standards as physics has been explained, the scientific method.

The Distinction between Scientific
vs. Volitional Predictions

Proof is defined in science as observational corroborability of a predicted event. If we can predict where Mars will be in its orbit at any given moment, at a given moment next Thursday evening as viewed from a certain place on Earth or where it will be a thousand years from now at a precise moment or any time, and you wait out that period and you will find that it's there. Well, it's either the most amazing coincidence, or there must be something behind what gave you the tool to compute this with. It beats any Nostradamus or Jeane Dixon [*some chuckling in the audience and AJG drinks*] in certainty of prediction, without ambiguity of representation. Please note that most of these soothsayers cloak their prognostications in major ambiguities and they don't have a very good track record for the most part. If they make thirty predictions, maybe two might come out approximately right. They'll harp on the rightness, skip the approximation and forget the other twenty-eight.

Same goes in the stock market. Somebody recommends twenty different stocks, three do well, twenty-seven do not. Guess which ones his future advertisements will stress? The ones he failed on or the ones he succeeded on? I'm not even discussing the fact that in the stock market you have direct influence on the market by—the prediction itself affects the market. Whereas the prediction of where Mars will be has no effect upon its future location. [*Audience laughter and AJG chuckles.*] Whereas if you predict the market will crash and you have a big enough readership, this will help it crash. This is a very certain fact. It's based on the supply and demand mechanism. And you are, by predicting a crash, you are essentially scaring people who read what you say or hear what you say, scaring them to say, "Well, if there's gonna be a crash I better quickly sell before it's too late!"

And a lot of people get on this scare track and they dump their stocks. That produces the very crash! Then a lot of people say, "Oh, he's right. The market is crashing! The sky is falling!" So a whole bunch of other jerks jump in and they sell. And so the very prediction produces the effect. That's called a self-fulfilling prophecy. That can only work with volition. You cannot do that with a planet.

Zero Failures of Predictions: A Tougher Criterion

A thousand people can predict where Mars will be differently than the gravitational law and they're out of luck. The gravitational prediction correctly calculated pursuant to Newtonian techniques—that you can put your life on. You can stake your life on it as long as the person who did the computation knows what he's doing and applies Newtonian principles correctly. All right. Now, if Newtonian principles have produced corroborability again and again and again without failure and the only times when there have been failures, it can be shown that the failures to corroborate were caused by incorrect computation which you can check the calculation or the input observations. If either the input observations or the calculations are wrong, the prediction will also be wrong. It has not happened yet once that any incorrect calculation which led to a faulty prediction was not identifiable as a faulty calculation.

You can do it over again, you know, and find somebody else's mistakes. The prescribed method of an orbit computation is, I won't say it's easy but it's certainly an identifiable procedure. And if you do it right, you can calculate this and calculate it right, without failure. How many failures can you tolerate when I say without failure? [*Audience responds, "Zero."*] Zero. That's a tougher criterion than any other aspect of human conduct requires.

329

One Way to Eliminate Politics:
Adopt the Rules of Physics

If a man, to become president, had to get 100% of the votes of everybody who's in the election, no one would ever be elected president. As a matter of fact, that would cure politics immediately, if you adopted the rules of physics. You say, "No one can become president or congressman or governor or alderman or city councilman or any other political job holder, he cannot obtain his job unless 100% of the people vote for him, with zero exception." That would be one way to eliminate politics except they don't have these rules. They can't afford such rules. They can't accept such a criterion. It's too harsh and too stringent for their flimsy mechanism.

Yet, we in physics have to accept that. That's the only criterion we can live with and still have a science. My science, the one I developed, volition, and occasional extensions of volition like this one beyond, will stand on the scientific method's criterion. It cannot tolerate any failures of generalization and to the extent that any failure of generalization occurs, then that part of what was said is not right.

Confidence Is Earned, Faith Is Credit Advanced

Now I'd like to point out that when you have had such an experience, such as we've had with Newtonian gravitation for instance, for centuries there has been no failure. And I'll make the statement, I believe next time we're going to get a calculation it's also going to come out predictably true. I have confidence in it. That's not the same as faith. This has had now a long history of success and the probability of there being an exception to it now is decidedly so small that you could more easily expect to be able to break the bank of Monte Carlo or of Las Vegas, for that matter, than to expect an exception to a successful generalization of physics to be found.

The probability of breaking the bank of Monte Carlo would be much easier to assume and to expect. Then you call that confidence. You see, the confidence is earned, the faith is credit advanced. The difference between confidence and faith, is that with faith you have no advance payment. You make the payment without having had any return yet, and it's a lousy market transaction.

Pursuing Happiness through Faith and Delusion

I'm still waiting for a favorable return to me over the belief in the Immaculate Conception. What has it helped my life with? In what way am I happier? In what way can I pursue happiness better because of believing in this? I don't, but let's say I did. Let's say I were one of the millions who did. How does that make it possible for me to pursue happiness? The only way is by a process which in psychology by myself has been called delusion. *Delusion* is the belief in something being true, which in fact is not true because it is not corroborable. Now let's say I did believe in the Immaculate Conception, and I believed in the belief in Jesus as the Son of God, the teacher of man, the savior of man, that he, his death, washes away any sins that I might commit.

If I believed this, you may say, "Well, it helps me pursue happiness because, boy, no matter what I can do all I have to do is to go and repent, confess to the church what I have done, ask for absolution, ask for, with certain little donations of time, money and effort [*AJG chuckles*] I can get absolution from even murder, and I can still enter the kingdom heaven if I take Christ the Lord into my heart as my savior. And his death two thousand years ago which he obtained his life through Immaculate Conception through the Holy Ghost having visited the Virgin Mary, and I believe this with all my heart and with all my soul with no question, no reservation, no thought possibly creeping into my mind that there might be something wrong with this [*audience laughter with AJG*] I accept this on faith, and I will enter into the kingdom of heaven. I'm thereby pursuing happiness because even though I committed a murder, robbed a bank, raped, assaulted, committed arson, treason, whatever. All of this can be washed away because I believe in Christ as my savior, and he got into this world by Immaculate Conception, I question it not. [*Chuckles in the audience.*] So I am pursuing happiness successfully."

Okay. I do not deny that this is a pursuit of happiness. This is the way some people do pursue happiness. I do hope I'm not hurting anyone's feelings, that is not my intention. [*Audience laughter.*] I have seen you before in other courses and I think you've had certain, if you pardon the expression, baptism [*loud audience laughter*] of rationality, baptism of rationality before so I hope some has registered. If I am hurting someone's feelings, I apologize for your emotional upset but not for what I said. [*Chuckles from the audience.*]

Can Pursuit of Happiness through Faith Be Successful?

That is a pursuit of happiness. It's an acknowledgeable pursuit of happiness. Is it successful though? Everyone pursues happiness. There's no question about that. Hitler pursued happiness. Christ pursued happiness when he did what he did. Those who believe this way pursue happiness. The question is, what have you gained thereby that's of durable consequence? What I just described, which I hope you will agree is some reasonable facsimile of the Christian faith, at least a small portion of it. I would say a substantial part of it, however. Can you build a bridge with that? Can you reach the moon with that? Can you have fluorescent lights with it? Can you air condition with it? Or wash clothes with it? And you will of course have obviously at your tongue tips a response, "Yes, but you see, these are all materialistic goals. [*Audience chuckling*.] What about the higher things in life?" I'm glad you asked that. [*Light audience laughter*.] I happen to have that very dear to my personal thinking.

The Higher Things in Life Is the Proper Thing to Pursue

I always have thought about the higher things in life more than the material things myself and I recognize that there's something more important than fluorescent lights. The desire to know enough about electricity to develop it. There's something more important than going to the moon. The desire to go and the proper reason why. Which I had, by the way, at age nine—the proper reason why. Which I'm not going to put in this class. But I had the right reason to go to the moon when I was nine. That's higher than the actual going. It's not a dream of glory. There's a necessity for a man to go beyond his own planet for him to become capable of coexisting with the cosmos so that he does not have his existence linked to merely an accidental piece of the universe that he happens to be initially born on as a creature. The higher things in life is the proper thing to pursue.

Galambos' Concept of the Higher Things in Life

Now let's discuss my higher things in life, and the person who accepts the Immaculate Conception, the resurrection and the saviorship of Christ the Lord, in relationship to my concept, let's say, of going into the cosmos as man's destiny. They're both the higher things in life from

332

two different vantage points. Which one will make the species last? Which one will give man something in a domain that we know exists because we are observing it, now. I may not live—as a matter of fact it's highly unlikely I would live to the time which man reaches the stars. The only way I could live that long is if my own theory took root so fast that my lifetime could be expanded to centuries, and it's not too likely it'll happen that quickly.

But even though I will never live to see man going to the stars, I know that because of what I'm doing this capability is more than an accidental, minute possibility. It's an almost certainty based upon the confidence factor, not the faith. Because the confidence in this knowledge comes from the already corroborated back-flowstream of what has preceded this. From the caveman who invented the wheel through Newton through Einstein through Eddington through what I'm doing. The back-flowstream has had adequate corroboration that what I'm predicting is not faith, it's confidence by the distinction I just made. I would not care to put everything I've got on my delusion on heaven that I would get from the standard theological perspective.

V-30 and V-31 in Relation to V-50 and V-201

Once I blew it in V-31,[4] which doesn't even require V-50 to hear it. My wife and Snelson sometimes almost cover their faces when I say something out of context in the wrong audience. I sometimes don't know when to keep silent. That is one of my problems. [*AJG chuckles.*] And by the way, I'm capable of seeing this problem in myself. I'm not ego minoring my way out, "Oh, I don't know how to stop that." As a matter of fact, I probably suffer from the same problem Bruno did, talk too much in the wrong circles. [*Chuckles from the audience and AJG.*] It's not so much I talk too much, but it's to the wrong people. [*More chuckles.*]

Anyhow, in V-31 I came off with what is absolutely correct and I rarely said anything more profound than this, but it was a little premature for those who haven't had V-50 yet. Now most people, fortunately,

---

[4] V-31: "Chaos & Plunder vs. Investments & Insurance."

who've had V-31 have had 201 first. But there were some strays in there who did not yet have V-50 and there is no technical requirement to have V-50 to take V-31, because the only prerequisite V-31 has is V-30 and V-30 you can take without V-50. You won't get as much out of it, but you will get something, mainly in the P_2 domain.

The Discussion of Wild Speculators in V-31

And I said in V-31, we were talking about the dangers of wild speculators, talking about the silver swindlers, the touters of this and that speculation. By the way, there's nothing wrong in silver, per se. It's the getting people to go whole hog into a speculation which is misrepresented as an investment and to call a speculation an investment. To go on high margin and call it riskless. That's a swindle. The ownership of silver is not a swindle. The sale of silver is not a swindle, but the whole hog total commitment to silver, gold or any other thing without diversification, calling it a riskless investment instead of a highly risky speculation, that's a swindle.

The Supreme Speculation of All Time

So I was talking about speculation and some of the promulgators of poor-quality speculation and I said, "But all of this is nothing compared with the supreme speculation of all time. To commit everything you've got, all of your primordial property, your entire primary property and everything you have in secondary property," and, in short, your entire! [AJG shouts] natural estate—that much I didn't say. [Audience laughter and AJG laughs.] But in short, your entire natural estate—P_0, plus P_1, plus P_2—in the supreme speculation of all time!

That the whole of your life from birth to death has no meaning, save one. That this is preparatory to the next life and that you can invest your entire life into the foolish speculation that there might be something else coming later and you blow the whole life you've got in the favor of a speculation of another life you have no knowledge of and cannot have until you're dead. And what a shock you'll get when you're dead and there isn't and you won't know it. [Audience laughs along with AJG.] The only good I can say about this for the people who believe this is they'll never know that they were wrong. Because when they die, they won't know it because there's nothing after that. [AJG stops to take a drink.]

There Can Be Hope for an Afterlife—
But There's No Evidence for or against It

I didn't say that for a fact, I said that's a high probability. Like Paine, I do not know if there is no afterlife, but I just have no evidence for it. I didn't say I wouldn't hope for it. I never said that. You will note, I never said I don't hope for it. I just say I don't believe in it. There's no evidence for it. Why would I hope for it? Well, if there were an afterlife, I might have a chance to see my Father again. If there were an afterlife, I'd see my Mother again. If there were an afterlife after I'm dead I'd see my wife again. For all these reasons I'd hope for it. For seeing some other people, I'd just as soon [*AJG pauses followed by loud audience laughter*] pass. There'd be a possibility I could meet Newton or Archimedes or Paine. Find out the inventor of the wheel and thank him personally. [*AJG chuckles with the audience.*] All these, I hope for. For most people I've met, if I could have a segregated [*audience laughter*]—I decide who I meet and who I don't, and then I'd very much hope for it. But that's all idle speculation. I could hope from now till forever. That's more than eternity. *Forever* is infinite time; *eternity* is the duration of the cosmos. My separation of these two terms. I could hope from now till forever. Hope won't make it so. You might say, "Galambos, you might have a shock when you die. You might find there is an afterlife." That won't be a shock. I've allowed for that possibility. I just say that it is not a likely possibility.

A Strong Hypothesis That God Is Not a Volitional Being

By the way, if there is such a separation as heaven and hell, which I highly doubt, since we already have the hell here, we don't need a new one. [*Audience laughter and AJG chuckles.*] If there were a true such separation as heaven and hell, the one thing I'm 100% certain of, and I have zero qualms or hesitation on it, the fact that I did not take Christ as my savior will not keep me out of heaven and put me into hell. That I would stake everything on. In case you don't think so, I'm doing it! [*Loud audience laughter and AJG chuckles.*] The fact that I say that I do not believe god is a volitional being I don't think would keep me out of heaven and put me into hell. Even if I were wrong, and I admit this possibility, this is a hypothesis when I say god is not a volitional being.

By the way, I want to make that clear. That is a hypothesis at this time. It is not subject to observational corroboration. It is not yet corroborated although I think it can be someday. Right now, the crucial observations cannot be made. I have given evidence for it. I consider that a strong hypothesis, it's not a theory. So, lest you think I'm claiming more for it than there is.

You say, "Supposing you're wrong and it turns out God is a volitional being and he's sore at you for talking this way." [*Audience laughter.*] I admit it is possible, I am mistaken and God is a volitional being. I do not think that is the case, but let's say I am wrong and there is such a volitional concept of God that really exists. Then I am certain that he will not keep me out of heaven, for this reason. If I don't get into heaven, it will not be for this reason. Do you know why not? Because if that kept me out of heaven and put me into hell, then he belongs in hell for being so criminal.

That creature, that personage, that super personage that would be a volitional God would be a criminal. How can he hold against me for thinking the way I do when he created me? He should blame himself! [*AJG laughs with the audience.*] And if he doesn't, he's irrational and suffering from an ego minor complex. [*Loud audience laughter.*] He's putting on me the blame for his own achievement. [*More laughter.*] Now argue your way out of that rationally! [*Laughter continues and then slowly begins to fade.*] You see the risk is very low. [*More audience laughter.*] I have a high risk with my fellow human beings, but not with god. If there were a volitional God, he could not punish someone who seeks to know even if he made a mistake, because that mistake injures no one. It is not coercively foisted on anyone.

Nothing of Significance Endures into the Species Time Scale on a Positive Basis That Isn't Right

For example, if you don't want to believe what I believe—don't. I don't have a gun. Not with me, not anywhere. You don't have to believe this. If you think I'm wrong, don't accept it. I'm offering you a tube to look through. You have the right not to look. I'm sorry, you have the prerogative not to look. You can't be right because it's not rational. You have the prerogative not to look. You may look and see something differently. That's your problem. One of us is mistaken. [*Audience chuckles.*] Maybe both of us are. It's possible we may be both wrong. You see, I accept as a judge something that most people cannot accept. The only judge I accept for my views is the long term because it is an observationally corroborable fact, ladies and gentlemen, that nothing of significance endures into the long term—by that I mean the species time scale on a positive basis—that isn't right. Why? Because it won't work. It won't work. As a purely pragmatic reason, it fits the scientific method. It will fail of observational corroboration.

Belief in Wrong Ideas Is Not Coercive, Just Not Successful

The fact that somebody a long time ago believed that there are four elements—earth, fire, water and air—and that somebody is the entire Greek culture, including, but not limited to Aristotle. He didn't do the world any damage. That was a procedure of thinking, a belief he once had, a lot of people had, it didn't harm anybody. It was better than believing in nothing. And there were certain things about Greek science that did benefit man. If nothing more, it led to Galileo. He didn't injure anybody by it. The fact that people believed in perpetual motion machines at one time, that's not going to injure someone. It won't work. The only ones it might injure are those who invest in it. They don't have to. Or buy one. It won't work. You got yourself a real lemon. [*Audience laughs and AJG chuckles.*] It won't work. After a while it loses its credibility.

Clear and open market response to products however faulty they may be, does not injure anyone except those who buy it, and if they're not forced to buy it everyone has to take a risk in whatever they do. To shield a person from risk is to say he is not worthy of making a decision. He has to have his decisions made for him. We have had such people already, they're called dictators. "We must not let anybody make a decision because he'll make the wrong one. I'll make it for him." Let a person make his own decision. If he makes a mistake, he should suffer for it. That's called justice. He learns quicker that way, by the way. When you learn to suffer from the times you've been burned, you get burned less often, but if you can't learn from it you can't profit from it.

Only Rationality Should Dominate the Culture

Now, the general nature of the interrelationship between emotional and rational aspects of psychology then are that there has to be recognized that psychology covers both, not just one. It's neither saying that rationality is the only thing that should be tolerated and all emotional outbursts must be suppressed, because then you would be suppressing the living nature of the creature, the volitional nature of the creature. On the other hand, to say that emotional behavior should dominate the species is to invite political and theological nonsense. They both exist but only rationality should dominate the culture because that's the only durable component.

Emotions Can Inspire Rational Thinking

Emotionalism does not produce durable consequences. It might inspire rational thinking. When emotions are properly harnessed, properly co-ordinated, properly and carefully looked upon as partially an incentive mechanism and partially a therapy mechanism and you recognize that from a rational point of view, I make no effort whatsoever to be unemotional. But I challenge anyone to tell me when I'm emotional, that leads to irrational conduct since I have grown up—which is not defined as when I stopped growing biologically. It is defined as when I assume self-responsibility, which is an evolutionary development. But I certainly have attained it long ago where I'm capable of taking a responsibility for everything that I do.

Temper Outbursts in Defense of Property
vs. Temper Outbursts in Defense of Error

Now I wouldn't make the same claim about myself as a child. I would not ever say that I never had a temper tantrum as a child. A tantrum being defined as more than temper. It's an irrational outburst where you lash out in vengeance. Short-term vengeance maybe. That's not the same thing as defense of property. Temper outburst in defense of property is not the same thing as temper outburst in defense of error. You may say, "Wait a minute, wait a minute. Error is property too. You have a property in your error." All right, let's clarify that.

Temper is frustration to defend positive property. Error is negative property. To defend one's emotional sanity, keep yourself on an emotionally satisfying course, let's say, so you don't bottle up—which could injure you more—in frustration over the damage to your positive property, without attacking somebody else's, is healthy. A temper manifestation which is coming from a defense of one's own error, whether that error is incompetence which is failure to deliver satisfactorily on a contract, or whether it is due to an intentional attack on someone else's property which is coercion—and to defend that and say it was right and have temper outbursts on that—that is entirely different in its social and volitional implication. The latter is injurious to the stability not only of the person involved, but to the environment he is immersed in which is basically the structure of society around him.

Emotion: A Builder of Personal Stability So Long as Not Used as an Escape Hatch from Reality

It injures him because he's trying to justify and thereby continue an error propagation which cannot do anything except expand the damage to himself and to everybody he touches or has interaction with. And therefore, if not stopped, if not somehow self-limited, it will end in coercion. This, of course, is the sorry history of emotionalism which is why to the extent that it has been studied by scholars it is looked down upon. And yet when you recognize that this is a two-edged concept—I don't want to use the word *sword*—but two-edged, well, sword is a weapon, but a two-edged blade. Or two-sided coin. There is a constructive and a destructive aspect to emotion. Emotion is a builder of personal stability so long as you don't use it as an escape hatch from reality, as measured by rational behavior.

Emotion as a Tool towards Long-Term Rationality

The reality of the universe is observationally corroborable. If emotion is a tool towards long-term rationality and does not conflict with it, it is both healthy and desirable. It makes a person less dull. It makes a person less dull to know. It makes it easier for himself to express himself. It increases his capability to function and even to articulate if he links his emotional behavior to rational long-term goals. Unfortunately, that's the minimum case, that's the less frequent mechanism which is why emotion has a sorry history, which is why probably most of what is called psychiatry, psychoanalysis deals with defective conditions in people who have not had their emotions in their proper harness.

The Direct Benefit of PBV-273 for the Few Who Understand V-201

This course will deal, unlike what I imagine standard psychology or psychiatry to do, with both emotion and rationality. I will deal with them frequently, both of them separately and together and their interaction, with this to be remembered from what has gone in the last few sessions before this and the present session. Emotionalism and rationality are two separate things which are both possible to be working in unison towards a long-term goal or in disharmony, one pulling for the short term and the other pulling for the long term, and the short term wins. And that's the majority of people. For this reason, this theory of psychology can directly benefit only a few people which is presumably including you, potentially if not actually. It can benefit you if you understand what's in V-201 and want to do something more with it. It can benefit anyone who's capable of understanding V-201, subsequently to V-201.

V-201 a Prerequisite for PBV-273

As you can see, this course could not be presented without V-201 as a prerequisite. Anybody not notice that? [*Audience laughter.*] I was asked last year—by that I mean last academic year, let's say, in the spring—by several people, if they could take PBV-273 at the same time that V-201 they're taking. I said no, out of the question. First of all, by the time V-201 starts, PBV-273 will be about eight or nine sessions into it, and you will have had no inputs on 201. And 201 is a real slow course to get anything out of because in the beginning you haven't got the slightest idea of what the jigsaw puzzle is about, and you're only getting a few of the pieces. They're all alien to everything you previously experienced, and they don't at all reconcile with what you previously knew or believed. And so, it's not until the end of the course that the pieces start falling together and by that time this course is over. You can see that this theory is completely a derivative of V-201. Is it not clear to everyone, is that clear? [*"Yeses" from the audience.*] All right.

On that basis, this can only benefit people who understand and can function in resonance with the theory of primary property. You say, "Well that's a very limited field." Well, at the moment it is. [*AJG sighs.*] A better market though than Bruno had, I believe, the day he was burned. Well, I'm not sure it's better—it's larger. [*AJG drinks.*] Although, unknown to him, he had some friends who kept his work alive and in later centuries it came to fruition.

The Longer-Term Effect of This Theory of Psychology

I want to point out that the market for it grows with the market for V-201, which in turn depends on the market for V-50. That's why I can't urge you more emphatically that you have to start the funnel at the front end, not the back end. Well, I'd like to emphasize even more so, this has a longer-term effect, this theory of psychology. That's the short-term effect, in the beginning it can directly affect those who understand V-201 and more than understand it, more than claim to understand it, but they do, and more than doing it, they live it. Not as an academic exercise in the ivory tower, but in the real, even, flatland world.

Rising Above the Flat Nature of the Society

The rest of the world may be flatland, you don't have to be flat with them. If you're not as flat as your society, you could be a towering giant even in your—you may not be comfortable, by the way. You might feel like Gulliver in Lilliput. Who was rather annoyed by those little twerps. [*Some chuckles in the audience.*] You know, he wasn't very well treated in Lilliput, you know that. That may be your situation, but you'd still be a giant in flatland. If you rise above the flat nature of the society. You can practice 201 even though no one else does around you. The only ones you need it with, by the way, are production associates. There, it's critical. There it is critical. All right.

Building the Integrity Machine

But just remember, that just as what's in V-201 in the long term will produce a production mechanism which will outproduce all other pro-duction mechanisms and produce a new culture based on the hot end, not the cold end. The cold end, which is the majority of the people, will inevitably be affected by this theory, not by direct study of it as you're asked to do and you have consented to be here. Not by studying it or hearing about it and then trying to apply it, but they will be living in a world that is structured via the integrity machine, which of course is automatically improved to the extent that the psychological attitude of those who build the integrity machine, the hot-enders, have made a stronger and more durable production mechanism wherein everyone else is immersed for the source of everything he gets that's positive, and he will conform not by coercion, but by custom because that's the only thing that will provide him the pursuit of happiness that naturally he seeks. And when the source of his property and his goals is automatic in terms of a mechanism he neither has built nor helped build nor un-derstands nor even knows exists, it still will benefit him.

341

The Leverage of the Few Who Understand and Properly Apply This Technology

How many people here prior to hearing my courses knew about the Maxwell theory of electromagnetic wave propagation? But nonetheless, you still have TVs. Do I make my point? [*"Yeses" from the audience.*] The majority of people who have TVs neither have heard of nor care, either about James Clerk Maxwell the person, nor about the theory of electromagnetic wave propagation or even the fact that there is such a thing as electromagnetic waves. What the hell do they care as long as the World Series is not interfered with as they're watching it? The average person who watches the World Series, if he were asked who is more important, the guy who hit the winning home run, or James Clerk Maxwell, what do you suppose he would answer? "James Who?!" [*Audience chuckles.*] But without James Who [*more chuckles*] he wouldn't even know there is a guy playing in the World Series till weeks later the Pony Express brings in the information! And even the Pony Express required a technology higher than he could master.

How many of the people who now consider the World Series to be the be all and end all of life would have been able to invent the horseshoe? [*Audience chuckling.*] Try having a Pony Express without a horseshoe on the poor horse that is the victim of this operation. [*Audience laughter.*] Horses in nature don't run around with metal on their feet and a rider on their back. Both are unnatural to horses. Do I make my point? [*"Yeses" in the audience.*] Good evening.

[*Applause from the audience for twenty-two seconds.*]

BIBLIOGRAPHY

Ardrey, Robert, *African Genesis*. New York: Dell Publishing Co., Inc. 1961.

——— , *The Territorial Imperative*. New York: Dell Publishing Co., Inc. 1966.

Conway, Moncure D., *The Life of Thomas Paine*. United States of America: Benjamin Blom, Inc. 1976.

Copeland, Lewis, *The World's Great Speeches*. New York: The Book League of America, 1942.

Eddington, Arthur, *The Philosophy of Physical Science*. New York: The Macmillan Company, 1939.

Galambos, Andrew J. & Smith, Jerome F., *V-120 Seminar on Money*.

Galambos, Andrew J., *Course 100 Capitalism – The Key to Survival!*.

——— , *F-201 This is the Original V-201: The Nature and Protection of Primary Property*.

——— , *φ-102 Selected Additional Topics in Physics*.

——— , *V-30 Investments & Insurance*.

——— , *V-31 Chaos & Plunder vs. Investments & Insurance*.

——— , *V-50 Capitalism—The Liberal Revolution*.

——— , *V-76 The Declaration of Independence, Thomas Paine, and Your Freedom*.

——— , *V-111X$_2$ (Witchcraft and Superstition)^2 Resulting in Permanent Crisismanship as the Residue Sludge Derived from Non-Proprietary Scientific Progress*.

——— , *V-113 National Socialism's Coercive Ecology and Posterity*.

——— , *V-201 The Nature and Protection of Primary Property*.

——— , *V-212 The Integrity Concepts of V-201 Applied as Personal Ethics in Flatland and Spaceland*.

——— , *V-215 Positive (Primary) Marriage: The Prime Personal Contract*.

——— , *PBVS-273 An Introduction to the Theory of Subvolition (Primary Psychology) – Part I*.

——— , *V-282 Positive History*.

——— , *VC-283 Positive Journalism (Communication)*.

Heath, Spencer, *Citadel, Market and Altar*. Baltimore, MD. The Science of Society Foundation, Inc. 1957.

Lewis, Joseph, *Thomas Paine: Author of the Declaration of Independence*. New York: Freethought Press, 1947.

Newton, Isaac, *Principia Mathematica*. Berkeley, Los Angeles, London: University of California Press, 1934.

Paine, Thomas, *Agrarian Justice*. Philadelphia: Bache edition, 1796.

——— , *Common Sense*. New York: Freethought Press Association, 1946.

——— , *Rights of Man*. New York: The Heritage Press, 1961.

——— , *The Age of Reason*. New York: The Truth Seeker Company, 1898.

Planck, Maxwell, *Black Body Radiation*. Berlin: 1900.

Plutarch, *Plutarch's Lives*. Great Britain: William Heinemann LTD, 1917.

Stoker, Bram, *Dracula*. United Kingdom: Archibald Constable and Company, 1897.

The Koran, London: Allen & Unwin, 1955.

The New Testament, Toronto-New York-Edinburgh: Thomas Nelson & Sons, 1953.

The Old Testament, Toronto-New York-Edinburgh: Thomas Nelson & Sons, 1953.

The Torah, Philadelphia: The Jewish Publication Society of America, 1962.

Woodward, W.E., *Tom Paine: America's Godfather.* New York: E.P. Dutton & Company, 1945.

COURSE DESIGNATIONS:

P = PHYSICS

B = BIOLOGY

V = VOLITION

S = SUBVOLITION

F = FREEDOM

C = COMMUNICATION

INDEX

Aaron
    as acting tribal chief getting
        tribe to have some pagan
        ritual, 290
    brother of Moses, 290
absolute
    a subjective concept, 324
absolute importance
    diagram of, 254
absolute knowledge
    as relative knowledge first, 324
    becomes uniformly, universally
        corroborable, 324
    distinguished from relative
        knowledge, 316
accidents
    having a psychological input,
        123
accomplishment
    you must not disintegrate if not
        attained, 93
accuracy
    a reduction to a well-identified
        quantity, 218
    always desirable to make a
        science stronger, 237
    desirable but not essential, 234
    difficult in volition, 220
    distinguished from precision,
        218, 219, 237
    not required to make a science,
        219
achievement
    comes from long-term time
        scale, 118

distinguished from dreams of
    glory, 255
dreams of glory not a substitute
    for, 92
flatland history classes do not
    glorify, 173
should be measured by money,
    106
value comes from, 65
acquaintance
    distinguished from friend, 77
Adams, John
    not for independence before
        Paine, 192
affluence
    as part of high production, 221
    not a natural condition but a
        manufactured one, 221
African Genesis, 183
afterlife
    hope for, 335
    no evidence for it, 335
Age of Reason, The
    dealing with fabulous and true
        theology, 290
    first book of Paine that
        Galambos read, 87
    written by Paine to show that
        atheism is wrong, 193
aging
    as cellular deterioration, 126
Agrarian Justice
    Galambos no longer agrees
        with, 87
    Galambos saying in V-76 it has
        unfortunate things in it, 265

airplane
  not being used for bombing
    cities, 251
  used for war, 51
Akhenaten
  a genius, 284
  at the fountainhead of our
    monotheistic civilization, 323
  example of nondurability of
    coercion, 284, 293
  first identifiable monotheist in
    history, 284
  Galambos' commendation and
    respect for, 293
  has remained through his
    primary property, 294
  Moses the intellectual disciple
    of, 284, 285
  Newton standing on shoulders
    of, 294
  personally unpopular even
    though pharaoh, 284
  was pharaoh only a short time,
    284, 293
albino
  definition of, 318
Alexander the Great, 164, 302
  actually not great, 171, 173,
    254
  having no effect upon 20th
    century civilization, 172
  Linus will be remembered much
    longer than, 254
  what was considered important
    in days of, 171
allergy
  vs. virus, 123
altruism
  falseness of, 31
  not claimed by Galambos, 253
  phoniness of, 47
American English
  vs. *English* English, 228, 229

American farmer
  self-esteem of, 311, 312
American republic
  based on wrong principles, 87
American Revolution, 55
  a partially short-term successful
    achievement which later
    derailed, 194
  as supreme cause of Paine, 248
  created by Thomas Paine, 192
  derailed because it did not
    include theory of primary
    property, 194
  durable and nondurable
    components of, 8
  Galambos' concepts basically
    the continuation of, 89
  leading to high production and
    low durability, 40
  Paine right on principal part of,
    88
  primary psychology course
    drawing from, 8
Americans
  Europeans eat more efficiently
    than, 269
  mostly monolingual, 214
ancient Egyptians
  noncontinuous culture of, 294
Andrade, Edward Neville da Costa
  biography of Newton, 252
anger
  release of having a personal
    therapeutic effect, 307
anti-cancer
  not a way to cure cancer, 6
anti-capitalists
  believe law of the jungle
    prevails in capitalism, 221
anti-communism
  a disease, 6
  vs. pro-capitalism, 6
Apollo 8, 171
Apollo 11, 93

apology
  a necessary minimum
    prerequisite, 160
  must be followed up with right
    behavior, 160
appreciation
  a weaker and shorter-term
    form of thankfulness, 175
  an inward aspect of
    thankfulness, 198
  short term, 198
Arab states
  blackmail of, 14–16
  have one of the lowest cultures
    in the world, 14
a-rational
  distinguished from irrational,
    185
  meaning of, 185
Archimedes, 39, 322
  at forefront of most important
    advances in history of
    mathematics, 217
  concept of the limit inherent in
    work of, 217
  dead but his work is alive, 46
  died without recognition but
    work survives, 248
  easy to get permission from, 36
  enormous expansion of
    civilization production due
    to, 199
  everyone in debt to, 176
  Galileo resurrected recognition
    of achievement of, 46
  had some public exposure, 250
  has Galambos' authority of
    respect, 86
  heroic stature of, 45
  ideological ancestry of, 44, 45
  if there were an afterlife,
    possibility for Galambos to
    meet, 335
  intellectual ancestor of
    Galambos, 44, 45
  large ego of, 46, 47
  mathematical applications of,
    216
  Newton and Leibniz depended
    on earlier work of, 217
  not called the Greatest, 171
  only a few manuscripts of
    survived, 250
  prevailing ideas the base of our
    present culture, 249
  providing and also receiving
    value, 45, 46
  providing Galambos greater
    value than 3.8 billion people
    combined, 45
  receiving market value from
    Galambos, 45
  remembered through *Plutarch's
    Lives*, 249
  size of natural estate, 45
  stands as towering giant of
    2,000 years ago, 45
  supreme example of durability
    of right ideas, 284
  theory of primary property
    applied to, 199
  totally ignored by his
    contemporaries, 249
  what makes him better than
    other Greeks?, 169
  zero-to-one transition of, 50
Archimedeses
  unimaginable how many
    potential ones existed whom
    we never even heard of, 249,
    251
ARD mechanism (automatic
  remoteness dilution)
  innovation released to when in
    public domain, 42
  not available for Newton, 38
  NRD the stage to get to, 34

release mechanism of, 33
safe to release to, 33
someday will be proper to have
   every product released to, 33
Ardrey, Robert
   recognized animals naturally
      guard their own territory,
      183
Aristarchus of Samos
   died without recognition but
      work survives, 248
   not valued adequately today,
      138
Aristotle
   beliefs of not harming anybody,
      337
Armstrong, Neil, 322
Arnot, F.L., 125
artificial disasters
   not properly controlled through
      lack of adequate technology,
      199
astrology
   definition of, 237
   distinguished from astronomy,
      237
astronautics
   a misnomer, 109
   Galambos intended to return
      to, 94
astronomy
   distinguished from astrology,
      237
astrophysics
   a larger subject than ordinary
      physics, 238
   Galambos' interest in, 238
atheism
   Paine writes *Age of Reason* to
      show it is wrong, 193
atomic bomb, 51, 60
atomic energy
   Newton didn't know anything
      about, 267

atomic time, 125, 128
attack
   distinguished from defense,
      271
Attila the Hun, 164
authority list
   same as Galambos' gratitude
      list, 86, 88
authority of respect
   a correct concept of authority
      and ideological followership,
      88
   as truest authority there is, 86
   expression coined by Alvin
      Lowi, 85, 86
   integrated with friendship and
      gratitude, 88
   others who are on Galambos'
      list, 86

back-flowstream
   adequate corroboration of, 333
bad
   a basic preference concept in
      volitional science, 219
   definition of, 219
   difficulty putting a number on,
      219, 220
bad reputation
   psychological defect of, 255
barbarism
   world turned around of from to
      civilization, 6
Barbary pirates
   cleaned out by the Marine
      Corps, 15
   demanding tribute from
      European nations, 15
   seizing merchant ships of
      Western countries, 15
baseball, 53, 58
Bastiat effect (significance)
   description of, 172

348

Bastiat, Frédéric
  major concept of, 172
Bible
  brainwashing people for
    thousands of years, 290
  legends of have some basis in
    fact, 286
biblical time scale
  preposterous, 57
bilingual thinking
  affects diversification capability
    of developing thought
    patterns, 213
biological maturity
  deals with biological growth, 10
  terminates around two
    decades, 30, 120
biology
  cure applied to, 3
  flowing from physics, 3
  included in Galambos'
    expansion of physics, 328
  includes living part of the
    universe, 152
  mathematics hard to apply in,
    216
  thermodynamics analogy
    shown to be right for, 148
  upstream of volition, 3
  will benefit when mathematics
    can be applied to, 217
blackbody
  an idealization, 258
  explained in Galambos' physics
    course, 258
blackbody radiation
  explanation of, 258
  phenomenon of favors
    corpuscular theory of light,
    258, 259
  phenomenon of not favoring
    wave theory of light, 258,
    259

Böhm-Bawerk, Eugen von
  a quality economist
    understanding market
    concept, 144
bolometers, 318
Braille
  invention of, 317
Branden, Barbara, 243
bridge to freedom
  a very difficult transition
    period, 36
British Empire
  vs. Genghis Khan's empire, 213
  was most world-round empire
    of history, 212
Bruno, Giordano, 333
  death of, 19
  died without recognition but
    work survives, 248
  listed in Index of Forbidden
    Books, 6
  long-term view of, 137
  murder of, 113
  not having a P_2 goal, 137
  not valued adequately today,
    138
  work of kept alive, 340
Bryan, William Jennings
  quotation of, 43
bureaucrats
  controlling Newtonian physics,
    94
business
  Galambos had close to
    socialistic attitude on, 247
  loathed by Galambos but he
    was still successful, 246
businesses
  examples of highest quality
    ones at this time, 142, 143
  honesty important for ones in
    the personal time scale, 142
  nature of in short-end of trivial
    time scale, 141, 142

349

none have lasted a thousand
years, 143
not lasting into species time
scale, 147
remained short-term in quality
and scope, 108
businessmen
cannot identify with anything
beyond their own lifespan,
22
examples of lowest quality of,
141
haven't got the dynamism and
imagination to carry into
species time scale, 22
majority are blowhards, 140
natural and easy linkage with
investors, 110
not on same time scale as
innovators, 22
not penetrating beyond
personal time scale, 22
not recognizing long-term time
scale, 108

Caesar, Julius
what was considered important
in days of, 171
calculus
invented by both Leibniz and
Newton, 192
invented by three men
independently of each other,
217
California
capital of rocket industry as
reason why Galambos moved
to, 93
cancer
not cured by being anti-cancer,
6
capibut
definition of, 14

capitalism
anti-capitalists believe law of
the jungle prevails in, 221
arbitrary word of, 17
economic origin of, 17
Galambos using term for a
larger concept, 17
quasi-political term of, 17
takes blame for all blunders
committed in socialism, 26
word turning off pseudo-
liberals, 6
"Capitalism—The Key to Survival!"
original title of V-50, 5, 6
"Capitalism—The Liberal
Revolution"
title of V-50, 5–7
capitalist
includes the investor, inventor
and entrepreneur, 98
innovator the third class of, 108
only class of people involved in
progress, 102
three types of, 102
Capone, Al, 254
Carnegie, Dale
Galambos could not have
attitude of, 319
Galambos not concealing
relationships with acts of, 83
one of biggest menaces against
civilization, 319
politicians easily influenced by
techniques of, 319
salesman philosophy of, 78
Carnot engine
an idealization, 258
cat
biologically mature at age one,
120
can twist like a pretzel, 181
cannot be enslaved, 197
Galambos' discussion of, 197,
198

Galambos never met on that
was ugly just funny looking,
186
hated by tyrants, 198
language and psychology of,
224, 225
man who cannot learn from is
an intellectual weakling, 197
naturally impatient, 114, 115
naturally possessive, 183
one-year-old can support
himself, 180
pleasant feeling of touching,
185
Catholic Church
abolished its Index of Forbidden
Books, 5
cause
all are lost, 248
Galambos dropping in favor of
proprietary productive
management, 248
centimeter
definition of, 234
"Chaos & Plunder vs. Investments
& Insurance"
title to V-31, 333n
Chaplin, Charlie, 12
childhood
ego minor a continuation of, 28
ego minor naturally begins in,
52
children
behavior of, 13
can be quite destructive, 51
don't know what long term is,
118
ego minor keeps most people
as when they shouldn't be,
48
formative years are exceedingly
important, 119
gradually acquiring self-
responsibility, 12

immaturity of, 11
impossible to be mature, 11
learn by generalizing, 118, 119,
124, 125, 129
natural for them to like
property, 52
natural short-termers, 129
naturally possessive, 183
normal to have ego minor, 28
not rational, 28
not understanding concept of
later, 126
parents having a proprietary
interest in, 28, 119
parents should be responsible
for misdeeds of, 29
start out as parasites, 119
taught what property is, 52
China
basically a feudal country, 311,
312
famine in the inability to adapt
to natural market
mechanism, 311
has produced famines, 309
not having healthy concept
towards property, 309, 310
whole culture of opposed to
preservation of ideas, 251
worship of Marx, 310
Chinese peasant
self-esteem of, 311, 312
chintziness
a manifestation of ego minor,
121
an abhorrent ego minor
characterisitc, 54
whole society succombing to,
54
Christianity
accepting the Ten
Commandments, 292
fabulous inputs of, 291

Galambos' psychology not
meant as substitute for, 313
integration with primary
psychology, 10
not a study in rationality but
mainly of emotions, 307
coercer
goes to his doom with his
death, 293
coercion
a crime when successful, 297
a malfunction, a disease, 2
accepted as a protection of
property instead of an attack
on property, 185
always emotional, 295
as source of all civilizational
collapse, 200
as source of flatland, 2
correlation with ego minor, 2,
200
destroying civilization, 60
disclosure barrier as source of
all of, 144
disclosure barrier the cause of,
112
discussion of in V-50, 297
ego minor behavior turning to if
successful at, 165
example where physical act can
represent or not represent,
306
fatal weakness of, 293
flatland teaches the
glorification of, 170
greater amount of leads to
greater envy of property, 55
idea durable without, 295
increases injustice, 135
irrationality of, 60
more primitive than
production, 165
nondurability of, 284, 293

not a healthy response to
emotional pain, 305
not involved with market
justice, 134
not part of a contract, 160
one of two possible ways to
interact, 112
only one meaning to word of,
234
people not conforming by, 341
relationship to property, 2
restitution due even in absence
of, 182
reversed when restitution is
made, 301
transfers focal point of injustice
to a single organization, 135
vs. contract, 112
when not intentional is not a
crime, 298
with an error is a crime, 159
working at counter purposes
with production, 55
coercive pressure
examples of, 133
producing state justice, 133
cold end
benefiting from a social
structure with stabler
surroundings, 204
busting up the hot end, 37
distinguished from hot end, 136
goes with short term, 78
ideological importance of, 111
improvement of, 204
inevitably affected by this
theory, 341
interaction needed with hot
end, 110
refers to consumption, 78
theory of psychology also has
value in, 279

cold end of ideological program
people in can get hotter with
effort, 279
cold-end relationship
not a friendship, 82
cold-enders
buy fanciest car to impress
customers, 206
natural society automatically
making integrity structure
for, 280
no self-improvement by direct
means, 231
reason there are more of, 79
very difficult to deal with hot-
enders, 111
coldness
a human deficiency and failing,
201
colds
elimination of with right
attitude, 123
Coleman, Ronald, 229
commodity concept of money
an unacceptable medium of
transmission of integrity, 144
distinguished from state
concept of money, 145
Common Sense
best-seller of a magnitude that
has never been equaled, 247
communication
and language required to
harnass knowledge of
nature, 222
characteristics of healthy form
of, 226
necessary for human beings,
210
needed for emotions, 240
communism
a disease, 6
based on complete defect in
observational reality, 183

communists
trying to make Paine their pet,
247
use of propaganda exceeding
Hitler, 26
compassion
a desirable emotion, 201
not reciprocated producing
frustration, 202
compensation
a return action for warmth, 201
and consent for use of ideas, 32
and gratitude, 198, 199
better ways of, 204
depends upon a proper concept
of gratitude, 174
law of logarithmic stimulation
in, 198
of positive value rendered to
others, 12
competence
produced by a long-term
attitude, 91
comprehension
necessary prerequisite for
testing of, 299
concept of the limit
basically inherent in
Archimedes' work, 217
confidence
acquired after proof, 325
as earned, 330
distinguished from faith, 327,
330, 333
replacing faith, 325
conservatives
turned off by word *liberal*, 6
consumption
as fourth step of ideological
program, 112, 136
cold end referring to, 78
maintains purpose for
production, 136, 137

356

deduction
distinguished from induction, 179
deeper
definition of, 176
defense
distinguished from attack, 271
delusion
absolutely no shortage of, 20
applicable to everybody, 27
definition of, 18, 20, 331
of accepting theft by the state, 35
DeMille, Cecil B., 286
Demosthenes
a great orator in Greek, 215
destroyers
ego minor of, 51
destruction
as negative property, 152
flatland teaches the glorification of, 170
more important to glorify production than, 174
preceded by production, 70
dictators
come to power and stay in power by emotional response, 278
shielding people from risk, 337
Dillinger, John, 253, 254
dinosaurs
found out how difficult it was to survive, 222
lasted much longer than man will have lasted, 58
lived by coercion too, 60
disaster
perpetuation of, 173
discipline
Galambos deciding if he's going to give lecture on, 71

disclosure barrier
a barrier to civilization to receive concepts of integrity, 144
as cause of all coercion and source of the state, 112
as source of all coercion, 144
can only be removed from side of second step of ideological program, 196
Galambos extending significance of depth and importance of, 144
in direction of both the innovator and rest of mankind, 146, 147
keeping innovator's higher standards and quality out of the marketplace, 158
not permitting those who develop ideas to live in a market concept, 144
preventing people to have capability of integrity, 144
removal of, 111, 251, 300
discoverer
as *cosmological innovator*, 100
discovery
von Mises not looking upon as property, 98
disease
of anti-communism, 6
of coercion, 2
of communism, 6
of ego minor, 29
prevention of with proper release mechanisms, 122
very closely related to psychology, 122, 123
dispute
how to determine who is right, 260, 261

distribution
  as third step of ideological
    program, 112, 136
dividend
  a morale booster in P_2 to keep
    people investing, 103
  a short-term concept, 103
  not strongest form of investing,
    104
  only important in the short run,
    104
Dixon, Jeane, 328
dog
  naturally possessive, 183
"Don't take name of the lord in
vain"
  psychological value of, 289
downstream
  not morally proper to criticize
    upstream, 264
  proper conduct of person in,
    271
Dracula
  means "little devil", 302
  nickname of Vlad the Impaler,
    302
dreams of glory
  a childish concept, 92
  a manifestation of ego minor,
    121
  a smaller variety of ego minor,
    53
  a very major form of ego minor,
    255
  an ego minor disease, 279
  and psychological disorders,
    259
  creation of, 159
  distinguished from
    achievement, 255
  not a substitute for
    achievement, 92
Du Pont de Nemours, I.E. Co., 142

duodecimal metric system
  ideal numerical system, 233
durability
  affecting quality, 152
  as subject of V-201, 7
  lack of from high coercion
    factor, 213
durable culture
  depends upon both morality
    and rationality, 79
dwarf machine
  as the ego minor, 48

Earth
  itself has a magnetic field, 154
  moon falling around, 177
ecological nuts
  as eco-nuts, 58
economics, 17, 67, 145
Eddington, Arthur Stanley, 260,
  333
  a giant who could afford to use
    colloquial terminology, 245
  delightful writings of, 244
  discussing most profound thing
    about the universe, 244
  excellent concept of subjective
    knowledge, 324
  Galambos' highest intellectual
    debt to, 324
  Galambos' major input by
    reading, 323
  Galambos not agreeing with on
    concept of faith, 326
  had much mysticism and faith
    in his total intellectual
    makeup, 324
  his own self-popularizer, 192
  of comparable stature to
    Newton, 268
  one of greatest physicists
    ignored in 20th century, 324

358

showing there's no possibility
of acquiring objective
knowledge, 323
terminology of, 316
ego
a form of property, 2
a principal term of psychology,
2
as a tensor, 155
as basic concept of psychology,
2
as only new term introduced by
Galambos, 2
as what you think of yourself,
176
called plus and minus, 155
called self-esteem, 312
ego minor explaining all
malfunctions of, 2
is to psychology what property
is to volition, 2
it takes more than one action
to build, 155
major and minor aspects of,
198
not a matter of self-vainglory,
90
ego major
builder of first step of
ideological program, 196
builder of second step of
ideological program through
theory of primary property,
196
by doing something which is
right, 137, 138
explains how hot end
mechanism affects
civilization, 199
nobody is born with, 52
of major achievers, 43
of producers, 51
positive base of, 11

ultimate maximum potential,
41, 42
ego minor
a continuation of childhood, 28
a delusion which leads to false
results, 2
a disease, 29
a false self-esteem, 199
a principal term of psychology,
2
a psychological disease, 161
a supreme simplifier explaining
all malfunctions of the ego, 2
all forms are childish, 92
always frustrated, 161, 162
as only disease in terms of
man's thinking processes,
165
as the dwarf machine, 48
basically a characteristic of an
action, 53
becomes dangerous in adults,
53
behavior of, 165
behavioral characteristic of, 161
concept as varied as number of
actions of all people in the
world, 53
correlation with coercion, 2,
200
dreams of glory a very major
form of, 255
enormous number of
characteristics, 54
everyone is born with, 52
has many manifestations, 121
having all kinds of effects, 165
Hitler a supreme case, 164, 165
Hitler one of biggest of all
history, 162
inseparably and inescapably
intertwined with short term,
121

is to the ego what coercion is to property, 2

keeps most people children, 48

naturally begins in childhood, 52

nearly universal disease, 200

normal in children, 28

not particularly harmful in most children, 53

of childish behavior, 11

of destroyers, 51

of dreams of glory, 279

of politicians, 165

offensive behavior of, 266, 267

producing short-term and unstable activities, 199

reduced by having long-term view, 124

reduction of is the stability mechanism, 96

self-image based on false inputs, 165

short term the source of behavior, 118

spectral range of, 53

the mechanism to have a lesser appreciation of reality, 200

the universal disease, 43

the universal psychological disease, 11, 161

totally a short-term phenomenon, 121

universal destroyer of stability, 196

vanishing with total maturity, 124

when a person is incapable of maturity, 11

ego minor mentality
applied to innovation, 30

egoism
distinguished from egotism, 162
highest form of, 47, 48

egotism
distinguished from egoism, 162

Ehrenhaft, Felix
a great but unaccepted scientist of 20th century, 67
Galambos had privilege of knowing, 67
Galambos helps put book into English, 89
would have accepted if his observations proved to be faulty, 89

eighteenth century
as Age of Enlightenment (Age of Reason), 56
only short-term violence in, 56

Einstein, Albert, 304, 333
an example of highest form of egoism, 47
contempt for P_2, 137
developer of theory of relativity, 259
farther account of Newton, 99
Galambos' pronunciation of, 223, 228
long-term view of, 137
Newton a shoulder account for, 328
not called the Great, 171
not having a P_2 goal, 137
of comparable stature to Newton, 268
owner of law of gravitation, 99
publishes photoelectric effect, 259
pursued happiness, 60
quotation of, 47

Einsteinian physics
Newtonian physics a first approximation to, 328

electrical charges
called plus and minus, 154
positive and negative, 153

360

more primitive than rational
concpets but not
undesirable, 242, 243
most not co-ordinated with
rationality, 315
most people cannot
subordinate to what's
necessary to be done, 206,
207
most volitional action
motivated by, 276
necessity for, 122, 314
needed for mass
communication, 189
not a bad thing per se, 280, 296
not a license to be irrational,
203
not all coercive, 295
not always stemming from
irrationality, 192
not as articulate as a rational
thought would be, 231
not in contradiction to
rationality, 10
not necessarily automatically
wrong, 297
not same as rational thinking,
241, 242
not the enemy of man, 308
number of reasons for not
demonstrating, 202, 203
of psychological feelings, 186
partially an incentive
mechanism and partially a
therapy mechanism, 338
positive factors of, 297
possible to be working in
unison with rationality, 340
proper expression of necessary
for human stability, 297
proper expression of necessary
to maintain stability and
health, 203

release of can be therapeutic,
297
repression of causing injury,
280
repression of leads to internal
frustration and psychological
problems, 203
responsible for thankfulness,
175
should not be scrapped, 188
some have positive, favorable
effects, 280
stability comes from harnessing
of to fit rational conduct, 314
subordination of to rationality,
201, 278
theory of psychology deals
with, 241
under better control when
language improves, 231
when properly harnessed, 338
when rational producing
progress, 308, 309
working in harmony with
rationality, 315
energy
as supreme unifier of physics, 2
energy crisis
repugnant concept of, 251
energy hoax
delusion of, 17
Galambos reads newsletter
showing there is no energy
shortage, 209, 210
leading to chintziness in
society, 54
English English
vs. American English, 228, 229
English language
a giant language with an
enormous vocabulary, 211,
212
a hodgepodge intellectually as
well as content-wise, 210

362

extrapolation
  as third step of scientific
    method, 295

F-201
  Galambos granted too much
    credit with, 24
  market rejection of, 164
fabulous
  meaning of, 290
fabulous theology
  distinguished from true
    theology, 291
faith
  a lousy market transaction, 330
  a substitute for thinking and
    corroboration, 102
  as credit advanced, 330
  considered a virtue, 325
  distinguished from confidence,
    327, 330, 333
  Galambos' definition of, 327
  in God, 325
  is pursuit of happiness
    successful through?, 332
  mankind has fallen for, for
    thousands of years, 326
  replaced by confidence, 325
  taken before proof, 325
  to believe something that is not
    corroborable, 326
  trying to explain things by, 292
  vs. truth, 324
false alternative
  characteristic of a coercive
    society, 62, 63
false modesty
  Galambos not suffering from,
    169
  not pretended by Newton, 48
  pretended by some people but
    not a true statement of their
    real convictions, 90

pretended for either social
    circumstances or market
    consequences, 96
Faraday, Michael
  close to Newtonian stature, 268
  discoverer of law of
    electromagnetic induction,
    111
  not called the Great, 171
farther account
  expression of coming from
    Newton's quote, 47
feelings
  an intellectual response to an
    external stimulus, 295
  from reactions to experiences
    one has been subjected to,
    242
  from the five senses, 185, 186,
    295
  moods based upon, 220
  nothing more dangerous than
    to repress, 207
fiat money
  as money by decree, 145
financial concepts
  subordinate to primary
    achievement, 53
  to be integrated properly in
    subvolitional domain, 53
financial stability
  Galambos giving courses on, 95
  integrated with emotional
    stability, 95
  unavoidably connected to
    emotional stability, 104
first moon landing
  Galambos' reaction to, 93, 94
  who got the credit for, 171
first step of ideological program
  as hottest part of the
    production cycle, 137
  as primary production, 147
  as production, 147

364

as source of knowledge and
integrity, 144
as source of progress, 144
ego major the builder of, 196
fourth step essential for, 111
having a continuous sustained
existence, 111
never been a market concept
before Galambos developed
theory of primary property,
137
never recognized in economic
sense, 144
no contract possible with other
steps unless of a time scale
alignment, 113
previously excluded from
marketplace, 158
relatively modest expenses in,
111
those in having capability of
integrity, 144
totally incompatible with other
three steps, 112
Five Books of Moses
called the Torah, 291
five senses
feelings from, 185, 186, 295
instinctive reaction to inputs
from, 242
sense of sight the only long-
range sense of, 317
source of our direct knowledge
from inputs of external
universe, 316, 317
flatland, 106, 175, 341
a coercive society, 63
a very powerful way of talking
about the world of the past,
17
all ideas are in public domain,
36, 37
business concept of, 195
coercion the source of, 2

concept of getting even in, 305
everyone has lived in, 36
full of irrational and immoral
producers, 79
haggling mechanism will perish
with, 34
harder to be truthful, 207
history classes in do not glorify
achievement or long term,
173
Hitler incapable of earning an
honest living in, 162
most people work for time
payments in, 74
most successful people in, 43
no proprietary handling of
primary property in, 36
people having no inputs on
gratitude, 168, 169
pretty low standard of time
scales and moral concepts,
107
produced by malfunctions of
handling property, 2
respect for authority a
prevalent expression in, 85
secrecy the only tool available
to innovators in, 35
teaches glorification of coercion
and destruction, not
production, 170
the universally politically
oriented, politically
dominated structure, 168
the world where property is not
considered sacrosanct, 36
truth a difficulty in, 208
usual concept of innovators in,
137
flatland base
prevented Galambos being first
on the moon, 92
someday will be improved with
spaceland base, 92

flatland businesses
    vast bulk are in trivial time
        scale, 140
flatland civilization
    real backbone of, 22
flatland culture
    restricts all of our output, 92
    scientists always the lackeys of
        various states, 170
flatland laws
    protect criminals not property
        holders, 37, 38
flies
    language and psychology of,
        225
forever
    definition of, 335
    distinguished from eternity,
        172, 335
fourth step of ideological program
    as consumption, 112, 136
    as maintenance, 111
    economic market conditions
        recognized in, 144
    essential for first step to exist,
        111
Franklin, Benjamin
    beautiful choice in naming
        electrical charges, 154
    not for independence before
        Paine, 192
fraud
    basic mechanism of, 18, 19
Free Enterprise Institute (FEI)
    all courses interrelated to each
        other, 4
    alternate name is University of
        Hard Knocks, 81
    as Galambos' detour to make
        world safe for physicists, 80
    in the true market, 81
    janitors made more on union
        wages than Galambos in first
        couple of years, 74
    main purpose of, 63, 64
    open-end course of, 4
    reason why Galambos founded,
        94
    stability a major course offered
        by, 199
freedom
    as subject of V-201, 7
    couldn't even be identified until
        now, 109
    depends on property, 101
    depends upon both morality
        and rationality, 79
    greatest tragedy in history of,
        246
French language
    irrational pronunciation in, 224
French Revolution, 277
Fresnel, Augustin-Jean, 257
Freud, Sigmund, 304
    recommended any
        psychoanalyst required to
        undergo psychoanalysis, 241
    saying Moses was an Egyptian,
        285
    sexual brand of psychology, 180
friend
    definition of, 76
    distinguished from
        acquaintance, 77
friendship
    a long-term concept, 80
    always a primary relationship,
        78
    between people who have
        never met, living in different
        times, 82
    concept of in species time
        scale, 82
    depends upon a primary
        concept, 78
    if it can terminate it means it
        never existed, 82

366

integrated with gratitude and authority of respect, 88

meaning of, 68

mimimum time it has to endure, 82

shortest-term concept of there can be, 81

frustration

a proper release mechanism to prevent disease, 122

causes of, 202

exaggeration a form of, 77, 78, 123

Galambos, Joseph B.

a giant of 20th century, a universal genius, 250

a smoker, 84

a unique input into Galambos' theory, 322

as Galambos' greatest teacher, 29

as Galambos' one friend in childhood, 82, 111

as reason Galambos doesn't smoke, 84

as supreme teacher of Galambos, 250

asking about English spelling on Galambos' report cards, 209, 214

asking Galambos as a child why he has no friends, 68

commemoration of, 320

concerned Galambos had no practical goals as a child, 320

could give lectures to English professors, 210

didn't like discipline of Prussianism, 85

Galambos glad he shaped him before it was too late, 84

Galambos partly getting his long-term concepts from, 169

Galambos' respect for, 84, 86

gives all kinds of Latin things for Galambos to study, 262

had the authority of respect, 85

had the mind of Leondardo da Vinci, 250

if there were an afterlife, Galambos would see again, 335

irritated Galambos only able to speak two languages, 210

letters of, 320

never forbid Galambos to smoke, 84

not wrong on any significant occasion, 86

provided fantastic insights to Galambos, 320

saying only a dog barks in one language, 211

superior character of, 250

teahces Galambos Latin phrase *Si duo faciunt idem, est non idem*, 261

telling Galambos gratitude is the rarest human characteristic of all, 175

tells Galambos he didn't bring him to America to look like a Prussian Junker, 85

the least coercive parent a man could have, 84

tolerated no insolence, 84

worried his son had dreams of glory to go to the moon, 93

Galambos, Margaret

if there were an afterlife, Galambos would see again, 335

quotation of, 281

German language
  also irrational, 223, 224
  ambiguities in, 230
  as poor compared to
    mathematics, 240
"getting even"
  a counterfeit for justice, 299
  a short-term emotion, 299
  concept in flatland, 305
Gibbs, Josiah Willard
  close to Newtonian stature, 268
goals
  don't set too low, 92
  mismatch between
    entrepreneurs and
    innovators, 109, 110
  must be realistic, 92
  what should they be?, 92
god
  as simply an explanation of
    natural phenomena, 327
  as source of natural
    phenomena, 285, 287
  can't be identified on cosmic
    scale with anthropomorphic
    concepts, 285
  concept of not hostile to but
    inseparable from science,
    287
  definition of, 285, 287
  God a false and foolish concept
    of, 266
  greatest merit to concept of is
    the universe is harmonious,
    287
  miracle an insult to concept of,
    286, 288
  not a volitional being, 61, 266,
    327, 335, 336
God
  a false and foolish concept of
    god, 266
  a mythological creature, 258
  as incompetent, 287, 288

as omnipotent, 287
being irrational and suffering
  from an ego minor complex,
  336
cannot be talked to, 327
chat with Moses, 289, 290
could not punish someone who
  seeks to know even if he
  made a mistake, 336
fables invented about, 323
faith in, 325
Galambos not thinking is god,
  266
if a volitional being, will not
  keep Galambos out of
  heaven, 336
needed by man to support his
  own fallibilities, 61
pursuit of happiness applies to,
  61
used as abdication of personal
  responsibility, 289
Goddard, Robert
  one of three major men who
    got man off Earth, 26
  work of harnessed to get man
    on the moon, 171
Goebbels, Joseph, 12
  looked like an emaciated rat, 21
Gogh, Vincent van
  market rejection of, 163
gold
  can stabilize an economy, 146
  cannot be easily counterfeited,
    146
  cannot have any production
    effect in long term, 146
  has been a traditional
    commodity concept of
    money, 144, 145
  necessary to keep politicians
    honest, 146
  will not attract major
    innovation, 146

gold standard
  can last longer than hot-air
    standard, 146
good
  a basic preference concept in
    volitional science, 219
  definition of, 219
  difficulty putting a number on,
    219, 220
Grand Canyon, 308
gratitude
  a rare characteristic, 166, 175
  an identifiable, articulable and
    durable concept, 175
  and compensation, 198, 199
  and debt, 176
  both an inner emotion and
    external reaction and action
    towards other people, 175
  closely identified with integrity,
    166
  compensation depends upon a
    proper concept of, 174
  correlation with integrity, 167,
    168
  dependent upon rationality,
    166
  determines *why* to pay, 174
  development of, 231
  distinguished from
    thankfulness, 175
  does not fade with time, 166
  grows qualitatively and
    quantitatively, 199
  integrated with friendship and
    authority of respect, 88
  lack of with majority of
    scientists, 138
  measured in P_2 in theory of
    primary property, 199
  not an emotional concept, 166
  not available to people with a
    short-term view, 166

people having no inputs on in
    flatland, 168, 169
  possibilities to accomplish a
    sensitivity to, 168
  primary and secondary
    payment of, 11
  rational base needed for, 176
  short-term facsimile of, 198
  source of, 176
  thankfulness a facsimile of, 202
  the rarest human characteristic
    of all, 175
  where it comes from, 195
gratitude list
  same as Galambos' authority
    list, 86, 88
gravitation
  cannot be owned, 99
  distinguished from law of
    gravitation, 99
  is not property, 99
  not invented by Newton, 177
Great Britain
  Norman Conquest of, 294
Great Dictator, The, 12
*Greatest Speeches in the English
    Language, The*, 215
grown up
  definition of, 338

habits
  formation of, 231, 232
haggling mechanism
  a very undignified way of
    conducting business, 34
  an ego minor disease, 34
  will perish with flatland, 34
Halley, Edmond
  a double great man, 39
  a giant compared with other
    people, 39
  got Newton to publish, 40
  only looks small next to
    Newton, 39

Hamilton, Alexander
  influenced by history of Roman
    Senate, 195
Hamilton, William Rowan
  close to Newtonian stature, 268
happiness
  moral quest for as only
    mechanism for man's
    survival, 17
  most people don't have
    articulable definition of, 188
  quality and quantity of, 3
  two different ways of
    measuring, 3, 4
hate
  distinguished from contempt,
    49
  implies a desire to counter-
    coerce, 49
  not compatible with rationality,
    281
health
  maintained by proper
    expression of one's
    emotions, 203
Heath, Spencer
  a genius, 273, 275
  absolutely no controversy with
    Galambos, 275
  attended all of Galambos' early
    lectures, 273, 274
  book of showing proprietary
    managmeent superior to
    political management, 275
  Galambos' first encounter with,
    272, 273
  Galambos preserving work of,
    275, 276
  gave inscribed copy of his book
    to Galambos, 273
  in attendance when Galambos
    started San Diego market,
    274

introduces Hoiles to Galambos,
  275
liked what he learned from
  Galambos, 273, 275
not having Linus-Newton type
  of interrelationship with
  Galambos, 276
talks with Galambos till 4 AM,
  274
tremendous ego major of, 275,
  276
heaven
  delusion from standard
    theological perspective, 333
Henry, Joseph
  close to Newtonian stature, 268
  discoverer of law of
    electromagnetic induction,
    111
"higher things in life"
  as proper thing to pursue, 332
  from two different vantage
    points, 333
  very dear to Galambos'
    personal thinking, 332
high-quality people
  probability of, 70, 71
Himmler, Heinrich
  not looking like a member of
    the master race, 22
hippies
  as short-termers, 83
  most conformist group of
    people today, 83
  mutual admiration society of,
    83
historians
  political interpretations not
    compatible with reality, 55
history
  as first step of scientific method
    applied to volitional science,
    173

if it persists, politics won't, 140
when dominated by emotional
behavior invites political and
theological nonsense, 338
humanitarian
Galambos not offended by
expression, when it's defined
right, 31
humanitarianism
highest form of, 51
nonexistent in raw nature, 222
true operation of, 47
Hungarian language
as poor compared to
mathematics, 240
diversification factor of, 213
more colorful than English, 212
not perfect, 225
phonetic pronunciation, 223,
225
Hungary
as center of an area which has
many different nationalities,
213
Hunyadi Mátyás
had Vlad the Impaler
imprisoned for twelve years,
303
one of few quality people who
ever occupied a throne, 303
Huygens, Christiaan
dispute with Newton a bona
fide major dispute between
two giants, 259
dispute with Newton on theory
of light, 256, 257, 259
not chasing Newton into the
mint, 260
hypothesis
extrapolation of producing
predictions, 187
not subject to observational
corroboration, 335

observation for corroboration
substantiating strength of,
187
hypothesis formulation
as second step of scientific
method, 295

IBM
as highest quality type of
business at this time, 142
at top of ladder as investment
vehicle for flatland, 107
built up by Thomas Watson,
107
Galambos reads biography of,
107
still a growing company but no
longer as spectacular as it
once was, 142
idea
an articulated thought, 149
as a communicated thought,
224
as property, 98
compensation and consent for
use of, 32
durability of, 250
durable without coercion, 295
moral usage of requires
consent from the owner, 34
must be rational to survive, 295
Paine very sensitive to strength
and durability of, 246
treated as property by moral
innovator, 38
whole culture of China opposed
to preservation of, 251
whole of culture depends on
cumulative impact of, 283,
284
idealism
psychological motivating
factors of, 174

induction
    distinguished from deduction,
        179
Industrial Revolution, 108
infant
    all are parasites, 119
    needing to make
        generalizations, 179–182
infinite pay for zero work
    ultimate goal of politicians and
        unions, 65
infinite price
    meaning something is not for
        sale, 34
inflation
    produced by state concept of
        money, 145
injustice
    Galambos very sensitive to,
        318, 319
    increased by coercion, 135
    meaning of, 299
innovation
    ego minor mentality applied to,
        30
    gross reduction in output by
        coercing most able
        producers, 113
    market for exists because of
        interaction of hot and cold
        end, 110
    not done for mankind, 42
    released through NRD, 42
    released to ARD when in public
        domain, 42
    theory of primary property
        preventing use of for harmful
        purposes, 251
innovator
    distinguished from *inventor*, 98,
        100
innovators
    believe they do it for mankind,
        42

can be in any time scale
    depending on quality and
        magnitude of innovation, 108
disclsoure barrier keeping
    higher quality and standards
    of out of marketplace, 158
disparity in time scales with
    entrepreneurs, 109, 110
don't like to be plundered, 35
have never been recognized as
    a market mechanism, 137
having an intellectual product,
    32
how can they benefit
    themselves?, 42
impossible to adopt short-term
    time scale attitude, 113
looked upon sort of as a natural
    resource you dip into, 137
mismatch of goals with
    entrepreneurs, 109, 110
most never really try to reach
    masses, 194
motivation and moral behavior
    of, 42
not on same time scale as
    businessmen, 22
owes to upstream flow
    derivatives, 36
owing their intellectual
    ancestors, upstream
    shoulder accounts, 42
previously not making terms to
    mankind, 51
secrecy the only tool available
    to in flatland, 35
set apart from rest of people
    and treated harshly, 147
third, final and highest class of
    capitalist, 108
time scale difference with
    entrepreneurs, 97

insanity
as irrationality on a permanent
continuing basis, 86
insurance
definition of, 237
insurance industry
based upon statistical
application of profit
mechanism, 235–237
integrity
a long duration concept, 144
a rare commodity, 133
as self-honesty, 136, 156, 280
concept of gratitude closely
identified with, 166
connection with long term, 131
correlation with gratitude, 167,
168
defined in V-201, 136, 168
development of, 231
disclosure barrier preventing
people to have capability of,
144
distinguished from honesty, 76,
133
distinguished from non-term
integrity, 132
doesn't exist on short-term
basis, 131
emanating from the hot end,
160
innate form of, 131
needed to discover laws of
nature, 137
no such thing with a short-
termer, 131
not imposed upon a person,
132
not needed for all people, 136
only source of true justice, 136
significance and importance of
stressed in V-212, 8
source of, 144

integrity machine
built by hot-enders, 341
of V-201, 195
strengthening of understanding
of, 8
intellectual products
as longest term products of all,
79
intellectuals
emotion capable of having a
rational impact on, 192
primary property comes first
with, 248
intelligence
as basis for generalization
capability, 129
mainly hereditary, 129
intent
most people cannot and do not
define, 297
most sensible way of defining,
298
originally defined in V-201, 298
internal frustration
caused by inconsistencies in
oneself, 159
caused by repressing emotions,
203
distinguished from external
frustration, 156, 159
examples of, 159
produced by ego minus, 156
invention of the wheel
as important to mankind as
major law of nature, 108
in species time scale, 108
of same magnitude as flying,
108
inventor
a kind of capitalist, 98
a primary capitalist, 98
as *technological innovator*, 100
distinguished from *innovator*,
98, 100

inventor of the wheel, 96, 108,
333
cartoons of, 166, 167
connection between
nonrecognition of and our
present plight, 167
everyone in debt to, 176
Galambos worried about as a
little kid, 58
how many people are grateful
to?, 166
if there were an afterlife,
possibility for Galambos to
meet, 335
investing
distinguished from speculating,
103
distinguished from spending, 71
dividend not strong form, 104
principle of, 105, 106
whole concept of, 104
investment
an example of self-discipline, 71
an expenditure with a longer
time scale, 75, 76
distinguished from expenditure,
74
distinguished from sacrifice,
255
distinguished from speculation,
334
short-term value usually called
an expenditure, 71, 72
two time scales involved in, 102
investor
a kind of capitalist, 98
a secondary capitalist, 100
natural and easy linkage with
businessmen, 110
irrational
distinguished from a-rational,
185
irrationality
always emotional, 192

becoming a non-successful
pursuit of happiness, 221
leads to error and ultimately to
failure, 188
irresponsibility
as what most people want, 101

"JBG Centennial"
a two-day commemoration
celebrating birth of Joseph
Galambos, 320n
Jefferson, Thomas
a great man but inferior to
Paine, 273
influenced by history of Roman
Senate, 195
not for independence before
Paine, 192
not paying tribute to Barbary
pirates, 15
Jesus Christ, 304
asked to believe was born of
Immaculate Conception, 325
belief in as Son of God, 331
belief in resurrection and
saviorship of, 332
case where turn the other
cheek would work, 271
Galambos compares his work
with philosophy of, 89
Galambos not taking as his
savior, 335
pursued happiness, 60, 332
Jews
had no major culture previous
to Moses, 286
Oberth said destruction of is
Allied propaganda, 25
Ten Commandments converting
from primitive tribe to high
civilization, 289
Joan of Arc
a popular hero of France, 277,
278

John Birch Society
  keeping people out of
    Galambos' course, 5
Judaism
  as most durable religion in
    history, 284
justice
  Galambos very sensitive to,
    318, 319
  getting even a counterfeit for,
    299
  integrity the only source of, 136
  market justice becoming
    prevalent form of, 134
  of people suffering from
    making a mistake, 337
  only one meaning to word of,
    234
  proper compensation for any
    form of market action or
    behavior, 10
  quest for by political means,
    305
  return flow of, 10, 11
  with market pressure, 134
justice mechanism
  maturity of abiding by instead
    of opposing it, 11

Kennedy, John F.
  getting credit for landing on the
    moon, 171
Khrushchev, Nikita, 293
kibbutz
  survive only by external
    handouts, 312
kindness
  a desirable emotion, 201
  not reciprocated producing
    frustration, 202
Kirchhoff, Gustav, 257
knowledge
  application of called
    production, 147

as anything that you can
    corroborate by observation,
    219
  definition of, 315
  none today that is non-
    Newtonian, 267
  of seeing connection between
    seemingly unrelated and
    dissimilar things, 325
  secondary production cannot
    exist without, 147
  should dominate the
    production world, 94
  source of, 144
  successful application of
    needed to have ego major,
    11
Kossuth, Louis
  gives speech before joint
    session of Congress, 215
  Hungarian patriot-liberal, 214
  probably the greatest orator of
    all time, 215
  spoke thirteen languages
    fluently, 214, 215

Lafayette, Marquis de
  granted permanent American
    citizenship, 215
  helps America during War for
    Independence, 215
Lagrange, Joseph-Louis
  close to Newtonian stature, 268
  continued Newton's work, 192
  in popular polls not greatest
    Frenchman, 277
  not called the Great, 171
Lange, Matt, 262
language
  a supremely important
    environmental factor for
    psychological domain, 232

predictions correctly calculated pursuant to Newtonian techniques, 329

you can't rule with, 310

law of logarithmic stimulation, 119, 128, 166

a volitional extension of Weber's law of physiology and psychophysiology, 174

determines *how much* to pay, 174

properly in compensation, 198

laws of nature

a volitional idea that fits the reality about nature, 98

application of to volitional affairs, 80

as perfect, 287

cannot be discovered without integrity, 137

cannot be violated, 41

discoveries of as property, 98

distinguished from natural phenomenon, 98, 99

do not have to be rediscovered, 283

inductive generalizations of history, 179

not protectable, 37

nothing more imporant than, 37

learning process

whole point of, 116, 117

legality

seldom parallel with morality, 134

Leibniz, Gottfried Wilhelm

depended on Archimedes' earlier work, 217

independent of Newton on the calculus, 217

inventor of the calculus, 192

scientific rival of Newton, 192

leisure time

one of troubles of civilization, 64, 65

Lemuel Gulliver

in Lilliput, 341

Leondardo da Vinci

Joseph Galambos had the mind of, 250

Lewis, John L., 59

Lewis, Joseph

an atheist and welfare-stater, 193

writes book proving Paine is author of Declaration of Independence, 193

liberal

word turning off conservatives, 6

Liberty Amendment

fallacy of, 265

life insurance

is truly death insurance, 235

light

absorbed and emitted in discrete particles or chunks, 258

Lincoln, Abraham

quotation of, 103

Linus, Francis

a criminal who does not fit on absolute importance diagram, 254

a parasite off of Newton's achievement, 256

a primary murderer, 39

acrimony and viciousness of, 260

attacking Newton's discoveries, 260

attained immortality as a generic type of criminal, 253

better for him not to be remembered at all, 39

causing emotional stress to Newton, 252, 253

dispute with Newton an example of *Si duo faciunt idem, est non idem*, 261

dispute with Newton not based on bona fide scientific difference of opinion, 259

has now become a generic name for a leech who seeks glory at expense of a genius, 253

injuring Newton, 299

might not care how he is remembered, 253

not probable he would be injured by Newton, 270

permanent damage done by, 254

will be more famous as a result of Galambos than Newton, 299

will be remembered longer than all other criminals in history, 254

Locke, John

Galambos bored by style of, 189

logic

facsimile form of, 150

long term

always more important than short term, 314

as defined in V-30, 148

concept of cannot be ignored for major producers, 89

connection with integrity, 131

counts more for primary production, 148

favored over the short term, 114

flatland history classes do not glorify, 173

gold cannot have production effect in, 146

nothing of significance endures into that isn't right, 337

primary property automatically strong in, 248

production method of outproducing all other production mechanisms, 341

shor term winning over for majority of people, 340

significance and importance of stressed in V-212, 8

species time scale considered as, 148

stability coming from concept of, 196

longer-term value

examples of, 72, 73

long-term attitude

a person's outlook on life affected by, 124

counterfeit of, 91

goes with long-term success, 90, 91

impossible for everyone to adopt, 113

produced by competence, 91

long-term integrity

distinguished from non-term integrity, 132, 133

long-term investor

distinguished from short-term investor, 103

long-term optimist

Galambos an example of, 81

long-term profit

significance for, 196

long-term success

goes with a long-term attitude, 90, 91

long-term time scale

all achievement comes from, 118

long-term view
  actuated by subtrivial time
    scale thoughts, 196
  as hardest of all intellectual
    feats, 124
  development of, 124
  needed to reduce ego minor,
    124
  rationality dependent upon,
    166
long-termer
  developing emotional
    disorders, 252
  doesn't have recognition and
    incentive in short time he
    lives, 249
  has a conscious goal as to what
    he wants to do, 133
  needs short-termers for a
    market, 252
  not possible for everyone to be,
    69, 251
  products of needed by short-
    termers for survival and
    civilization, 252
  qualities of, 114
  rational base of, 252
  theory of primary property
    becoming of greater benefit
    to, 251
looking through the tube
  offered by Galambos, 336
looting
  can be primary or secondary or
    both, 64
Lord Kelvin
  not called the Great, 171
Lowi, Alvin
  coins expression "authority of
    respect", 85, 86
luck
  distinguished from being right,
    117

Madison, James
  a great man but inferior to
    Paine, 273
  influenced by history of Roman
    Senate, 195
magnetism, 154
maintenance
  proprietary profit mechanism,
    111
major achievers
  having a positive ego, 43
  large scale ego majors of, 43
  motivation for, 43
major innovation
  not attracted by gold, 146
majority
  comprehension and acceptance
    of, 66
  not accepting responsibility for
    their errors, 158
majority rule
  preposterous example of
    property dispute settled by,
    261
majority view
  almost always wrong, especially
    when ideas are new, 19
  when it's the prevalent view, 19
man
  destiny to go into the cosmos,
    332
  dinosaurs lasting much longer
    than, 58
  duodecimal metric system the
    ideal numerical system for,
    233
  ego minor as only disease in
    terms of thinking processes
    of, 165
  emotional component of here
    to stay, 315
  emotions necessary for, 314
  emotions not the enemy of,
    308

Mao Tse-tung
  leading exponent of Marxism in
    the world, 310
Marine Corps
  cleaned out Barbary pirates, 15
market function
  goes with market value, 29
market justice
  as reason for higher quality
    people being honest, 141
  as reason for person to make
    amends, 300, 301
  becoming prevalent form of
    justice, 134
  causing the state to wither
    away, 135
  distinguished from state justice,
    136
  does not involve coercion, 134
  replacing state justice, 135
  stronger than state justice, 136
  ultimately will survive and
    supersede state justice, 300
  understanding of primary
    property needed to establish
    mechanism of, 136
market pressure
  a stronger form of justice, 134
market risk
  totally disregarded by
    scientists, 187
market value
  goes with market function, 29
marketplace
  and language, 220
  disclosure barrier keeping
    innovator's higher standards
    and quality out of, 158
  first step of ideological program
    previously excluded from,
    158
  has always been inferior and
    short term, 158

Marquise du Châtelet
  translated Newton into French
    from Latin, 191, 192
  Voltaire's mistress, 191
Mars
  predictions of orbit, 328, 329
Marx, Karl
  worshiped by China, 310
Marxian doctrine
  need crap influencing state
    policy, 312
Marxism
  can be used to destroy, using
    Newtonian technology, 310
  justice mechanism for is
    starvation, 312
  Mao Tse-tung the leading
    exponent of, 310
  Russia had dosage of in 1920s,
    312
Marxists-Leninists, 135
mass
  transformed into high amounts
    of energy and used for
    atomic bombs, 51
mass communication
  emotions needed for, 189
mass popularity
  people who do things for
    always dangerous, 20, 21
mass, length and time, 2
  fundamental entities of all
    knowledge, 127
masses
  believe the printed word, 290
  delusions of coming from
    propaganda, 20
  have market function in
    ideological program, 110
  less competent, less
    productive, 55
  most innovators never really try
    to reach, 194

no direct way to improve,
either in volition or
psychology, 231
reached indirectly through
second step of ideological
program, 194
reaching of the easiest thing in
the world to do, 22
still not understanding Earth
not center of the universe,
19, 20
survival of species not
depending on, 48
tyrant cannot rule without
concurrence of, 55
where they are right, 19
who do you reach them with?,
22
will believe anything that's told
to them often enough, 23
mathematics
applied to volition, 236, 237
as supreme form of
communication, 239, 240
hard to apply in biology, 216
single supreme language of
mankind, 215
strongest advances done by
physicists who needed new
tools, 217
superior language of, 222, 240
the language of physics, 216
used in physics where precision
is possible and accuracy
normally not possible, 234
usefulness as communications
tool, 217
maturity
comes when you are self-
responsible, totally, 320
definition of, 182
description of, 10, 11
ego minor vanishes with, 124

no single definition of in all
subjects, 10
nobody is born with, 11
Maxwell, James Clerk
majority of people not caring
about, 342
never tried to reach the
masses, 194
not called the Great, 171
of Newtonian stature, 268
Maxwell's theory
explanation of not the way to
sell television sets, 188, 189
megalomania
a manifestation of ego minor,
121
can be an enormous disaster of
ego minor, 53
megalomaniac
definition of, 162
memory
poor system of, 179
time the greatest dimmer of,
198
Menger, Carl
a quality economist
understanding market
concept, 144
metric system
Galambos making comments
about in physics course, 233
microcosm
subject of needing a new
Newton, 267
Milne, E.A., 125
minus ego
a negative ego, 155
called ego minus, 155
description of, 155
producing internal frustration,
156
producing psychological
disorders, 156

mysticism
  abandonment of, 143

Nader, Ralph
  probably is sincere, 20, 21
Napoléon I, 164
  as most popular Frenchman in
    history, 277
  no rationality in respecting, 278
  not even French, 277
  paying tribute to Barbary
    pirates, 15
national socialism
  fraud of, 9
  German application of a
    worldwide volitional disease,
    8
  letting you think you're free
    when you're enslaved, 9
  not a German invention, 9
  PSC a form of, 9
  the most sophisticated form of
    attack on property ever
    designed, 9
nationalization
  definition of, 14
natural disasters
  not properly controlled through
    lack of adequate technology,
    199
natural estate mechanism
  through trusteeship, 36
natural integrity
  stems from studying of nature,
    137
natural law
  defied by a miracle, 286
natural phenomenon
  distinguished from laws of
    nature, 98, 99
natural republic
  concept not available to
    President Jefferson, 15
  described in V-50, 134

developed in V-201, 134
  hastening of, 279
natural society
  automatically making integrity
    structure for cold-enders,
    280
  hastening of, 279
  if already existing Galambos
    wouldn't even be in this field,
    320, 321
  spaceland a stronger word
    than, 17
nature
  communication and language
    required to harnass
    knowledge of nature, 222
  cruel and harsh to living beings,
    222
  natural integrity stems from
    studying of, 137
  no paradise for lunacy in, 221
  not humanitarian, 222
  technology coming from large
    understanding of, 222
"Nature and Protection of Primary
  Property"
  title of V-201, 7
negative ego
  can produce suicide, 156
  could produce a lifetime of
    morbidity, 156
  not recognized by majority of
    people, 156
negative history
  as history of politics and
    militarism, 257
  examples of, 170
negative output
  distinguished from positive
    output, 151–153
negative posture
  an unhealthy way of handling
    anything, 6

long-term view of, 137
loss of primary property a
    greater loss to mankind, 39
magnitude and grandeur of
    mind of, 327
mathematical applications of,
    216
motivated to stop disclosing
    and become a hermit from
    science, 50
never tried to reach the
    masses, 194
no mechanism to protect ideas
    when publishing his book, 38
not always exciting in his
    writings, 191
not called the Great, 171
not chased into the mint by
    Huygens, 260
not having a P_2 goal, 137
not infallible, 266
not knowing anything about
    atomic energy, 267
not picked by Mao Tse-tung,
    310
not probable he would injure a
    Linus, 270
not valued adequately today,
    138
on ideological flow chart, 264
only a few successors are of his
    stature, 268
other people popularizing work
    of, 191, 192
owner of law of gravitation, 99
Paine an in-between point on
    ideological flowstream
    between him and Galambos,
    194
Paine's stature in history not
    close to, 195
personality of, 49
preserved from oblivion by
    Halley, 39

pursued happiness, 60
quotation of, 47, 48
receiving value from
    Archimedes, 46
reputation of will last longer
    than Hitler, 254
respected by Paine, 88
scientific rival of Leibniz, 192
scientific work the principal
    input for Paine and
    Galambos, 190
spent forty years at the mint
    and Royal Society, 40
stated he was standing on
    shoulders of giants, 294
stature of, 251
successors standing on
    shoulders of, 267
tired of being plundered, 35
very reluctantly published his
    book, 38
what he didn't publish is
    probably enormously larger,
    39, 40
what makes him excel over
    other people born in
    Woolsthorpe?, 169
when succeeds in pursuing
    happiness, rest of human
    species will find it easier, 300
will be remembered, 254
work of harnessed to get man
    on the moon, 171
wouldn't have written this
    much but for Halley, 40
writings of not easy to read,
    190
zero-to-one transition of
    Archimedes, 50
Newtonian gravitation
    long history of success, 330
    Paine wrote favorably about, 88
    understood by Paine, 190

391

Newtonian integration
  influence of monotheism on,
    285, 294
Newtonian physics, 195
  a first approximation to
    Einsteinian physics, 328
  Paine only person who did
    something significant with,
    194
  works for bureaucrats too, 94
Niagara Falls, 308
  a very expensive operation, 111
Nixon, Richard M.
  getting credit for landing on the
    moon, 171
non-term integrity
  concept of, 131
  distinguished from integrity,
    132
  distinguished from long-term
    integrity, 132, 133
  not really anything productive,
    132
  of Sam (office cat), 131, 132
non-termer
  might have honesty but not
    integrity, 132
  no thought-out purpose with,
    133
  not producing anything, 132
normality
  meaning of, 241
  subject of psychology to
    understand, 241
North Pole
  in Southern Hemisphere, 23
  usually called plus, 154
nosiness
  the facsimile of curiosity, 202
Nostradamus, 328
NRD mechanism (negotiated
  remoteness dilution)
  innovation released through, 42

the stage we have to go
    through to get to ARD, 34
nuclear energy, 16, 60, 63
  Galambos writes essay on when
    sixteen years old, 59
  not being used to wipe out the
    population but to produce
    cheap and abundant energy,
    251

Oberth, Hermann
  astronautics pioneer, 24
  believed Hilter was defending
    Germany, 24–26
  Galambos thinks he swallowed
    Nazi propaganda, 31
  meeting with Professor
    Galambos, 24–26
  one of three major men who
    got man off Earth, 26
  work of harnessed to get man
    on the moon, 171
objective knowledge
  distinguished from subjective
    knowledge, 324
  not possible, 315, 316, 318, 323
objectivity
  scientists talking about, 324
oblivion
  more desirable than to be
    remembered as a criminal,
    253
observation
  as first step of scientific
    method, 295
  role of in scientific method, 244
  universe knowable through,
    323
observation for corroboration
  as basis for proof, 327, 328
  as fourth step of scientific
    method, 295, 296
  as observational reality, 296
  producing knowledge, 219

Galambos considers a true
friend though he never met
him, 81
Galambos has deepest respect
for, 87, 272
Galambos' love for, 87
Galambos not tolerating
criticism of political
conclusions, 265, 266, 272
Galambos' personal affection
for, 195
Galambos reads complete
works of at age twenty, 87
Galambos receives offensive
letter from woman who took
course on, 265, 266
Galambos recognizing heroism
of, 272
Galambos standing on
shoulders of, 245
had two greatest best-sellers of
the period, 247
has Galambos' authority of
respect, 86
if there were an afterlife,
possibility for Galambos to
meet, 335
in context of theory of primary
property, 193
looked down upon as just a
political pamphleteer, 190
major significance in world
history, 193, 194
making a rational impact on the
intellectuals, 192
making an emotional impact on
the masses, 192
mistaken belief in
representative democracy,
272
more of a scholar than
intellectual snobs that think
they are scholars, 189

Newton's scientific work the
principal input for, 190
not a socialist, 247
not correlating primary and
secondary property, 248
not infallible, 266
not recognizing beauty of profit
mechanism, 246
one of finest human beings
who ever lived, 87
original iron bridge builder, 191
part of what was wrong with
American republic, 87
personality of, 49
poor in P_2, 247
respected same authority as
Galambos, 88
right on principal part of
American Revolution, 88
Rights of Man the second best-
seller of, 248
stature in history not close to
Newton's, 195
thought about constructing a
constitution with same
precision as an engineer
builds a bridge, 191
use of word *fabulous*, 290
would concur that political
philosophy has failed, 89
writes *Age of Reason* to show
atheism is wrong, 193
writings of easy to read, 190
wrote favorably about
Copernican astronomy and
Newtonian gravitation, 88
wrote with strong, witty and
emotional terminology, 245
parasite
either mooching or stealing,
101
parents
bringing up children poorly, 13

first person an infant is likely to
  encounter, 179, 180
growing up in the present era
  making for lousy parents in
  next generation, 141
having a proprietary interest in
  their children, 28, 119
most are emotional children, 29
most responsible for what
  children think, 119
must assume liability for child's
  immaturity, 11
not knowing much about
  property, 13
should be responsible for
  misdeeds of their children,
  29
unable to explain concept of
  later, 126
partial secondary capitalism (PSC)
a form of national socialism, 9
Roman Empire an example of,
  107
Pasteur, Louis
in popular polls not greatest
  Frenchman, 277
patents
relatively short expiration
  periods in terms of species
  time scale, 37, 38
patience
as necessary, 114
cultivated and not innate, 115
once you learn, it's learned,
  115, 116
takes character and courage,
  115
PBV-273
a course on emotional stability,
  95
bringing together many inputs
  and coming out with a single
  unified subject, 28

deals with both emotion and
  rationality, 339
digging deeper into
  subvolitional domain of
  secondary production, 139
drawing from a number of
  areas covered in other
  courses, 28
drawing from all other
  volitional science courses, 8
drawing from American
  Revolution, 8
first major course born in OEC,
  9
Galambos explains his teaching
  method, 260
Galambos has to discuss both
  hot and cold end, 110
Galambos integrating various
  human characteristics in, 169
Galambos never expected to
  give course, 66–68
purpose of to improve hot end
  of ideological program, 204
reaching mainly hot-enders,
  204
slopover effect of in V-201, 298
stability a large part of, 96
V-201 a prerequisite for, 340
perpetual motion machine
belief in not injuring anyone,
  337
personal stability
development of, 196
emotion can be a builder of,
  339
personal time scale
as short term, 148
businessmen not penetrating
  beyond, 22
from point of view gold can
  stabilize an economy, 146
highest quality companies
  operating in, 143

property protection
  in seven of Ten
    Commandments, 289
  inherent in Ten
    Commandments, 292
  mechanism of, 17
proposition
  definition of, 179
proprietary management
  Galambos' based on primary
    property and flowstream of
    knowledge, 275
  Heath's real estate
    interpretation of, 275
  superior to political or coercive
    management, 275
proprietary mechanism
  as property protecting and
    property expanding, 18
  explained in V-50 and V-201, 18
  only thing capable of providing
    propulsion for volitional
    apparatuses, 18
  path to man's survival, 95
  subvolitional motivations
    turning people off against, 18
Prussian army
  a substitute for stability, 197
Prussian Junker, 85
pseudo long-termer
  definition of, 256
  will not care whether he earned
    his reputation, 256
pseudo-humanitarians
  believe affluence should be
    distributed, 221
pseudo-hypothesis
  examples of, 150
pseudo-liberals
  turned off by word *capitalism*, 6
psychiatrists
  don't have too much occasion
    to deal with normal people,
    241

many in need of their own
    services, 241
psychiatry
  a luxury man can afford when
    he has relative tranquility,
    314
  deals with defective conditions
    in people, 339
psychoanalysis
  deals with defective conditions
    in people, 339
psychological attitudes
  unstable for people, 220
psychological diseases
  beginning of, 188
psychological disorders
  and dreams of glory, 259
  examples of, 156
  produced by ego minus, 156
psychological feelings
  as emotions, 186
  examples of, 186
psychological intent
  distinguished from operational
    intent, 298, 301
  in a person's thoughts prior to
    articulation, 298
psychological internal disaster
  of not accepting what is, 91
psychological maturity
  deals with emotional growth,
    10
  not reached by everyone, 182
  time for depends, 120
psychology
  a subvolitional subject that fits
    the universe, 328
  application of plus and minus
    to, 153
  concept of incentive in, 4
  connections with disease, 122,
    123
  cure applied to, 3

pursuit of happiness
  applies to God, 61
  as automatic, 61
  as simply the selection of goods
    and rejection of bads, 75
  as what one lives for, not to
    service his teeth or
    fingernails, 205
  depends on time scales, 75, 76
  done by everyone, 332
  how is helped by believing
    Immaculate Conception, 331
  is successful through faith?, 332
  no exceptions to, 60–62
  not interfering with other
    people's property, 204
  requires self-discipline not to
    take something that favors
    you immediately, 114, 115
  sought naturally, 341
  unsuccessful if irrational, 188,
    221
  via profit, 17
Pythagoras
  Galambos' pronunciation of,
    228, 240
Pythagorean theorem
  triangle of, 239, 240

quackery
  as the rule and not the
    exception, 238
quacks
  amongst psychologists and
    psychiatrists but doesn't
    make the profession false, 1
quality
  affected by both quantity and
    durability, 152
  depends upon a person's time
    scale attitude, 76
quantum
  definition of, 258

quantum theory, 266
  origin of, 258
  wave mechanics making a hell
    out of, 259
Quod licet Jovis, non licet bovis
  in Galambos' childhood
    notebook as Latin phrase
    #33, 262

rabble rousers
  as hooligans and hoodlums, 6
  not revolutionaries, 6
  not turned off by word
    revolution, 6
Rand, Ayn
  exposed falseness of altruism,
    31
  Galambos personally meeting
    with, 243
  Galambos visits three lectures
    of in New York, 243
  heroes of are psychological
    degenerates, 243
  robot-like followers of, 243
  showing phoniness of altruism,
    47
  wooden characters of, 122, 243
Rathbone, Basil, 229
rational
  distinguished from emotional,
    187, 188
  meaning of, 187
rational people
  not always dull, 189, 192
rational thinking
  involves logical analysis, 241
  not same as emotions, 241, 242
rationality
  and their proper juxtaposition
    in human motivation, 283
  as too rare a characteristic, 201
  baptism of, 331
  books of Newton a monument
    to, 191, 194

classical psychology not a study
in, 307
classical psychology not mainly
dealing with, 241
comes before morality, 269
concept of gratitude dependent
upon, 166
dependent on highest and most
precise form of
communication, 239
dependent upon a long-term
view, 166
deteriorates at terminal stages
of civilization, 64
dominant in hot-enders, 281
emotions can be compatible
with, 192
emotions compatible with, 281
emotions not in contradiction
to, 10
emotions subordinated to, 201,
278
hate not compatible with, 281
hot-enders reached with, 239
influenced by time scale
attitudes, 64
integration of certain emotions
with, 276
morality a derivative of, 79
most emotions not co-
ordinated with, 315
most volitional action not
motivated by, 276
needed to have anyting that
endures for a long time, 322
no systematic way to improve
behavior of a person who
does not possess, 220, 221
one of characteristics of, 260
possible to be working in
unison with emotions, 340
responsible for gratitude, 175
scientific method the birthplace
of, 216

should dominate the culture,
338
theory of primary property
increased the mechanism of,
195
theory of psychology deals
with, 241
truth and validity involved in
development of, 241, 242
working in harmony with
emotions, 315
Read, Leonard
quotation of, 263
real world
as observational, 295
realism and idealism
connected with short- and long-
term concepts, 198
goes with pessimism and
optimism, 81
weaved into stability, 96
reality
definition of, 170
lesser appreciation of ego
minor, 200
measured by rational behavior,
339
not recognized by a psychotic,
159
psychological motivating
factors of, 174
reason
as your guide, 143
Red Sea
easy explanation for so-called
parting of, 286, 288
redistribution of wealth
socialist concept of, 221
redshift of the nebulae
without an expanding universe,
125
Reformation
produced more religious wars
amongst Christians, 292

regular definition
  distinguished from operational
    definition, 127
reincarnation, 48, 238
relative
  a subjective concept, 324
relative knowledge
  distinguished from absolute
    knowledge, 316
religion
  conditioned man to be a
    nonthinker, 102
  reducing man's psychological
    capabilities, 102
religious hell
  invented to pacify men living in
    real hell, 102
Renaissance
  beginning of, 46, 250
representative democracy
  Paine's mistaken belief in, 272
respect for authority
  prevalent expression in
    flatland, 85
responsibility
  for making mistakes, 188
  not possible without risk, 182
restitution
  as negative compensation, 11
  due even in absence of
    coercion or error, 182
  erasing crime, so long as it's
    reversible, 301
  need for, 160
  of damaging someone else's
    property, 12
  when made undoing coercion,
    301
  when voluntarily made reverses
    a crime, 298
revolution
  word turning off everybody
    except rabble rousers, 6

revolutionary
  a clean, sweet, beautiful word,
    6
  meaning of, 6
right triangle
  definition of, 240
rightness
  defined in V-50, 138, 149
  distinguished from luck, 117
  does not involve someone
    else's property but your own,
    263
  most people do not know how
    to evaluate, 203
  necessary to be a producer of
    durable consequence, 79
  of something still being around
    thousands of years later, 89
Rights of Man
  Galambos no longer agrees
    with, 87
  has unfortunate things in it, 265
  second best-seller of Paine, 248
risk
  dictators shielding people from,
    337
  needed to have responsibility,
    182
riskless investment
  swindle of, 334
Rogers, Will
  quotation of, 76, 77
Roman Empire
  basically a PSC country, 107
  recognized what a merchant's
    profit is, 108
  significance of their P_2
    enterprises has disappeared,
    108
  you cannot name richest
    businessman in, 80
Roman Senate
  influenced Madison, Hamilton
    and Jefferson, 195

Roosevelt, Franklin D., 107
Royal Society
    Newton president of, 40
Russia
    had a dosage of Marxism in
        1920s and got famine as
        their just reward, 312

sacrifice
    distinguished from investment,
        255
Salk, Jonas, 97
Sam (office cat), 120, 131, 132
Sanders, George, 229
Santa Claus
    a harmful fable, 52
    paternal attitude of, 66
    principle of, 102
Saracens
    principal major value in history,
        250
scholar
    definition of, 189
science
    accuracy desirable to make
        stronger, 237
    accuracy not required for, 219
    based on precision, 234
    beauty of, 287
    big discoveries of, 177
    concept of god inseparable
        from, 287
    concept of synonym contrary
        to, 234
    content of explaining the
        universe, 323
    definition of cure applicable to
        all of, 3
    harnessed to be lackey of the
        state, 310
    history of, 173
    innate superiority of positive
        production of, 170
    means *knowledge*, 219

monotheism essential to, 323
precision required for, 219, 237
should dominate the
    production world, 94
world's attitude towards, 30
scientific method, 337
    application in determining self-
        comprehension, 299
    birthplace of rationality, 216
    difference between ordinary
        use of and scientist's use of,
        117, 118
    easier to make excuses than to
        apply, 157
    explaining things by ultimately
        producing all value, 293
    first step of applied to volitional
        science, 173
    four steps of, 295
    Galambos discussing in all his
        courses, 187
    Galambos' extension of to
        things over than classical
        physics, 216
    learning process of, 117, 157
    not used by most people to
        adopt conclusions, 18
    produces greatest advances in
        science and success in
        business and personal living,
        157
    role of observation in, 244
    step one involving historical
        inputs, 65
    the technique of arriving at
        conclusions, 187
    use of as basis for proof, 327
    use of showing everything has a
        reason, 143
    used by physics to explain the
        universe, 328
    volitional science standing on
        criterion of, 330

Paine not correlating with
  primary property, 248
secrecy
  the only tool available to
  innovators in flatland, 35
security
  only one that man can ever
  have, 184
  political state arises as an
  answer to, 184, 185
self-discipline
  external discipline an
  unacceptable alternative to,
  197
  lack of as cause of ego minor,
  121
  needed for stability, 197
  needed to have patience, 115
  required not to take something
  that favors you immediately,
  114
  the solution for high quality, 71
self-esteem
  called the ego, 312
  of having a primary goal, 137
self-honesty
  as integrity, 136, 156, 280
self-rule
  as only rule man needs, 88
Semmelweis, Ignaz
  died when forty-five years old,
  121
  pushed into early death, 113
sense of sight
  the only long-range sense of
  five senses, 317
  without, how would we even
  know there's a universe
  beyond Earth?, 316–318
sensitivity
  deals with primary matters
  more than to secondary, 202
  definition of, 202
  discussion of in V-212, 202

lack of with majority of
  scientists, 138
significance and importance of
  stressed in V-212, 8
touchiness the facsimile of, 202
sensory inputs
  source of everything we know,
  316
Sermon on the Mount, 304
Shakespeare, William, 214
short selling
  a pessimistic concept, 104
  Galambos long ago detached
  from, 105
  mechanism used by
  speculators, 104
  no upper limit to what you can
  lose, 105
  speculating based on
  pessimism, 104
short term
  always less important than long
  term, 314
  as source of ego minor
  behavior, 118
  counts more for secondary
  production, 148
  goes with cold end, 78
  inseparably and inescapably
  intertwined with ego minor,
  121
  long term favored over, 114
  meaning either trivial or
  personal time scale, 148
  scientists living in, 138
  strengthening of, 248
  winning over long term for
  majority of people, 340
short-term attitude
  a person's outlook on life
  affected by, 124
short-term investor
  distinguished from long-term
  investor, 103

408

short-term pessimist
  Galambos an example of, 81
short-term quick-buck artists
  cannot enter hot end of
    ideological program, 197
short-term value
  examples of, 72, 73
short-termer
  best you can hope from, 131
  children an example of, 129
  hippies an example of, 83
  impatience of, 114, 115
  needed by long-termers for a
    market, 252
  needs products of long-termers
    for survival and civilization,
    252
  no such thing as having
    integrity, 131
  reason there are more of, 79
  theory of primary property
    becoming of greater benefit
    to, 251
shoulder account
  expression of coming from
    Newton's quote, 47
Si duo faciunt idem, est non idem
  in Galambos' childhood
    notebook as Latin phrase
    #32, 262
  Latin phrase Galambos learned
    from his Father, 261
sidereal
  definition of, 128
sidereal day
  distinguished from solar day,
    128
slavery
  a provably wrong concept, 191
Smithsonian Institute, 113
Snelson, Jay, 298, 333
  changes title to "The V-50
    Lectures", 7

hands message to Mrs.
  Galambos, 96
heard V-201 five times, 114
social science
  anti-social and anti-science, 218
  as pseudo-science, 237
  confusion of concepts in, 2
  Galambos changes name to
    volitional science, 1, 237
  Galambos considers
    antecedents as total quacks,
    1
  Galambos' contempt of people
    engaging in, 237
  neither social nor science, 218
Social Security
  as wrong, 246
  belief in, 23
social structure
  almost all totally dominated by
    emotional attributes, 312
  cold end benefiting from with
    stabler surroundings, 204
socialism
  all blunders committed in called
    capitalism, 26
  as reason everything going to
    hell in this country, 26
  based on complete defect in
    observational reality, 183
  Galambos' deviation from, 247
  standard thing that leads to,
    320
socialists
  essentially meddling in
    someone else's property, 263
  hatred of people who have
    property, 247
  trying to make Paine their pet,
    247
society
  ability to have systematic stable
    form of, 222
  bad guys in today's form of, 13

direct correlation with
  psychology, 313
  rising above flat nature of, 341
solar day
  distinguished from sidereal day,
  128
solar energy, 16
something came up (SCU)
  of God, 287
  used as an excuse for breaking
  contracts, 287n
South Pole
  in Northern Hemisphere, 23
  usually called minus, 154
spaceland
  a stronger word than *natural
  society*, 17
  a stronger word than quasi-
  political term *capitalism*, 17
  as where we're going, 17
  easier to be truthful, 207
spaceland base
  someday will improve flatland
  base, 92
species time scale
  affected by long-term concepts,
  169
  all cosmological discoveries are
  in, 109
  businessmen haven't got the
  dynamism and imagination
  to carry into, 22
  can endure as long as the
  species can, 57
  concept of friendship in, 82
  considered long term, 148
  cosmological innovator
  naturally in, 109
  entrepreneurialship capable of
  entering into, 195
  highest quality businesses not
  entering into, 143
  in terms of unrecorded
  prehistory, 57

innovator can be in depending
  on quality and magnitude of
  innovation, 108
  invention of the wheel in, 108
  never any monumental
  achievement of secondary
  production in, 139
  no durable businesses in, 147
  not everybody's in, 81
  nothing of significance endures
  into that isn't right, 337
  of investment, 102
  only in domain of you can have
  significance to living a life,
  144
  patents and copyrights of
  relatively short duration in,
  37, 38
  recorded period of, 57
speculating
  distinguished from investing,
  103
speculation
  distinguished from investment,
  334
spending
  distinguished from investing, 71
Spinoza, Baruch
  died when forty-five years old,
  121
stability
  a large part of PBV-273, 96
  comes from harnessing of
  emotions to fit rational
  conduct, 314
  coming from concept of long
  term, 196
  dependent upon capability to
  generalize, 184
  ego minor the universal
  destroyer of, 196
  emotions have a positive aspect
  to, 281

410

follows from Galambos'
discussion of psychology, 199
Galambos giving a course on,
95
maintained by proper
expression of one's
emotions, 203
major course offered by FEI,
199
needed to accept bad results
and survive, 95
needs to be developed above
all and before everything, 95
self-discipline needed for, 197
totally unknown to average
person, 199
weaved into realism and
idealism, 96
stability mechanism
reduction of the ego minor, 96
Stalin, Joseph, 293
Stalingrad, 277
standard restrictions
when releasing property for any
use, 33
stars
Galambos' detour to, 321
Galambos not living to see man
going to, 333
starvation
as justice mechanism for
Marxism, 312
state
as only source of major crime,
135
concept of eminent domain, 34,
35
confiscating property, 135
control of individual transfered
to with national socialism, 9
controlling producers, 136
demands respect for authority,
85

disclosure barrier the source of,
112
expending enormous amount
of tax money to get to the
moon, 93
having a monopoly in murder
and theft, 292
having a monopoly of theft, 135
loses its image in declining
culture, 135
losing its potency to cope with
criminals, 135, 136
rise of comes from man's
infantile attitude on
property, 185
same paternal attitude as Santa
Claus, 66
science harnessed to be lackey
of, 310
scientists traditionally the
lackeys of, 170
ultimate credit went to for
landing on moon, 171
wants a monopoly to steal, 134
withering away with market
justice, 135
state concept of money
as the hot-air standard, 145
distinguished from commodity
concept of money, 145
example of, 145
produces the inflation effect,
145
the nondurable, trivial time
scale money that
disintegrates in purchasing
power, 145
state justice
a contradiction, 133
as reason for higher quality
people being honest, 141
as reason for person to make
amends, 300, 301
coercive pressure, 133

411

therapeutic if no one else's
property invaded, 304
temper tantrum
becomes dangerous when
injuring someone else's
property, 301
distinguished from temper
outburst, 304
not dangerous when
noncoercive, 301
Ten Commandments
accepted by Christianity, 292
accepted by Mohammedanism,
292
converting Jews from primitive
tribe to high civilization, 289
essentially written by Moses,
289
inseparable from monotheism,
292
property protection inherent in,
292
represent durability of Hebrew
culture, 290
seven of are clearly property
protection, 289
used by Christian countries, 304
Ten Commandments, The, 286,
290
Tennyson, Alfred, 326
tensor
ego a form of, 155
Territorial Imperative, 183
Tesla, Nikola
immobilization of, 113
tired of being plundered, 35
thankfulness
a short-term emotional factor,
168
appreciation a weaker and
shorter-term form of, 175
appreciation an inward aspect
of, 198

distinguished from gratitude,
175
fades with time, 166
the facsimile of gratitude, 202
theft
a monopoly reserved to the
state, 135
distinguished from mooching,
65
immoral, 65
little of in rising cultures, 135
theory
meaning of, 187
theory of electromagnetic wave
propagation
majority of people not caring
about, 342
theory of freedom
V-113 partially a historical
summary of what led to, 8
theory of gravitation
example of a big discovery of
science, 177
theory of light
Newton-Huygens dispute, 256,
257, 259
theory of primary property
a natural integration of two
entirely different forms of
property, 251
alternate name of V-201, 7
application to Archimedes, 199
covering concepts of debt and
gratitude, 176
enabling first step of ideological
program to be a market
concept, 137
function of secondary property
in, 248
Galambos doesn't claim
infallibility for, 89
Galambos' exceedingly great
sensitivity needed to
develop, 318, 319

thieves
  most so stupid they don't even
    know they have stolen, 30,
    31
thinking
  faith a substitute for, 102
  has to be done in some form of
    communicative or
    abstractive manner, 224
Third Reich
  made possible by mass
    propaganda, 26
third step of ideological program
  as distribution, 112, 136
  economic market conditions
    recognized in, 144
thought
  an emotion, 239
  as prior to articulation, 298
  becomes an idea when
    communicated, 224
  either rational or irrational,
    articulable or non-
    articulable, 149
  expressed in language, 212
  harnessing of to longer-term
    objectives, 196
  higher priority ones are longer
    time scale ones, 197
  higher quality produces
    greatest yield, 149, 150
  more articulate with stronger
    language, 324
  most carried out in subtrivial
    time scale, 150, 151
thought processes
  inherent in language structure,
    232
  takes time when rational, 295
thoughtprints
  no two are alike, 69
  the most fundamental
    characteristic of man, 69, 70

thoughtsteps
  better than footsteps, 108
thousand-year investment
  no such thing as, 73
Thrust for Freedom—No. 2
  originally printed in Course 100,
    99
  property defined in, 99
time
  concept of the supreme
    generalization of history, 182
  definition of through
    measurement techniques,
    128
  developing operational
    definition of, 126, 127
  not possible to go backward in,
    158, 159
  perhaps the greatest dimmer of
    memory, 198
  technical evaluations of, 128
  the most intuitively obvious
    and most scientifically
    difficult to identify, 182
  the single most difficult entity
    in physics to define, 125–127
  various definitions of, 125
time investment
  more important than money
    investment, 74
time payment
  basis of payment of people, 74,
    75
time scales
  a new one (political) between
    personal and species, 55, 57,
    140
  almost everything really
    depends on, 75
  disparity in between
    entrepreneurs and
    innovators, 109, 110
  introduced in V-201, 97

key to understanding human
motivation, 60
morality and rationality
influenced by attitudes in, 64
very major to theory of
psychology, 118
Torah
as Five Books of Moses, 291
touchiness
the facsimile of sensitivity, 202
usually deals with trivia and
superficialities, 202
Toynbee, Arnold J., 60
transcendentalism, 238
trespassing
a crime, 270, 306
tribal chief, 9, 168
tribal chief state
letting you know you're
enslaved, 9
witchcraft state worse than, 8
Trinity
a three-in-one God, 291
trivial time scale
as short term, 148
cheap, two-bit promoters living
in, 139
concept of money in, 146
is long term in subtrivial time
scale, 139
of most people, 43
secondary production in, 22
vast bulk of businesses are in,
140
true integrity
lies only in hot end of
ideological program, 138
source of, 139
true theology
distinguished from fabulous
theology, 291
Truman, Harry S., 107
truth
a difficulty in flatland, 208

an absolute, 161
involved in development of
rationality, 241, 242
Occam's Razor speaks for, 207
vs. faith, 324
when corroborated is
knowledge, 219
two pigs story
showing everybody loves
property when he's on the
receiving end, 30
tyrants
cannot rule without
concurrence of the masses,
55
hate cats, 198

U.S. Congress
Kossuth gives speech before
joint session of, 215
unification of all knowledge
beginning of, 2, 3
unions
insane irrational attitudes of, 65
ultimate goal of, 65
United States of America
approaching its termination, 40,
194
based upon Marxian doctrine
*To each according to his
need*, 312
death of, 41
decadent educational system
in, 203
has produced abundance of
food, 309, 310
import of Asiatic mysticism
into, 238
kowtows to 12th century
barbarians, 14, 15
mainly urban today, 310
political policy to subsidize
nonproduction of farmers,
310

teaching incidental to
monitoring windows and
aisles, 208
terminal stages of, 56
unbelievably short life cycle, 55
victor of WW II, 14, 15
United States republic
political catastrophe following
from, 245, 246
universal psychological disease
of the ego minor, 11, 161
universe
beauty and grandeur of, 308
biology includes living part of,
152
Eddington discussing most
profound thing about, 244
entropy the measure of the
disorder of, 3
explained by use of scientific
method, 328
harmonious, 287
heat death of, 126
man's expansion into, 322
more disorganized means *later*,
126
new domain of, 143
not knowable except by
observation, 323
only one source of explanation
for, 285
physics deals with content and
phenomena of, 323
physics includes all of, 152
reality of is observationally
corroborable, 339
requires an explanation, 323
scientists harnessing their own
conduct to fit nature of, 137
volition includes choice-making
capability with living part of,
152
wouldn't have any significance
if we didn't exist, 80

University of Minnesota
Galambos attending when
twenty-four years old, 85
upstream
may criticize downstream, 264
to attack is immoral, 270
you can't drift but have to have
strength to swim, 279
urgency
the rate of change of property,
151

V-2 rocket, 24
V-30, 206
a much deeper course than it
seems at first blush, 157
deals with investments and
insurance for defective
flatland products, 8
definition of insurance in, 237
describes present world of
defective insurances and
investments, 148
discussed long term in terms of
investments, 148
discussing corroboration of
your achievement, 54
discussing you must take
responsibility for mistakes,
188
explaining difference between
long-term and short-term
investor, 103
flunked by many people, 157
Galambos describes how he got
out of the boondoggle, 115
not needed to take V-50, 334
only investment course
compatible with theory of
primary property, 157
pointing out dividend is not
strongest form of investing,
104
prerequisite of V-31, 334

417

volitional science
    a new branch of physics, 80
    a true science, 237
    as Galambos' name for social
        science, 1, 237
    Galambos' creation of, 226
    good and *bad* the basic
        preference concept, 219
    history as first step of scientific
        method applied to, 173
    precise definitions in, 219, 220
    standing on scientific method's
        criterion, 330
    theory of psychology
        dependent upon, 152
    V-50 the basic course of, 5
Voltaire
    brought Newton to European
        continent, 191
    in popular polls not greatest
        Frenchman, 277
    popularized Newton's writings
        into French, 191, 192

War for Independence
    Lafayette helps American side
        during, 215
warmth
    a desirable emotion, 201
    not reciprocated producing
        frustration, 202
    requires a return action called
        compensation, 201
Washington, D.C.
    as the local Mecca, 146
Washington, George
    not for independence before
        Paine, 192
Watson, Thomas
    an absolute ass on politics, 107
    built up IBM, 107
wave mechanics
    atom explained as a probability
        wave, 259

wave theory of light, 260
    always successful in its
        application in transmission of
        light, 259
    not favored by phenomenon of
        blackbody radiation, 258, 259
    of Huygens, 256, 257, 259
Weber's law
    volitional extension of, 174
welfare state
    a mixture of capitalistic
        production and welfare state
        parasitism, 221
    concept of is anti-natural, 221,
        222
    general, whining philosophy of,
        13
    insidious philosophy of, 26
    paradise for lunacy in, 221
Western civilization
    much came through Christian
        countries, 304
Western culture
    characteristic of at this phase of
        degeneracy, 135, 136
Western religions
    durable portion of, 292
Western world
    as Judaic in culture because of
        its monotheism, 309
    knowledge of Greco-Roman
        period brought to, 250
witch doctor, 168
witchcraft
    highest form of, 26
witchcraft state
    nondurable components of
        American Revolution leading
        to, 8
    worse than tribal chief state, 8
women's liberation, 61
Woodward, W.E., 190
Woolsthorpe, 169

422

www.ingramcontent.com/pod-product-compliance
Lightning Source LLC
Chambersburg PA
CBHW050555270326
41926CB00012B/2064